Benchmark Papers in Geology

Series Editor: Rhodes W. Fairbridge
Columbia University

Volume

1 ENVIRONMENTAL GEOMORPHOLOGY AND LANDSCAPE CONSERVATION, Volume I: Prior to 1900 / *Donald R. Coates*
2 RIVER MORPHOLOGY / *Stanley A. Schumm*
3 SPITS AND BARS / *Maurice L. Schwartz*
4 TEKTITES / *Vigil E. Barnes and Mildred A. Barnes*
5 GEOCHRONOLOGY: Radiometric Dating of Rocks and Minerals / *C. T. Harper*
6 SLOPE MORPHOLOGY / *Stanley A. Schumm and M. Paul Mosley*
7 MARINE EVAPORITES: Orgin, Diagenesis, and Geochemistry / *Douglas W. Kirkland and Robert Evans*
8 ENVIRONMENTAL GEOMORPHOLOGY AND LANDSCAPE CONSERVATION, Volume III: Non-Urban / *Donald R. Coates*
9 BARRIER ISLANDS / *Maurice L. Schwartz*
10 GLACIAL ISOSTASY / *John T. Andrews*
11 GEOCHEMISTRY OF GERMANIUM / *Jon N. Weber*
12 ENVIRONMENTAL GEOMORPHOLOGY AND LANDSCAPE CONSERVATION, Volume II: Urban Areas / *Donald R. Coates*
13 PHILOSOPHY OF GEOHISTORY: 1785–1970 / *Claude C. Albritton, Jr.*
14 GEOCHEMISTRY AND THE ORGIN OF LIFE / *Keith A Kvenvolden*
15 SEDIMENTARY ROCKS: Concepts and History / *Albert V. Carozzi*
16 GEOCHEMISTRY OF WATER / *Yasushi Kitano*
17 METAMORPHISM AND PLATE TECTONIC REGIMES / *W. G. Ernst*
18 GEOCHEMISTRY OF IRON / *Henry Lepp*
19 SUBDUCTION ZONE METAMORPHISM / *W. G. Ernst*
20 PLAYAS AND DRIED LAKES: Occurrence and Development / *James T. Neal*
21 GLACIAL DEPOSITS / *Richard P. Goldthwait*
22 PLANATION SURFACES: Peneplains, Pediplains, and Etchplains / *George F. Adams*
23 GEOCHEMISTRY OF BORON / *C. T. Walker*
24 SUBMARINE CANYONS AND DEEP-SEA FANS: Modern and Ancient / *J. H. McD. Whitaker*
25 ENVIRONMENTAL GEOLOGY / *Frederick Betz, Jr.*
26 LOESS: Lithology and Genesis / *Ian J. Smalley*

27 PERIGLACIAL PROCESSES / *Cuchlaine A. M. King*
28 LANDFORMS AND GEOMORPHOLOGY: Concepts and History / *Cuchlaine A. M. King*
29 METALLOGENY AND GLOBAL TECTONICS / *Wilfred Walker*
30 HOLOCENE TIDAL SEDIMENTATION / *George deVries Klein*
31 PALEOBIOGEOGRAPHY / *Charles A. Ross*
32 MECHANICS OF THRUST FAULTS AND DÉCOLLEMENT / *Barry Voight*
33 WEST INDIES ISLAND ARCS / *Peter H. Mattson*
34 CRYSTAL FORM AND STRUCTURE / *Cecil J. Schneer*
35 OCEANOGRAPHY: Concepts and History / *Margaret B. Deacon*
36 METEORITE CRATERS / *G. J. H. McCall*
37 STATISTICAL ANALYSIS IN GEOLOGY / *John M. Cubitt and Stephen Henley*
38 AIR PHOTOGRAPHY AND COASTAL PROBLEMS / *Mohamed T. El-Ashry*
39 BEACH PROCESSES AND COASTAL HYDRODYNAMICS / *John S. Fisher and Robert Dolan*
40 DIAGENESIS OF DEEP-SEA BIOGENIC SEDIMENTS / *Gerrit J. van der Lingen*
41 DRAINAGE BASIN MORPHOLOGY / *Stanley A. Schumm*
42 COASTAL SEDIMENTATION / *Donald J. P. Swift and Harold D. Palmer*
43 ANCIENT CONTINENTAL DEPOSITS / *Franklyn B. Van Houten*
44 MINERAL DEPOSITS, CONTINENTAL DRIFT AND PLATE TECTONICS / *J. B. Wright*
45 SEA WATER: Cycles of the Major Elements / *James I. Drever*
46 PALYNOLOGY, PART I: Spores and Pollen / *Marjorie D. Muir and William A. S. Sarjeant*
47 PALYNOLOGY, PART II: Dinoflagellates, Acritarchs, and Other Microfossils / *Marjorie D. Muir and William A. S. Sarjeant*
48 GEOLOGY OF THE PLANET MARS / *Vivien Gornitz*
49 GEOCHEMISTRY OF BISMUTH / *Ernest E. Angino and David T. Long*
50 ASTROBLEMES—CRYPTOEXPLOSION STRUCTURES / *G. J. H. McCall*
51 NORTH AMERICAN GEOLOGY: Early Writings / *Robert Hazen*
52 GEOCHEMISTRY OF ORGANIC MOLECULES / *Keith A. Kvenvolden*
53 TETHYS: The Ancestral Mediterranean / *Peter Sonnenfeld*
54 MAGNETIC STRATIGRAPHY OF SEDIMENTS / *James P. Kennett*
55 CATASTROPHIC FLOODING: The Origin of the Channeled Scabland / *Victor R. Baker*

Benchmark Papers in Geology/55

A BENCHMARK® Books Series

CATASTROPHIC FLOODING
The Origin of the Channeled Scabland

Edited by

VICTOR R. BAKER
The University of Texas at Austin

Dowden, Hutchinson & Ross, Inc.
STROUDSBURG, PENNSYLVANIA

Benchmark Papers in Geology, Volume 55
Library of Congress Catalog Card Number: 79-9071
ISBN: 0-87933-360-X

83 82 81 1 2 3 4 5
Manufactured in the United States of America.

LIBRARY OF CONGRESS CATALOGING IN PUBLICATION DATA
Main entry under title:
Catastrophic flooding.
(Benchmark papers in geology; v. 55)
Includes bibliographical references and indexes.
1. Erosion—Columbia Plateau—Addresses, essays, lectures. 2. Geology—Columbia Plateau—Addresses, essays, lectures. 3. Bretz, J Harlen, 1882– I. Baker, Victor R. II. Title: Channeled Scabland.
QE581.C36 551.3'5 79-9071
ISBN 0-87933-360-X

Distributed world wide by Academic Press,
a subsidiary of Harcourt Brace Jovanovich,
Publishers.

SERIES EDITOR'S FOREWORD

The philosophy behind the Benchmark Papers in Geology is one of collection, sifting, and rediffusion. Scientific literature today is so vast, so dispersed, and, in the case of old papers, so inaccessible for readers not in the immediate neighborhood of major libraries that much valuable information has been ignored by default. It has become just so difficult, or so time consuming, to search out the key papers in any basic area of research that one can hardly blame a busy person for skimping on some of his or her "homework."

This series of volumes has been devised, therefore, as a practical solution to this critical problem. The geologist, perhaps even more than any other scientist, often suffers from twin difficulties—isolation from central library resources and immensely diffused sources of material. New colleges and industrial libraries simply cannot afford to purchase complete runs of all the world's earth science literature. Specialists simply cannot locate reprints or copies of all their principal reference materials. So it is that we are now making a concerted effort to gather into single volumes the critical materials needed to reconstruct the background of any and every major topic of our discipline.

We are interpreting "geology" in its broadest sense: the fundamental science of the planet Earth, its materials, its history, and its dynamics. Because of training in "earthy" materials, we also take in astrogeology, the corresponding aspect of the planetary sciences. Besides the classical core disciplines such as mineralogy, petrology, structure, geomorphology, paleontology, and stratigraphy, we embrace the newer fields of geophysics and geochemistry, applied also to oceanography, geochronology, and paleoecology. We recognize the work of the mining geologists, the petroleum geologists, the hydrologists, and the engineering and environmental geologists. Each specialist needs a working library. We are endeavoring to make the task of compiling such a library a little easier.

Each volume in the series contains an introduction prepared by a specialist (the volume editor)—a "state of the art" opening or a summary of the object and content of the volume. The articles, usually some twenty to fifty reproduced either in their entirety or in significant extracts, are selected in an attempt to cover the field, from the key papers of the last century to fairly recent work. Where the original works are in foreign languages, we have endeavored to locate or commission translations. Geologists, because of their global subject, are often acutely aware of the oneness of our world. The selections cannot therefore be restricted to any one country, and whenever possible an attempt is made to scan the world literature.

To each article, or group of kindred articles, some sort of "highlight commentary" is usually supplied by the volume editor. This commentary should serve to bring that article into historical perspective and to emphasize its particular role in the growth of the field. References, or citations, wherever possible, will be reproduced in their entirety—for by this means the observant reader can assess the background material available to that particular author, or, if desired, he or she too can double check the earlier sources.

A "benchmark," in surveyor's terminology, is an established point on the ground that is recorded on our maps. It is usually anything that is a vantage point, from a modest hill to a mountain peak. From the historical viewpoint, these benchmarks are the bricks of our scientific edifice.

RHODES W. FAIRBRIDGE

PREFACE

In a brilliant series of papers written between 1923 and 1932, J Harlen Bretz shocked the geological community with his studies of an enormous plexus of proglacial channels eroded into the loess and basalt of the Columbia Plateau, eastern Washington. This region, which he named the "Channeled Scabland," contained erosional and depositional features that were unique among fluvial phenomena. With painstaking fieldwork, before the advent of aerial photography and modern topographic mapping, Bretz documented the field relationships of the region. He argued that the landforms could be explained only as the product of a relatively brief but enormous flood, which he called the "Spokane Flood." Considering the nature and vehemence of the opposition to this "outrageous hypothesis," the eventual triumph of that idea constitutes one of the most fascinating episodes in the history of modern geology.

Simultaneously, the Spokane Flood hypothesis established a conflict between two important cornerstones of geological philosophy: (1) the triumph of the glacial theory over diluvian myth, and (2) the scientific tolerance of outrageous hypotheses. The recounting of the Spokane Flood controversy is thus something more than a discourse on the topic of catastrophic flooding. The papers by Bretz and others represent a marvelous exposition of the scientific method in action. One cannot but be amazed at the spectacle of otherwise objective scientists twisting hypotheses to give a uniformitarian explanation to the Channeled Scabland. Undoubtedly these men thought they were upholding the very framework of geology as it had been established in the writings of Hutton, Lyell, and Agassiz. The irony may be that Bretz's critics never really appreciated the scientific implications of Agassiz's famous dictum, "Study nature, not books." Perhaps no geologist has understood and lived the spirit of those words more enthusiastically than J Harlen Bretz.

The most amazing twist to the studies of the Channeled Scabland is the most recent. As the Viking spacecraft were orbiting Mars in the summer of 1976, their cameras were trained on the great Martian channel systems. The photographs revealed uplands streamlined by fluid flow, eroded scabland on the channel floor, and many other features that we now know to be diagnostic of bedrock erosion by catastrophic flooding. Fifty years after J Harlen Bretz's theory of scabland erosion on the Columbia Plateau was denounced at an infamous meeting of the Washington Academy of Science, astrogeologists were using Bretz's well-documented studies of the Channeled Scabland as the major earth-analog to Martian channel erosion. The origin of the great Martian outflow channels is a new controversy of which Bretz is not a part. Nevertheless, the fact remains that few geological concepts have continued to have such relevance to modern geology as Bretz's theory of catastrophic flood origin of the Channeled Scabland, born amid bitter controversy over a half century ago.

This book will use the original papers to recount the development, resolution, and implications of the controversy over Bretz's catastrophic flood hypothesis. Because the personalities of the participants form a fascinating part of the story, the editor's comments for Papers 1–11 will integrate them with the development of ideas. Papers 12–15 will present some of the modern research that derives from Bretz's inspiration.

This book would not have been possible without the aid of Dr. J Harlen Bretz. Through his painstaking fieldwork, scientific insight, and firm conviction, Dr. Bretz proved the catastrophic flood origin of the Channeled Scabland. I thank him for sharing his personal recollections on the Spokane Flood controversy and for his encouragement of my own studies of the Channeled Scabland.

VICTOR R. BAKER

CONTENTS

CONTENTS BY AUTHOR

CATASTROPHIC FLOODING

INTRODUCTION

In one sense this book represents the narrowest field of inquiry addressed thus far by the Benchmark series. The concern is with geologic events that encompass a small fraction of Pleistocene time in a portion of the northwestern United States. Indeed, in one of its earlier editions, the standard textbook on Quaternary geology (Flint 1957) did not even mention the subject of this Benchmark volume. The updated version of that textbook devotes but one sentence to the flood origin of the Channeled Scabland (Flint 1971).

On the other hand, the Channeled Scabland and its origin exemplify the very essence of what is implied by the word "catastrophe." Many geologists would probably accept the following American Geological Institute (1972) definition: "*catastrophe*—A sudden, violent disturbance of nature, ascribed to exceptional or supernatural causes, affecting the physical conditions and the inhabitants of the Earth's surface; e.g. the Noachian flood, or an extinction of an entire fauna." The definition is unfortunate. The phrase "exceptional or supernatural causes" implies that the study of catastrophes is something beyond the bounds of science. This implication explains part of the initial reaction by the scientific community to Bretz's flood hypothesis for the Channeled Scabland. The debates over that hypothesis were not always marked by scientific objectivity, but their recounting here provides a fascinating example of the triumph of an outrageous hypothesis (Baker 1978). Only in the last two decades has the flood hypothesis gained general acceptance. It is a measure of scientific maturity that, in current studies of the Channeled Scabland, "the idea, but not the man, has become central" (Olson 1969).

CATASTROPHE AND REVOLUTION

Geology textbooks will tell us that landforms result from forces acting on resistant rock and regolith at the interface between the lithosphere and the atmosphere/hydrosphere. Students of landforms have long been impressed with the capability of rivers to impose forces on the earth's surface. The power of Niagara, Victoria Falls, the Nile cataracts have all inspired scientific reflection. However, most geomorphologists today emphasize the observation that rivers do their work very slowly. Valley walls are weakened by the prolonged action of rainsplash, frost action, soil formation, and other processes. Rivers are mainly thought of as transport agents that remove debris during their occasional floods. The valley-cutting process generally takes place over millions of years. We are told that fluvial landform development, like science itself, appears to proceed progressively and with order.

Scientists like to view their activity as an orderly search for truth. The concept portrayed in scientific textbooks is that knowledge is gained by a steady, cumulative process. One philosophical view, however, holds that science is characterized by long periods of conformity to a commonly held theoretical framework (Kuhn 1962). This framework, which Kuhn calls the "paradigm," is the tradition in which new investigations proceed. The science that occurs during paradigm-governed periods is "normal science" and mainly consists of solving puzzles generated by the paradigm. It is "normal science" that appears in textbooks.

A corollary to Kuhn's paradigm concept is that the new hypotheses generated to explain phenomena in a particular scientific field will be most often conditioned by the paradigm. In essence, a set of unstated "ground rules" dictates the manner in which science will identify puzzles for its study. These rules will also act to exclude many possible explanations for natural phenomena if such explanations lie outside the bounds of the accepted or "normal" science of the day.

Kuhn's thesis holds that normal science eventually turns up puzzles that cannot be solved within the existing theoretical framework. Such anomalies are ignored, and attempts may even be made to suppress their study in order to prevent interference with the rapid progress that appears to occur in normal science. However, these anomalies hold the key to a superior kind of scientific progress. When their importance can no longer be ignored, a scientific field may experience crisis, pronounced change, and eventually a completely new synthesis. This is the "scientific revolution."

The concept of scientific revolutions has been applied mainly to chemistry, biology, and physics. Some have suggested that a revolution in geology occurred recently with the advent of the plate-tectonic model. Another view is that the plate-tectonic concept arose progressively, even though its acceptance had some aspects of a Kuhnian revolution (Kitts 1974).

Geology has experienced one undeniable scientific revolution. During the eighteenth century various philosophers had reconciled the rock record with the concept of great cataclysms, such as the biblical flood. Within this framework of normal science, however, anomalies appeared that led to crisis. The resulting revolution produced the grand synthesis in nineteenth-century geology that Hubbert (1967) termed "Huttonian-Lyellian-Darwinian." Within this new paradigm geology has advanced rapidly as a science and has solved the new puzzles posed by the commonly held faith in progressive change and natural order.

Bretz's flood hypothesis was a decided crisis for the normal geological science of the 1920s. The idea was at first thought simply to require ignoring. When Bretz pressed his point, attempts were made to change his mind on the matter. Eventually several studies were made explicitly to disprove the hypothesis. When the anomaly was finally verified, it was introduced to the normal science of the textbooks in the role of a novelty. Geology textbooks are nevertheless still written according to the Huttonian-Lyellian-Darwinian synthesis.

CATASTROPHISM OR UNIFORMITARIANISM

The principle of uniformity is usually expressed through Archibald Geikie's maxim, "The present is the key to the past." However, the principle is also often taken to imply that geologic processes have operated at approximately uniform rates and that past processes were identical to those operating today. The many other ramifications of uniformitarianism are beyond the scope of this brief review.

A convincing argument can be made for an inherent ambiguity in the principle of uniformity. Gould (1965) stressed this point by distinguishing two types of uniformitarianism: (1) methodologic uniformitarianism—the principle that the basic laws of nature are invariant with time; and (2) substantive uniformitarianism—the statement that similar processes and rates prevailed in the geologic past as at present. Hubbert (1967, p. 29) points out that methodologic uniformitarianism is not merely a fundamental principle of geology: "Instead of being an assumption or an *ad hoc* hypothesis, it is simply a succinct summation of the totality of all experimental and observational evidence." In short, methodologic uniformitarianism is a direct consequence of the inductive method that can be traced back to Sir Francis Bacon.

Substantive uniformitarianism is a much more complex concept, and it requires an appraisal in historical terms. It is indeed a hypothesis, because of the concept, again stated by Hubbert (1967, p. 30): "History, human or geological, represents our hypothesis, couched in terms of past events, devised to explain our present-day observations." As a hypothesis, substantive uniformitarianism evolved as an alternative geological explanation to assumptions of special creations and interferences by Divine Providence.

The concept of a catastrophe is difficult to extricate from the above philosophical-historical background. Albritton (1967, p. 2) stated the problem as follows: "Those who oppose substantive uniformity but advocate the methodological variety are sometimes denounced as neocatastrophists. This accusation is based on the queer assumption that whoever is not a gradualist must of necessity be a Victorian catastrophist."

As will become clear in this volume, Bretz's catastrophic flood hypothesis was posed in the best tradition of inductive scientific investigation. His critics, who thought they were upholding the concept of uniformitarianism, were actually violating methodological uniformity to save the substantive variety. Geologists have since concluded that substantive uniformitarianism is an anachronism that was useful only in the theologically based debates of the last century but is unnecessary today (Hubbert 1967). Methodological uniformity remains an essential concept. However, it is really no more than the well-known principle that among competing hypotheses, the simpler hypothesis will generally prevail. This is the famous principle of scientific parsimony, or Occam's razor. Its application to his scabland research is well stated by Bretz (written communication, 1978): "I have always used Chamberlin's method of multiple working hypotheses. I applied Occam's razor to select the most appropriate hypothesis, but always with due regard for possible dull places in the tool."

The first expressions of substantive uniformitarianism are generally ascribed to James Hutton. Hutton developed his ideas over several decades, but their major publication came in 1795. Most appropriate to the present discussion is Hutton's (1795, vol.

2, pp. 418–432) argument that river valleys are produced by the rivers running in them. This idea was developed more fully by Playfair (1802, pp. 102–115), who categorically denied the role of debacles in creating complex valley systems.

The "fluvialist" school of Hutton and Playfair contrasted with the general adherence to diluvialist concepts by British geomorphologists of the early nineteenth century. Many of these catastrophists were not motivated merely by religio-scientific dogma. They pointed to landscape features that could not be explained by Hutton's fluvialism: erratic boulders, alpine lakes, hanging valleys, cross-axial drainage, dry valleys, misfit streams, pervasive modern aggradation, waterfalls, and rapids. For such geologists as William Buckland, Adam Sedgwick, William Conybeare, and Sir James Hall, these anomalies seemed better explained by one or more debacles than by a gradual and progressive fluvialism.

The diluvialist school was terminated by the glacial theory of Louis Agassiz. During his British tour of 1840, Agassiz demonstrated that the features cited as incompatible with fluvialist doctrine were produced by the phenomenon of glaciation. In a sense, glaciation was a catastrophe, but it constituted an anomaly that could not be denied. Glaciation was embraced by the fluvialist school. Lyell's famous textbook (Lyell 1830) had already made the case for our planet's orderly sequence of development through time, and orderly development became the paradigm of modern geology.

In his history of British geomorphology, Davies (1969) clearly shows that Hutton's fluvialist doctrine was not his unique discovery, nor was it even adequately justified by field evidence in the glacially modified terrains to which it was often applied. In essence Hutton was correct, but for the wrong reasons. His much-maligned critics, the nineteenth-century advocates of diluvianism, were wrong, but their scientific methology was, in many cases, quite proper.

Had the fluvial-diluvial debates of the early nineteenth century occurred in the setting of the semiarid American West, a controversy would not have arisen. Fluvialist principles govern the landscape. In regard to the Henry Mountains of Utah, G. K. Gilbert (1877) was able to formulate the fluvial "laws" of slope, structure, and divides. The landscape was easily explicable in terms of physical theory.

Had the fluvial-diluvial debates occurred in the Channeled Scabland, geology today could well be a very different science. The landform relationships in that region indicate a system in which the existing order was overthrown. The Channeled Scabland emphatically did not form by a continuous progression of processes that were similar to those we observe in action today.

CATASTROPHE AND ORDER

Webster's dictionary cites one definition of catastrophe as follows: "any event that disturbs or overthrows the existing order of things." This definition points to the heart of the problem of philosophically resolving catastrophism and uniformitarianism. The order of nature has been confused with the principle of uniformitarianism. What was not perceived by Bretz's critics is that catastrophes may be a part of a process that maintains a larger order.

My colleague at The University of Texas at Austin, Ilya Prigogine (1977) tells the following story:

> In the Museum of Vienna, at the beginning of the 19th century, there was a marvelous collection of meteorites. And then a new curator appeared, and he said meteorites are obviously the product of superstition; meteorites do not exist, because there is no place for them to exist in the Newtonian view of the planetary system. And therefore he [threw] them all out. Then the French Academy was asked to give its opinion. Some meteorites . . . [had fallen] not very far from Paris. They sent a few people there—very famous people like Lavoisier—and they came back with the message: meteorites do not exist. They are artifacts due to, perhaps, old industry or iron, or I don't know what. But then it happened that there was a little later a real, I would say a real *rain* of meteorites, smashing . . . the windows of the French Academy. And at this point they could not avoid to think that, after all, there were meteorites.

Much of the current research in process geomorphology focuses on fluctuations that can be represented by mean values. There are two rationalizations for this emphasis. The first is operational; such fluctuations occur on short enough time scales that they can be easily measured. The second is theoretical; relatively more work in modifying landscapes appears to be accomplished by the many geomorphic events of moderate frequency than by rare catastrophic events, such as large floods (Wolman and Miller 1960). In essence, the Wolman-Miller principle derives from the so-called law of large numbers: in systems with many degrees of freedom, fluctuations occur, but their magnitudes are very small. The fluvial system returns to an equilibrium state because many factors usually act to damp out large fluctuations. Order in river morphology appears to derive from mean conditions. Fluctuations serve merely as corrections to the average trend.

Were the catastrophic floods of the Channeled Scabland isolated events, no more worthy of scrutiny than the thousands of flood peaks that we have observed on thousands of streams that dissect our plantet? Most floods do very little to the landscape. Like most human ideas, they have little influence on the larger system of which they are a part. Nevertheless, a small number of ideas may lead to revolution and to a new order. Biological evolution involves mutations, most of which have no effect on the ecosystem. Given sufficient time, however, a mutant may appear that will have a profound effect and will change the entire operation of the ecosystem. The sequence of flood events is no different. A finite possibility exists that fluvial systems may react in the same manner when perturbed by a phenomenal excess of input that exceeds a certain threshold limit. The usual factors for damping out such a fluctuation are overwhelmed, and the system enters a high-energy state of disequilibrium.

How does one study systems that have been driven far from equilibrium? Does disequilibrium simply lead to chaos? Such questions are fundamental in thermodynamics. Prigogine (1977) offers the following approach:

> Instead of . . . looking [in physics for] simplicity, you can now also go the other way, and try to find the laws of complexity.
>
> How can order be generated out of chaos? . . . The usual theory was to calculate the average number of collisions . . . which would produce new species . . . but these new structures arise from violations of the laws of large numbers. This breakdown of the average . . . leads to the new structures.

The new structures that develop from nonequilibrium are called "dissipative structures" (Prigogine 1978). They are giant fluctuations stabilized by exchanges of mass and energy. The Channeled Scabland is a landscape that was generated by an immense fluctuation in fluvial system behavior. As will be seen in this volume, the Channeled Scabland comprises an orderly assemblage of landforms consistent with the high-energy state of its formation.

REFERENCES

Albritton, C. C. 1967. Uniformity, the ambiguous principle. In Uniformity and simplicity, Albritton, C. C., ed., *Geol. Soc. America Spec. Paper* **89**:1–2.

American Geological Institute. 1972. *Glossary of Geology*. Washington, D. C.: American Geological Inst., 805 pp.

Baker, V. R. 1978. The Spokane Flood controversy and the martian outflow channels. *Science* **202**:1249–1256.

Davies, G. L. 1969. *Earth in Decay: A History of British Geomorphology, 1578–1878*. New York: Amer. Elsevier Publ. Co., 390 pp.

Flint, R. F. 1957. *Glacial and Pleistocene Geology*. New York: John Wiley and Sons, 553 pp.

Flint, R. F. 1971. *Glacial and Quaternary Geology*. New York: John Wiley and Sons, 892 pp.

Gilbert, G. K. 1877. *Geology of the Henry Mountains*. U. S. Geog. Geol. Survey of the Rocky Mountain Region, 160 pp.

Gould, S. J. 1965. Is uniformitarianism necessary? *Am. Jour. Sci.* **263**:223–228.

Hubbert, M. K. 1967. Critique of the principle of uniformity. In Uniformity and simplicity, Albritton, C. C., ed., *Geol. Soc. America Spec. Paper* **89**:3–33.

Hutton, James. 1795. *Theory of the Earth*, Vol. 2. Edinburgh Cadwell, Junior, and Davies, 568 pp.

Kitts, D. B. 1974. Continental drift and scientific revolution. *Am. Assoc. Petroleum, Geologists Bull.* **58**:2490–2496.

Kuhn, T. S. 1962. *The Structure of Scientific Revolutions*. Chicago: University of Chicago Press, 172 pp.

Lyell, Charles. 1830. *Principles of Geology, Being an Attempt to Explain the Former Changes of the Earth's Surface, by Reference to Causes Now in Operation*, vol. 1. London: John Murray, 511 pp.

Olson, E. C. 1969. Introduction to J Harlen Bretz's paper, The Lake Missoula floods and the Channeled Scabland. *Jour. Geology* **77**:503–504.

Playfair, John. 1802. *Illustrations of the Huttonian Theory of the Earth*. Edinburgh: Wm. Creech, 528 pp.

Prigogine, Ilya. 1977. Order out of chaos. Public lecture, Nov. 18, 1977 (printed in the *Texas Times*, Univ. Texas System, Dec. 1977, pp. 6–8).

Prigogine, Ilya. 1978. Time, structure, and fluctuations. *Science* **201**:777–785.

Wolman, M. G., and Miller, J. P. 1960. Magnitude and frequency of forces in geomorphic processes. *Jour. Geology* **68**:54–74.

Part I

AN OUTRAGEOUS HYPOTHESIS

Editor's Comments on Papers 1, 2, and 3

1 **BRETZ**
Introduction

2 **BRETZ**
The Channeled Scablands of the Columbia Plateau

3 **BRETZ**
Excerpts from *The Spokane Flood Beyond the Channeled Scablands*

The term "channeled scabland" was introduced by Bretz (1923b) to describe a spectacular type of erosional and depositional complex best developed in the loess and basalt of the Columbia Plateau in eastern Washington. The formal physiographic region named the "Channeled Scabland" comprises the northern portion of the plateau, whose surface gently dips southwestward from the high cliffs facing the Spokane and Columbia rivers to the north. Lying in the rain shadow of the Cascade Mountains, this land of loess-mantled uplands and jagged rock channelways has intrigued all who have studied it scientifically.

The unique character of the dry river courses ("coulees") of the Channeled Scabland was appreciated by the first scientific observers of the region. Rev. Samuel Parker (1838) provided the first published statement on the Grand Coulee: "[It] was indubitably the former channel of the river [Columbia]." Lieutenant T. W. Symons (1882) of the U. S. Army traversed the Grand Coulee; he "went north through the coulee, its perpendicular walls forming a vista like some grand old ruined roofless hall, down which we travelled hour after hour." Symons (1882) initiated the widely held notion that during glacial episodes of the Pleistocene the Columbia had simply been diverted across the Columbia Plateau (Symons 1882). Variations on this general theme were standard in the early literature (Russell 1893; Dawson 1898; Salisbury 1901; Calkins 1905).

The Grand Coulee gained international fame in 1912, when it was traversed by the American Geographical Society's Transcontinental Excursion. Karl Oestreich (1915) of the University of Utrecht described the coulee as "eines mächtigen Flusses Bett . . . ohne jede Spur von Zerfall der frischen Form." He provided an excellent description of significant features that must have had a special origin: exhumed granite hills, perpendicular walls, and the hanging valleys marginal to the upper coulee. He ascribed these hanging valleys to glacial erosion and to deepening of the coulee by the glacial Columbia River. Moreover, he recognized that the upper Grand Coulee was carved through a preglacial divide, which he correctly located just north of Coulee City.

Another foreign observer on the American Geographical Society excursion was H. Baulig, Université de Rennes. Baulig (1913) described the loess, coulees, dry falls ("cataracte dessechéé de la Columbia"), rock basins, and plunge pools. The origin of these features was ascribed to a glacial diversion of the Columbia. Nevertheless, he marveled at the scale of erosion (Baulig 1913, p. 159): "peut-être unique du relief terrestre,—unique par ses dimensions, sinon par son origine."

Dr. O. E. Meinzer, the eminent hydrologist of the U. S. Geological Survey, took an early interest in the western part of the Channeled Scabland. He observed (1918) that the glacially diverted Columbia at Grande Coulee "cut precipitous gorges several hundred feet deep, developed three cataracts, at least one of which was higher than Niagara . . . and performed an almost incredible amount of work in carrying boulders many miles and gouging out holes as much as two hundred feet deep." He implied that the great erosion occurred because the Columbia River was diverted across the steeply dipping basalt surface of the northern Columbia Plateau.

J HARLEN BRETZ

Scientific study of the Channeled Scabland did not begin in earnest until the studies of J Harlen Bretz. Bretz (Papers 1, 2, and 3) interpreted the erosional and depositional features of the region as the product of a brief but enormous flood, which he called the "Spokane Flood." For geology in the 1920s this was clearly an outrageous hypothesis. Olson (1969, p. 503) has described the reception of the idea:

> During its not always calm history, the story of the development of the Channeled Scabland was thought by some to have brushed beyond the dividing line in flaunting catastrophe too

> vividly in the face of the uniformity that had lent scientific dignity to interpretation of the history of the earth.

The reaction of the scientific community to Bretz's hypothesis was simply stated: "This heresy must be gently but firmly stamped out." (Bretz et al. 1956, p. 961).

J Harlen Bretz was born on September 2, 1882, in Saranac, Michigan. He attended Albion College, where he studied biology and met classmate Fanny Challis, whom he married in 1906. As a high school teacher in Flint, Michigan and later in Seattle, Washington, he developed an avid interest in geology. His spare time was consumed by studies of glacial geology in the Puget Sound region. During their first four years in Seattle, he and Fanny explored much of that area on foot and kept careful field notes on the Pleistocene geology. Bretz supplemented his field observations with frequent visits to the Seattle Public Library and the University of Washington Library. He also made many visits to the Department of Geology at the university.

Bretz (personal communication, 1978) recalls looking at the newly published Quincy topographic sheet of the U. S. Geological Survey on one of his university visits. The map clearly showed the great Potholes cataract of the western Quincy Basin. He recalls his curiosity about the origin of this immense waterfall, with its lip now perched over 200 m above the modern Columbia River.

With his savings and the aid of a fellowship, Bretz entered the University of Chicago Department of Geology as a graduate student. His extensive field notes from Washington served as the basis for his dissertation. The entire thesis was actually in press (Bretz 1913) when he submitted it to his professors, T. C. Chamberlin and R. D. Salisbury. Bretz received his Ph.D. summa cum laude in 1913.

Bretz's three years in graduate school were followed by an appointment as assistant professor of geology at the University of Washington (1913–14). During this short appointment Bretz made one especially noteworthy trip to the great gorge of the Columbia River across the Cascade Range. The trip impressed him with evidence for the immense quantities of water necessary to float ice and thereby distribute erratics in this region. Unfortunately, however, some of Bretz's colleagues at the University of Washington did not share his zeal for fieldwork, and he accepted an invitation from R. D. Salisbury to return to the University of Chicago. Nevertheless, the geology that he had glimpsed in the Columbia Gorge continued to intrigue him. Within two years after coming to the University of Chicago, Bretz organized a summer field course centered in the Columbia Gorge. He repeated this course in three subsequent summers, each time expanding his knowledge about Pleistocene events in the northwestern United States.

Bretz (1919) published a preliminary discussion of the extensively distributed berg-drifted boulders of the lower Columbia drainage. He reasoned correctly that the composition of these boulders required derivation from outcrops in east-central Washington, north of the Channeled Scabland. However, the boulders were distributed in obviously Pleistocene contexts throughout the lower Columbia River system and far from the Pleistocene glaciers that occurred north of the Channeled Scabland. Bretz tried to explain this enigma by invoking the concept of a late Pleistocene submergence. In doing so he was following a prevailing theory about the existence of extensive inundation in the Willamette Valley of Oregon. The idea had been popularized by Thomas Condon, who described the setting as follows (Condon 1902, p. 62): ". . . that fine old Williamette sound . . . may in the days of the Mammoth and the Broad Faced Ox have welcomed to its scores of sheltered harbors, the ancient hunter who, in his canoe, if he had one, floated one hundred feet or more above the present altitude of the church spires of Portland and Salem."

Bretz (1919) realized that the inundation had to have been fresh water and not marine. He reasoned that the immense discharge of the Columbia maintained fresh-water conditions during a period of late Pleistocene submergence induced by diastrophic movements. However, he was clearly troubled by this explanation and noted several disturbing qualities: (1) The duration of ponding had been so brief that it did not create shore features or lacustrine plains. Only the berg deposits recorded its extent. (2) Bretz could not find evidence of similar submergence in southwestern Washington, where he would have expected the unknown diastrophic origin to have also operated. (3) Finally, he was dissatisfied with the postulated origin of the submergence. Other world examples of late Pleistocene submergence involved the invasion of marine waters into terrains from which a thick glacial load had recently been removed. In the Columbia Gorge no such glacial load ever existed, yet the region was clearly invaded by immense volumes of fresh water during the late Pleistocene.

THE FLOOD HYPOTHESIS

In the summer of 1922, Bretz made the decision to take his small field parties of advanced students into the Channeled Scabland. His scabland studies continued over the next seven field seasons. During those years Bretz traversed the entire region first on foot and later in his trusty Dodge 4, an early enclosed-body car. He did this with parties of students, his wife, son, daughter, and collie dog. The results of the research of those years constitute one of the most remarkable achievements in modern geology. Aaron Waters (written communication, 1978) has stated the accomplishment as follows:

> With Landsat photographs and detailed aerial photos it is easy for us now to see the nature of the scablands . . . the giant ripples and other small-scale features "prove the case." But Bretz did all this with the aid of only an agile and well-disciplined mind, and a pair of legs that were pushed over thousands of weary miles.

Bretz (personal communication, 1978) relates one field experience that reveals the rigor of his explorations. Accompanied by three students, he set out to view the dune field that blocks Crab Creek to form Moses Lake in the Quincy Basin. At nightfall the party, expecting an easy journey across the sandy floor of the Quincy Basin, continued southward toward Othello (no topographic maps existed at the time). Bretz describes what happened:

> Darkness found us well past the black sand dunes that dam back the lake, but open country still had enough night sky light to travel by. Then we encountered some ragged rocky outcrops, around which Crab Creek became almost lost. Drumheller Ranch was somewhere among the craggy black basalt knobs, blades, pinnacles, rock basins and dry channels, but we missed it in the darkness.
>
> However, I knew that we were crossing the tract where Frenchman Springs anticline pitched its crest downward, for all this wild rocky country we were encountering was made by that part of the Spokane Flood that had surged southward from Grand Coulee.
>
> Huge volume and adequate gradient for the escaping water had demanded the excessive erosion of the nose of Frenchman Springs anticline. Why not a single erosional gash, why all this butte-and-basin labyrinthine complex? I learned in later scabland exploration that the crests of anticlinal bulges were the most fractured and most vulnerable to the huge-volume, high-gradient flood discharges across the Columbia basalt plateau.
>
> Well, we stumbled, climbed, descended in the darkness for an hour or more, heartily sick of having attempted the traverse at night. I have since seen a topographic survey and aerial views of Drumheller Channels and unhesitantly give this area the palm for complexity of all flood-made topography.

> Fairly close to midnight, we saw a single light a mile or so ahead and found easier going. We were escaping that tangle of short channels, rock basins, basalt buttes; that is, getting across the backbone of the Frenchman Springs anticline. Of course, we aimed for the light. It proved to be a farm house. Reaching it, we saw through the unshaded windows a scantily clad woman fussing over two young children bedded down on the floor of bare boards. This wouldn't do, this Peeping Tom behavior. We shortly found the barn and a wagon loaded with newly threshed wheat. Sleeping on the wheat was the farmer. We awakened him, introduced ourselves and craved a refuge for the remainder of the night. We could sleep on just-threshed wheat with him.
>
> The next morning he invited the four of us to breakfast. All that his daughter could provide was a carton of breakfast flakes. Glimpses of the interior of the house showed a table and a few straight chairs. Probably a bed or so that we didn't see. Otherwise, that house showed the direst rural poverty.
>
> We rode to Warden and a R.R. elevator. Enroute, the farmer told a bit of his misfortunes and stated firmly that, once he "got a crop," he was abandoning the farm. Years later, I saw his "farm" again. The house and barn were gone, the field was gone, the site was now an active gravel pit. Yes, he had been trying to make a scabland gravel *bar* into a farm.

Bretz's first paper on the Channeled Scabland was the text of an oral presentation to the Geological Society of America (Bretz 1923a). In that paper he took special care not to call upon cataclysmic orgins. The paper provided a detailed description of physiographic relationships in the region. He did note, however, that the indicated channel erosion required prodigious quantities of water. Referring to the three outlets at the south end of the Hartline Basin (Dry Coulee, Lenore Canyon, and Long Lake Canyon), Bretz (1923a, pp. 593–594) stated, " . . . these are truly distributary canyons. They mark a distributive or braided course of the Spokane glacial flood over a basalt surface which possessed no adequate pre-Spokane valleys."

The idea of a truly catastrophic flood appeared in Bretz's second scabland paper (Bretz 1923b). This paper is reprinted here as Paper 2. Note especially that Bretz's interpretation of the mounded scabland gravel deposits as subfluvial bars led directly to the requirement for great water depths. This paper also included the first detailed geomorphic map of the entire Channeled Scabland, showing the overall anastomosing pattern assumed by a great flood of water.

In a subsequent paper Bretz (1924) argued that the rugged

scabland of anastomosing channels and rock basins cut into the basalt was the product of subfluvial quarrying. He described this process for the modern Columbia River near The Dalles, Oregon (Bretz 1924). Moreover, he asserted that only large, vigorous streams could produce such forms. The eventual conclusion from these varying lines of evidence was that so much glacial meltwater occupied the preexisting valleys on the Columbia Plateau that it must have constituted a vast but short-lived flood, the Spokane Flood (Bretz 1925, p. 98). The flood spilled across preglacial stream divides, eroding the maturely dissected loess topography to form linear channels and leaving a legacy of scoured loess scarps, hanging distributary valleys, and high-level fluvial deposits. The flood also built the huge constructional bars of gravel and then subsided so quickly that these bedforms were left almost unmodified by the subsiding water (Bretz 1925, p. 105).

Bretz's 1925 paper is here reprinted in part as Paper 3. Note that his earlier Columbia Gorge studies allowed him to trace the path of the great flood downstream through the Columbia Gorge to its debouchure into the Willamette Lowland, where it built the "Portland Delta." Bretz (1925) even made the first estimate of the flood discharge. The phenomenal quantities of water required by his hypothesis led Bretz to speculate about the origin of the Spokane Flood. He could think of only two possible explanations: (1) a very rapid and short-lived climatic amelioration, and (2) a gigantic jokullhlaup produced by volcanic activity beneath an ice cap. He noted severe objections to either hypothesis but held that the great flood had occurred, despite the problems in accounting for its source.

REFERENCES

Baulig, H. 1913. Le plateaux de lave du Washington central et la Grand 'Coulee.' *Annales Geographie* **22**:149–159.

Bretz, J H. 1913. Glaciation of the Puget Sound region. Washington Div. *Mines and Geology Bull,* **8**:244 pp.

Bretz, J H. 1919. The late Pleistocene submergence in the Columbia Valley of Oregon and Washington. *Jour. Geology* **27**:489–505.

Bretz, J H. 1923a. Glacial drainage on the Columbia Plateau. *Geol. Soc. America Bull.* **34**:573–608.

Bretz, J H. 1923b. The Channeled Scabland of the Columbia Plateau. *Jour. Geology* **31**:617–649.

Bretz, J H. 1924. The Dalles type of river channel. *Jour. Geology* **32**;139–149.

Bretz, J H. 1925. The Spokane Flood beyond the Channeled Scabland. *Jour. Geology* **33**:97–115, 236–259.

Bretz, J H., Smith, H. T. U., and Neff, G. E. 1956. Channeled Scabland of Washington; new data and interpretations. *Geol. Soc. America Bull.* **67**:957–1049.

Calkins, F. C. 1905. Geology and water resources of a portion of east-central Washington. *U.S. Geol. Survey Water-Supply Paper 118,* 96 pp.

Condon, T. 1902. *The Two Islands and What Came of Them*. Portland, Ore.: The J. K. Gill Co., 211 pp.

Dawson, W. L. 1898. Glacial phenomena in Okanogan County, Washington. *Am. Geol.* **22**:203–217.

Meinzer, O. E. 1918. The glacial history of the Columbia River in the Big Bend region. *Washington Acad. Sci. Jour.* **8**:411–412.

Oestreich, K. 1915. Die Grand Coulee. American Geog. Soc. Memorial Volume of the Transcontinental Excursion of 1912. *Am. Geog. Soc. Spec. Pub.* **1**:259–273.

Olson, E. C. 1969. Introduction to J Harlen Bretz's paper, The Lake Missoula floods and the Channeled Scabland. *Jour. Geology* **77**:503–504.

Parker, S. 1838. *Journal of an Exploring Tour beyond the Rocky Mountains.* Auburn, N. Y.: privately published.

Russell, I. C. 1893. Geologic reconnaissance in central Washington. *U.S. Geol. Survey Bull. 108,* pp. 1–108.

Salisbury, R. D. 1901. Glacial work in the western mountians in 1901. *Jour. Geology* **9**:721–724.

Symons, T. W. 1882. Report of an examination of the upper Columbia River. U.S., Congress, Senate, 47th Cong., 1st sess., 1882, S. Doc. 186, 135 pp.

1

Reprinted from pages 1–2 of *The Channeled Scabland,* V. R. Baker and D. Nummedal, eds., NASA Office of Space Science, Planetary Geology Program, 1978, 186 pp.

Introduction

J HARLEN BRETZ
Professor Emeritus of Geology
The University of Chicago
Chicago, Illinois 60637

The publication in 1923 of a geomorphic study of "The Channeled Scabland of the Columbia Plateau" in southeastern Washington state launched a controversy that lasted for decades. A map accompanying the paper depicted a pattern of abandoned erosional waterways, many of them streamless canyons (coulees) with former cataract cliffs and plunge basins, potholes and deep rock basins, all eroded in the underlying basalt of the gently southwestward dipping slope of that part of the Columbia Plateau. The pattern of dry stream ways was described as at network, a plexus, an anastomosis; totally unlike any other drainage pattern on earth. A debacle was asked for, the volume of which filled existing normal stream valleys to overflowing. This great flood spilled over former divides, eroding their summits to complete the network. Associated with the enormously enlarged drainage ways in favorable places were similarly huge mounds of stream gravel which the writer called great river bars. Huge stream-rolled boulders occurred in these bars. The boulders were obviously plucked from the columnar basalt bedrock by the postulated high-velocity currents.

The term valley would no longer suffice. The abandoned rock-bound former waterways were called channels, and the entire composite was named "Channeled Scabland." The total area involved was 18 townships wide by 22 townships long (approximately 40,000 km^2).

The Cordilleran ice sheet had advanced to make contact with the heads of the dominant channel ways. Obviously the plexus of channels was in some way genetically related to the northern ice sheet. But where on our planet had glacial wastage produced a record of flooding and the total remaking of existing topography? The writer called for a catastrophic event.

Catastrophism had virtually vanished from geologic thinking when Hutton's concept of "the Present is the key to the Past" was accepted and Uniformitarianism was born. Was not this debacle that had been deduced from the Channeled Scabland simply a return, a retreat to catastrophism, to the dark ages of geology? It could not, it must not be tolerated.

This, the writer of the 1923 article learned when, in 1927, he was invited to lecture on his finding and thinkings before the Geological Society of Washington, D.C. an organization heavily manned by the staff of the United States Geological Survey. A discussion followed the lecture, and six elders spoke their prepared rebuttals. They demanded, in effect, a return to sanity and Uniformitarianism.

The upstart theorist was not upset nor silenced. Despite his knowledge that the country was full of other dissenters to his flood theory, he proceeded to publish more papers on his favorite topic, now named the Spokane Flood. He described other features of the afflicted plateau which he claimed were inexplicable without his flood of glacially derived meltwater. His apostacy would not be corrected as advised by the elders. The one-man rebellion was still alive.

1924 saw two new papers, 1925 saw one more and 1927, three more. One of the three contained the lecture and discussion already noted. Another paper in 1928 was the writer's reply to all alternative hypotheses thus far suggested. Also in 1928, he traced his flood down the Columbia as far as Portland, Oregon, adding a 200 square mile delta in the Willamette Valley.

By 1930, he had found a source for that immense discharge across the plateau. Clark Fork of the Columbia River, draining a large mountainous region of western Montana, had been dammed by the Cordilleran ice sheet at its traverse of the Idaho panhandle. This formed an immense glacial lake with an estimated volume of 500 cubic miles. The lake had been named some years before as Glacial Lake Missoula. The first geologist to describe the lake ironically was one of the six challenging elders in Washington in 1927 and the author of a short paper on problematical features, perhaps glacial in origin, in what came to be known as the Channeled Scabland.

If Lake Missoula had a properly located place for its ice dam and a clear route thence to the Channeled Scabland, then presto, we would have the big problem solved. Missoula's depth at the dam was known from its shorelines to have been 2000 feet, and there was a clear route to Spokane and the Scabland. A catastrophic failure of the dam would release 500 cubic miles of glacially derived water with adequate gradient to Spokane.

Late in the field study, a criterion of undeniable validity for the occurrence of a flood, or several such, came to light. Hidden largely by sagebrush were numerous occurrences of current ripple marks. They were discovered because the U.S. Bureau of Reclamation had taken aerial photographs of the area to be irrigated with Grand Coulee water. Then it became clear that some gravel surfaces, curiously humpy, were covered with giant current ripples. An investigator, standing between two humps, could not see over either one. Indeed, the size of these ripple ridges made them really small hills. Finally came the discovery of giant current ripples in parts of Lake Missoula where, in a catastrophic emptying, strong currents were formed. When investigated by a specialist in fluvial hydraulics and hydrodynamics there could be no question whatever of their origin, no other explanation for their rhythmic patterns than that of bedform development by amazingly deep, swift flood water.

Measurements of records for depths of water and gradients of water surfaces in channels with proper cross sections also yielded amazing figures for velocities. Plucking of channel bedrock to yield and transport huge boulders became understandable when this 1973 paleohydraulic and hydrodynamic study was made.

Back in 1927 geologic opinion still wavered. One man accepted flooding but did it with local iceberg dams in different channels. Another was a die-hard Uniformitarian who in 1938 still argued that no flood had occurred; that the erosional complex on the plateau was made by "leisurely streams no larger than the Snake in flood today." (But that was before the giant ripple marks were discovered.)

"Meticulously detailed" criticisms of the gross errors and assumptions in the 1938 non-flood interpretation appeared when three geologists spent much of the summer of 1952 (published in 1956) in studying chiefly the excavations made by the U.S. Bureau of Reclamation in diverting Columbia River water behind Grand Coulee dam. This water irrigates from a million acres of the Quincy Basin, an extensive fill made by scabland floods. There were repeated burstings of Lake Missoula dams and refillings when later advances of the ice front made other episodes in the lake's history. How many is not yet definitely determined, despite recent and continuing research on this problem.

When the Geological Society of America held its annual meeting in Seattle in November, 1977, and field trips were made to the flood-ravaged Columbia Plateau and valley, guidebooks made clear that the *catastrophic* history of plateau and master river valley is being pursued with enthusiasm.

2

Reprinted from *Jour. Geology* **31**:617–649 (1923)

THE CHANNELED SCABLANDS OF THE COLUMBIA PLATEAU

J HARLEN BRETZ
University of Chicago

[*Editor's Note:* Plate III showing the channeled scablands of the Columbia Plateau and their associated features could not be reproduced because of problems in reproduction and size reduction. Figure 5.2 in "Large-Scale Erosional and Depositional Features of the Channeled Scabland" (Paper 13) illustrates the overall pattern of the scabland channels discussed by Bretz.]

OUTLINE

DEFINITION OF "SCABLAND"

The terms "scabland" and "scabrock" are used in the Pacific Northwest to describe areas where denudation has removed or prevented the accumulation of a mantle of soil, and the underlying rock is exposed or covered largely with its own coarse, angular débris. The largest areas of scabland are on the Columbia Plateau in Washington, north of Snake River. These scablands have a history which is believed to be unique. The prevailing feature of their topography is indicated in the term here used: channeled scablands.[1] They are scored by thousands of channels eroded in the underlying rock. The plateau in Washington, north of Snake River, has a total area of about 12,750 square miles, of which at least 2,000 square miles is channeled scabland. The scabland is widely distributed over the region in linear tracts among maturely dissected hills which bear the loessial soil (wheat lands) of the plateau.

PHYSIOGRAPHIC RELATIONS OF THE CHANNELED SCABLANDS

The following features and relations of the scablands exist in all tracts. They must form the basis of any interpretation for the origin of channeled scabland. The map should be examined as this list is read.

1. Scabland tracts are developed invariably on or in the Columbia basalt formation.
2. Scabland tracts are invariably lower than the immediately adjacent soil-covered areas.
3. Scabland tracts are invariably elongate.
4. The elongation of scabland tracts is with the dip slope of the underlying basalt flows. There are eight known exceptions to this rule,[2] all minor affairs so far as length is concerned.
5. Scabland tracts, considered as units, invariably have continuous gradients.
6. Scabland tracts are invariably bounded by maturely eroded topography.

[1] An earlier paper on this subject was published by the writer in the *Bulletin of the Geological Society of America*, Vol. XXXIV (1923), pp. 573–608. The study on which it was based involved about a 1,000-mile traverse of the plateau. Since then, the writer has studied the plateau more thoroughly, having added more than 2,000 miles to the previous total traverse. Much more detailed information and several modifications of the earlier interpretations justify the appearance of a second paper on the subject. The accompanying map (Plate III) is based on a field examination of every scabland there indicated. In a few places the boundaries are inferred (dashed lines) but future work will hardly do more than make minor changes or additions.

[2] A part of Othello Channels, a part of Drumheller Channels, at Palisades and near Spencer in Moses Coulee, at Soap Lake and near Bacon in Grand Coulee, 6 miles south of Almira on Wilson Creek, and at Long Lake in Spring Coulee.

7. Scabland tracts are developed in pre-existing drainage lines of the mature topography.[1]

8. Scabland tracts are connected with each other.[2]

9. (*a*) The areas surrounded by scabland invariably have the dendritic drainage pattern, mature topography and loessial soil of the plateau. (*b*) They are almost invariably elongate with the scabland tracts. (*c*) They commonly have steep marginal slopes descending from 50 to 200 feet to the scabland. These slopes are almost invariably in loess. Slopes of 30° to 33° are not uncommon. They are much younger topographically than the slopes of the valleys among these mature hills.

10. Scabland tracts with steep gradient are narrow, while those with gentle gradient are wide.

11. The pattern of scabland tracts, where hills of the older topography are isolated in them, is anastomosing or "braided."

12. Scabland tracts invariably contain "channels." These are gorges or canyons or elongated basins eroded in the basalt. The channels are invariably elongate in parallelism with the tract as a whole and, in most cases, the channel pattern is anastomosing or braided.

13. (*a*) Scabland tracts invariably bear discontinuous deposits of basaltic stream gravel. (*b*) These deposits invariably contain a small proportion of pebbles and cobbles of rock foreign to the plateau. (*c*) These deposits invariably rest on an eroded, scabland surface of the basalt. (*d*) They commonly lie on the down-gradient side of eminences in the scabland.

14. (*a*) Scabland tracts invariably bear scattered bowlders of foreign rock. (*b*) The proportion of foreign débris, either the fragments in the gravel or the scattered bowlders, is invariably smaller with increasing distance down-gradient.

15. Scabland tracts are invariably without a mantle of residual soil.

16. Scabland tracts are traceable up-gradient to a narrow basalt plain bordering the south side of Spokane River in the northern

[1] There are many exceptions to this rule, occurring where scabland tracts cross pre-existing divides, but the total length of such is only a small fraction of the aggregate length of all scabland tracts.

[2] Moses Coulee is the only exception.

part of the plateau.[1] This basalt plain bears many glaciated erratic bowlders and some patches of till, but no channeled scabland, no mature topography, and no loessial soil.

17. Only where the minor valleys of the mature topography adjacent to this basalt plain open northward on to the plain are any glacial erratics found in them.[2]

18. There are but ten scabland openings to this basalt plain to the north.

19. Scabland tracts are invariably traceable down-gradient to Snake River on the south or to Columbia River on the west. There are nine places where scabland tracts enter these two streams Only three of them were drainage ways before the scablands were formed.

20. There is no channeled scabland on the plateau in western Idaho or south of Snake River or west of Columbia River.

21. Nowhere in the scablands or the maturely dissected country, during ten weeks of field study, has a till been found, or any deposit of doubtful genesis which could be interpreted more satisfactorily as till than as non-glacial in origin.[3]

GENERALIZED STATEMENT OF THE ORIGIN OF THE CHANNELED SCABLANDS

This unique combination of topographic features of the Columbia Plateau in Washington has only one interpretation consistent with all the foregoing items. The channeled scablands are the erosive record of large, high-gradient, glacier-born streams. The basalt plain records the southern limit reached by the ice sheet from which these streams took origin. Before this glaciation occurred, the entire plateau of Washington was covered with a loessial soil, varying in depth from a few feet to 200 feet. This and the underlying basalt had been eroded to maturity and a network of drainage lines covered it. The major water courses of this mature topography were consequent on the warped surface of the

[1] Grand Coulee and Moses Coulee are exceptions. Grand Coulee opens upgradient into Columbia River Valley and the upper end of Moses Coulee is obliterated by the drift of a later glaciation.

[2] Exceptions due to a later episode in the history of the region are noted later.

[3] One exception, noted later.

plateau which descends in a general way from the northeast to the southwest.

The ice sheet approached and invaded the plateau from its northern high margin. It barely crossed to the headwaters of the consequent drainage. In places, it did not cross, but by blocking all other escapeways, its waters were forced to cross. By about a dozen different routes, at different altitudes and distributed along more than 150 miles of the ice front, water entered the mature drainage system. The capacity of the pre-existing valleys was wholly inadequate for the volume of most of these streams. Furthermore, gradients were high and the glacial waters eroded enormously, sweeping away the overlying loessial material, crossing low divides and isolating many groups of the maturely eroded hills to produce the anastomosing pattern of the scablands, biting deeply into the basalt to make the canyons and rock basins, and spilling into the Snake and Columbia in three times as many places as the pre-existing drainage had used.

This procedure of glacial streams was unique, so far as the writer is aware. It was unorthodox, at any rate, for no valley trains and but two outwash plains[1] were built on the plateau south of the basalt plain. The stream gravel of the scablands is almost wholly in separate bars.

The conception above outlined is amply sustained by every feature and relationship of the scablands. All other hypotheses meet fatal objections. Yet the reader of the following more detailed descriptions, if now accepting the writer's interpretation, is likely to pause repeatedly and question that interpretation. The magnitude of the erosive changes wrought by these glacial streams is nothing short of amazing. The writer confesses that during ten weeks' study of the region, each newly examined scabland tract reawakened a feeling of amazement that such huge streams could take origin from such small marginal tracts of an ice sheet, or that such an enormous amount of erosion, despite high gradients, could have resulted in the very brief time these streams existed. Not River Warren, nor the Chicago outlet, nor the Mohawk channel,

[1] The Hartline gravel flat and the Quincy basin fill, both in structural depressions in the plateau.

nor even Niagara falls and gorge itself approach the proportions of some of these scabland tracts and their canyons. From one of these canyons alone[1] 10 cubic miles of basalt was eroded by its glacial stream.

THE BASALT PLAIN, NORTH OF THE SCABLANDS AND MATURE TOPOGRAPHY

This physiographic feature extends westward from Spangle for 50 miles along the south side of Spokane River and varies from 3 to 12 miles in width. It is determined by the upper surface of the Columbia basalt formation. It is interrupted in places by short valleys tributary to Spokane River, and the different portions are known as prairies.[2] This plain is bounded on the south by channeled scabland and maturely eroded loessial hills. The differences between it and adjacent broader scablands are not marked, but the loessial hills are in striking contrast with it. These hills which, elsewhere on the plateau, come right to the edge of Snake and Columbia valleys,[3] nowhere overlook the Spokane Valley. They terminate abruptly on the southern margin of this plain. On the plain there is no mature topography and no channeled scabland. There is no area on the plateau like it. The nearest approach to it is the northern portion of Douglas County, back of the Wisconsin terminal moraine. This narrow plain must be the result of conditions which prevailed no farther south.

These conditions can be summed up in one word—glaciation. Deposits of till and many striated erratics have been found in every township examined on it. The till is patchy in distribution. No moraine margins the southern edge of the plain and no good moraine ridges occur on it, though here and there is morainic topography.

The genesis of the plain thus established, the questions of its character before glaciation and the method of its development arise. These are answered clearly when the adjoining mature loessial topography is studied. The larger valleys of this topog-

[1] Upper Grand Coulee.

[2] Paradise Prairie, Sunset Prairie, Indian Prairie, Four Mound Prairie, etc. On the north side of Spokane River are Pleasant Prairie and Five Mile Prairie, also parts of this plain.

[3] With the exception of northern Douglas County.

raphy have been eroded to varying depths into the basalt. The bottoms of such as lead out across the basalt plain to the north are lower among the hills than the general surface of the plain. The profile of the plain, extended back among the hills, cuts their slopes somewhere between hill summits and valley bottoms. And this transection is at the contact of loess on basalt. The ice sheet which covered the plain, therefore, simply removed the upper, weaker formation, and only to a minor extent altered the surface of the basalt formation.

THE MATURE TOPOGRAPHY

This is the dominant type of topography of the Columbia Plateau. The major drainage lines are structurally controlled, but the minor ones constitute a dendritic network. The pattern is eroded largely in a weak sedimentary deposit, chiefly loessial, which overlies the basalt. Maturity is expressed in the complete development of the drainage system, in the reduction of the original surface to valley slopes[1] and in the concavity of the lower part of many of these slopes. This maturity has been developed with reference to the underlying basalt as a base level, for progress of the cycle of erosion in the loess has been very much more rapid. Neglecting the loessial cover, the basalt plateau is in early youth, and will still be when the loess has been entirely removed. Nevertheless, the absence of deep trenches in basalt in the interior of the plateau, similar to Spokane, Columbia, and Snake River valleys about its margin, and the cutting through of the loess has allowed the development of shallow mature valleys in the upper part of the basalt.

The loessial deposit varies in thickness, in some places being only a soil, and in others being 200 feet or more in thickness. It is not all loess. There are places where it is chiefly a residual soil from the basalt, and others where it is a waterlaid sediment.[2] But many widely distributed sections show a succession of loessial deposits,

[1] A few broad, undissected divides are still left. Michigan Prairie, south of Lind, is a good example.

[2] Probably the Ellensburg formation. The Pleistocene Ringold formation, in Franklin County, is younger than the mature topography.

with abundant root and rootlet casts throughout and with reddened upper surfaces of each deposit, aggregating 50 feet or more.[1]

The mature topography is older than the scablands and the basalt plain. Literally hundreds of isolated groups of maturely eroded hills of loess stand in the scablands. Their gentle interior slopes are identical with those far from the scabland tracts. But their marginal slopes, descending to the scablands, commonly are very steep, over large areas amounting to 30° and even 35° (Figs. 1, 2, and 3). These steep slopes are seldom even gullied, except

FIG. 1.—One of a group of loessial hills in the scablands a few miles southwest of Rock Lake. One of the steepened slopes and its alignment are shown. Photo by O. C. Clifford.

where a drainage line leads out from the hill group to the scabland. Where the minor valleys transected by the steep slopes lead backward into the interior of a hill group they are simply hanging valleys.

There are few places where basalt occurs in these steepened slopes. Where present, it is always restricted to the lower part and shows itself in conspicuous ledges.

A very striking and significant feature of the steepened slopes is their convergence at the northern ends of the groups to form great

[1] Near Harrington and near Kahlotus are two excellent cuts which show this very well.

prows, pointing up the scabland's gradient (Figs. 2 and 3). The nose of a prow may extend as a sharp ridge from the scabland to the very summit of the hill. It is impossible to study these prow-

FIG. 2.—An isolated loessial hill on the scabland south of Hooper. The prow of the hill is pointed at the observer. The hill is 180 feet high, more than half a mile long, has a very narrow crest, and sides which slope 35°. It is entirely surrounded by scrubbed basalt. Half a mile to the west is a canyon in the scabland 75 feet deep, with an abandoned waterfall at the head.

FIG. 3.—The same hill as shown in Figure 2. The prow is at the right. The corresponding steepened slope at the tail of the hill shows at the extreme left. Most of the apparent left slope of the crest is a matter of perspective. Photo by O. C. Clifford.

pointed loessial hills, surrounded by the scarred and channeled basalt scablands, without seeing in them the result of a powerful eroding agent which attacked them about their bases and most effectively from the scabland's up-gradient direction.

DETAILS OF A SCABLAND SURFACE

All scablands are channeled to a greater or lesser extent. These channels are eroded in basalt to depths varying from a few feet to

FIG. 4.—Devils Canyon at mid-length, looking south. A double fall existed here when the canyon was eroded. The island and the eastern part still remain.

hundreds of feet. Commonly there are many shallow channels on each tract. Most tracts also have a few deeper channels, of the proportions of canyons[1] (Figs. 4 and 5). All channels in a tract are intricately interlaced, resulting in a multitude of butte-like hills, knobs, and ridges among them. Few channels have

[1] Upper Grand Coulee (1,000 feet deep), Lower Moses Coulee (900 feet deep), Devils Canyon, Franklin County (500 feet deep), and Palouse Canyon (500 feet deep) are the most noteworthy examples.

accordant grades where they unite or diverge, the bottoms of the shallower ones hanging above the floors of the deeper ones. Many canyoned channels have abandoned cataracts and cascades in them or at their heads.[1] Most canyoned channels have elongated rock basins (see Fig. 6). Even in the shallow channels, basins or pockets in the rock are common. Some of these rock basins clearly were produced by recession of a cataract whose scarp still

FIG. 5.—Devils Canyon near mid-length, looking north. Note the scrubbed basalt ledges above the canyon rim, and the profile of the loessial bluffs, still higher and farther back.

exists.[2] Others were produced by plucking of the columnar basalt in the canyon floors where the gradient was high.[3]

These features of the channeled scablands on the Columbia basalt plateau do not closely resemble any other type of topography.

[1] Dry Falls (400 feet high) in Grand Coulee, The Potholes (350 feet high) south of Trinidad, Frenchman Springs (400 feet high) south of The Potholes, and The Three Devils (600 feet total descent) in Moses Coulee are especially noteworthy.

[2] Deep Lake, below one of the Grand Coulee abandoned falls, has many associated huge potholes, drilled into the basalt at the foot of the falls as they retreated. Each of the two cataracts of "The Potholes," south of Trinidad, has a single elongated rock basin at the foot (Fig. 7).

[3] Rock Lake in Whitman County, Goose Lake in Grant County, Washtucna Lake and Eagle Lake in Franklin County, Pacific Lake and Tule Lake in Lincoln County, Goose Lake in Grant County, Medical Lake, Silver Lake, and Farrington (Fish) Lake in Spokane County, and Big Swamp Lake and Cow Lake in Adams County are examples of hundreds of such basins.

The narrowness and elongation of channels and the continuous gradients of tracts as a whole suggest river valleys but these features are all inherited from pre-existing valleys. Furthermore, some tracts are nearly as broad as they are long. The pattern of channels on a tract, like the pattern of some tracts and their isolated loessial hills, is much like that of great braided streams (Fig. 8).

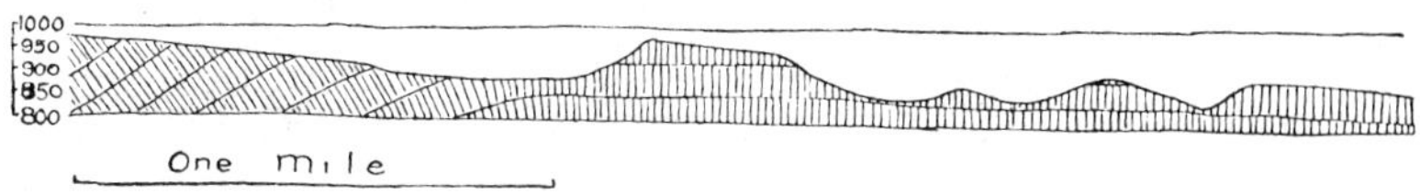

FIG. 6.—Longitudinal profile of the deepest of the Drumheller Channels across the nose of the anticline. Four rock basins are indicated, the largest of which is 75 feet deep.

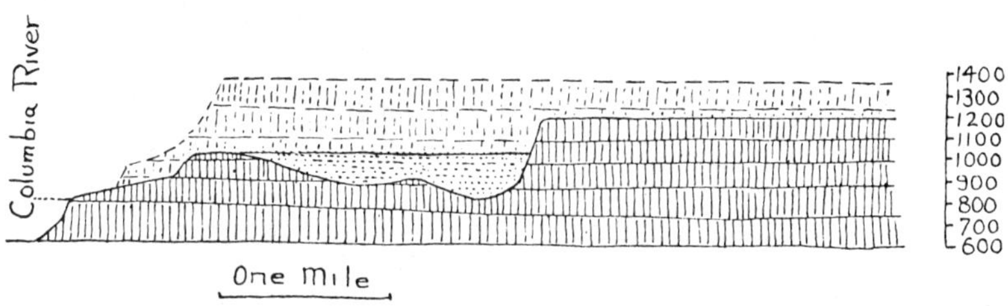

FIG. 7.—"The Potholes," longitudinal profile of the northern half, showing (1) cliff along northern side of the canyon, (2) amount of recession of the falls, (3) elongated rock basin below the falls, (4) great gravel bar along edge of the rock basin, and (5) approximate level of Columbia River when the cataract was formed.

But the scablands are erosional in origin, while the braided stream pattern is depositional.

The evidence for the origin of channeled scabland by stream erosion is overwhelming. The evidence of contemporaneity of action of all channels of a given tract, at least in its early history, is equally convincing. The only sequence indicated is that of development of the greater channels later in the epoch and consequent draining off of the shallower channels.[1] The scablands of the plateau in Washington are the beds of huge river courses in which

[1] Especially well shown in Lower Grand Coulee, in Moses Coulee between Palisades and Spencer, in The Potholes and Frenchman Springs south of Trinidad, and in Palouse Canyon south of Hooper, in all of which cataract recession in main channels cut off smaller channels alongside in the same scabland tract.

the streams once spread completely from side to side and only later became concentrated in the deeper canyons.

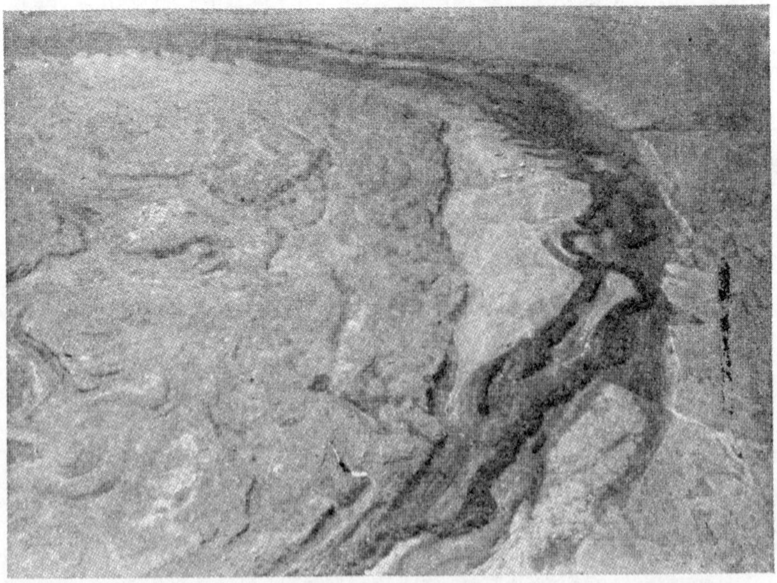

FIG. 8.—A part of Rock Creek (Lincoln County) and its associated scabland. The creek in the main channel is margined by vegetation. The scabland to the left is barren and the irregularity in lighting is due almost entirely to the cliffs which border the shallow anastomosing channels. (Aeroplane Photo by F. H. Frost.)

ALTITUDES AND GRADIENTS OF THE SCABLAND TRACTS

The heads of scabland tracts which are open to the basalt plain range in altitude between 2,350 and 2,500 feet above tide. There are six or eight of these, the number depending on just what is considered to constitute an individual scabland tract. At least the following should be recognized as unit tracts. The order in the list is from east to west.

1. Pine Creek channel. Altitude of head, 2,450 feet A.T.
   Gradient approximately 25 feet per mile
2. Cheney-Hooper tract
   Four heads:
   a) Marshall-Spangle, altitude, 2350±
   b) Four Lakes, altitude, ?
   c) Medical Lake, altitude, 2,425
   d) Deep Creek, altitude, 2,350±
   Total width of the group is 22 miles

Gradients:
Spangle to Hooper, 21.9 feet per mile
Cheney to Hooper, 22.6 feet per mile
Medical Lake to Sprague, 24 feet per mile
Medical Lake to Hooper, 22.5 feet per mile

3. Reardan channel. Altitude of head, 2,500+
Gradient, Reardan to Odessa, 20 feet per mile
4. Davenport-Harrington channel. Altitude of head, 2,450±
Gradient:
Davenport to Harrington, 27 feet per mile
Harrington to Odessa, 26 feet per mile
5. Telford tract
a) Eastern head. Altitude, 2,500±
Gradient:
Rocklyn to Odessa, 30 feet per mile
Rocklyn to Krupp (Marlin), 26 feet per mile
b) Western head. Altitude, 2,500±
Gradient:
Near Creston to Krupp (Marlin), 32 feet per mile
Near Creston to Wilson Creek, 30 feet per mile
Total width of heads is 17 miles
c) Wilbur branch
Gradient:
Creston to Wilson Creek, 32 feet per mile
Wilbur to Wilson Creek, 25 feet per mile
Almira to Wilson Creek, 26 feet per mile

The only scabland tracts which do not open on the basalt plain are Grand Coulee and Moses Coulee. For the head of Moses Coulee there are no altitude measurements. Grand Coulee has had a peculiar history, not yet fully deciphered, but the significant altitude at its head for present purposes is not the coulee floor (1,530 feet A.T.) but the scabland margining the brink of the canyon, about 2,500 feet A.T. The canyon has been cut subsequent to the first spilling over of glacial waters. The floor near Coulee City is 1,510 feet A.T. Most of this descent occurred within a few miles of Coulee City, the original slope being as steep as 20° in part and averaging perhaps 10° for 1,000 feet of descent. This is the chief reason for the great canyon across the divide. No other scabland head has been notably canyoned. None other had a gradient to exceed about 30 feet to the mile. All the canyons of the channeled scablands are located in places of exceptionally steep original gradients.

If the channeled scablands are the product of stream erosion, and if all tracts are of the same age, the fact that the scabland heads vary in altitude, though all but two are open to the same glaciated basalt plain, can have but one satisfactory explanation. Each glacial stream must have had a source of water which was unconnected with the others. This means that the ice sheet whose melting yielded these streams must have covered the basalt plain and in most places must have been in contact with the mature loessial hills which separate the scabland heads. It means, furthermore, that but a few miles of ice front supplied the water for streams so huge that they flooded over many preglacial divides of the mature topography even in the southern part of the plateau. It was this flooding across divides which produced the scabland plexus of the plateau.

About 40 miles of ice front in one case[1] yielded water sufficient to denude a non-elongated tract 250 square miles in area of a loessial cover about 100 feet in maximum thickness. This was done by lateral planation in the preglacial drainage lines of the tract. These lines were so shallowly intrenched in the basalt and the volume of the water was so great that, as the loessial hills were eroded away, the flood spread over the entire area, 13 miles wide. Gradients were low, however, and it did not develop canyoned channels. Steeper gradients farther from the edge of the ice, and greater capacity of the preglacial valleys, held the waters within the confines of these valleys, but six such[2] were necessary to contain the flood and they were all greatly eroded in the underlying basalt.

Another large scabland area[3] is 75 miles long and 15 miles in average width. Its total descent is 1,850 feet. Its altitude at the head is the lowest of all such tracts. It differs from the one above in its notable linear extent, in the possession of a large number of isolated groups of loessial hills of the older topography, in its greater

[1] The Telford scabland and its dependencies.

[2] Coal Creek, Duck Creek, Lake Creek, an unnamed creek, Connawai Creek and Wilson Creek.

[3] The Cheney-Hooper area, extending from Spangle and Cheney to Snake River, south of Hooper.

gradient and in the development of canyons.[1] Though a much greater volume of water passed through this tract, the gradients were steep enough to draw off the flood and prevent a complete spreading over the area. Much water came to it from the ice margin to the east in Idaho and no estimate can now be made of the length of ice front which contributed to it. There were five or six places of distributary discharge where this flood crossed preglacial divides and one of them[2] eventually obtained most of the discharge. Along this distributary route, the glacial flood swept away the loessial hills for a width of 10 to 15 miles and eroded 500 feet into the basalt.

The glacial drainage route which possessed the highest crossing of the preglacial surface of the plateau is Grand Coulee. It also found the steepest gradient[3] and its volume was sufficient with this gradient to cause the deepest erosion in the basalt. Upper Grand Coulee, across the preglacial divide, is 1,000 feet deep. But its floor, after the epoch had closed, was lower than that of any other glacial drainage route at the head. In its early history it drew water from about 40 miles of ice front, but never spread widely. The gradient, determined here by exceptional warping of the basalt, prevented that. There was no noteworthy preglacial stream along its course. Grand Coulee is, therefore, the simplest but grandest case of canyon-cutting by glacial streams on the plateau.

DEPTH OF GLACIAL STREAM EROSION IN THE SCABLANDS

Criteria.—The courses of the larger valleys in the mature drainage system were determined by the warped surface of the basalt. The dominant feature of this warping is the southwestward dip from Spokane River and Columbia River on the north to Snake River on the south and Columbia River on the west. Many minor folds are superposed on this dip slope, so recent geologically that anticlines determine divides and synclines contain stream valleys.[4] Commonly, only the major valleys are intrenched in the

[1] Cow Creek, Rock Lake, and Creek and lower Palouse River now occupy the most striking of these canyons.

[2] Palouse River Canyon below Hooper.

[3] In one place, 1,000 feet in about a mile.

[4] Examples are Moses Coulee east of Palisades, Wilson Creek above Almira, Crab Creek below Corfu, Washtucna Coulee, Lind Coulee below Lind, Snake River near Lewiston and Clarkston, Union Flat Creek, Rebel Flat Creek, and Palouse River above Winona.

basalt. The minor valleys, in general, are not eroded through the loessial mantle.

Except near the bounding canyons of Spokane, Columbia, and Snake Rivers, or in exceptionally upwarped parts of the plateau, the preglacial valleys in basalt which were unviolated by the glacial flood, have the same mature slopes as those in the loess. Where glacial streams found routes eroded but slightly in basalt, they commonly spread widely at first[1] and only later eroded canyons, if they did so at all. Such wide scablands commonly are bounded by steep bluffs of loess. Many of the mature valleys in basalt were sufficiently capacious to contain the glacial waters which entered them. In such cases[2] the scabland of the route lies on the sides and bottoms only, and unsteepened slopes of the bounding older topography may come to the edge of the scabland. Such a tract is narrow and instead of having a multitude of lateral shallow channels anastomosing with the main canyon, it consists of exceedingly roughened ledges of basalt outcropping on the slopes of the main valley. Shoulders in the curves of these valleys were treated with especial vigor, in some cases being wholly cut away to leave prominent cliffs on the valley walls.

By smoothing out these ledges, something of the cross-section of the preglacial valley may be obtained. Remnants of the old floors are recorded in isolated buttes of basalt on the present floor and in the lowest prominent rock terraces, below which is the canyoned channel eroded in mid-current by the huge glacial stream.

Instances.—In Cow Creek, southwest of Ralston, the remarkable number of knobs and buttes in the lower part of the valley indicates clearly that the preglacial floor must have been at least 75 feet higher than the present. In Crab Creek Valley, between Krupp (Marlin) and Stratford, there are prominent buttes, isolated or partially isolated on the floor of the canyon. The tops of these are remnants of the preglacial floor and their height (100 feet in places) is a minimal measure of the depth of canyon-cutting by the glacial waters.

[1] As in the Telford and the Cheney-Hooper areas.

[2] Washtucna Coulee, Esquatzel Coulee, Lower Moses Coulee, Lind Coulee, Pine Creek above Rock Lake, Rock Creek (Lincoln County) and Coal Creek are examples.

In Washtucna Coulee, numerous prominent rock terraces from 150 to 200 feet above the present floor are probably remnants of the earlier valley bottom. In Esquatzel Coulee, into which Washtucna opens, these terraces are 200 feet and more above the bottom.

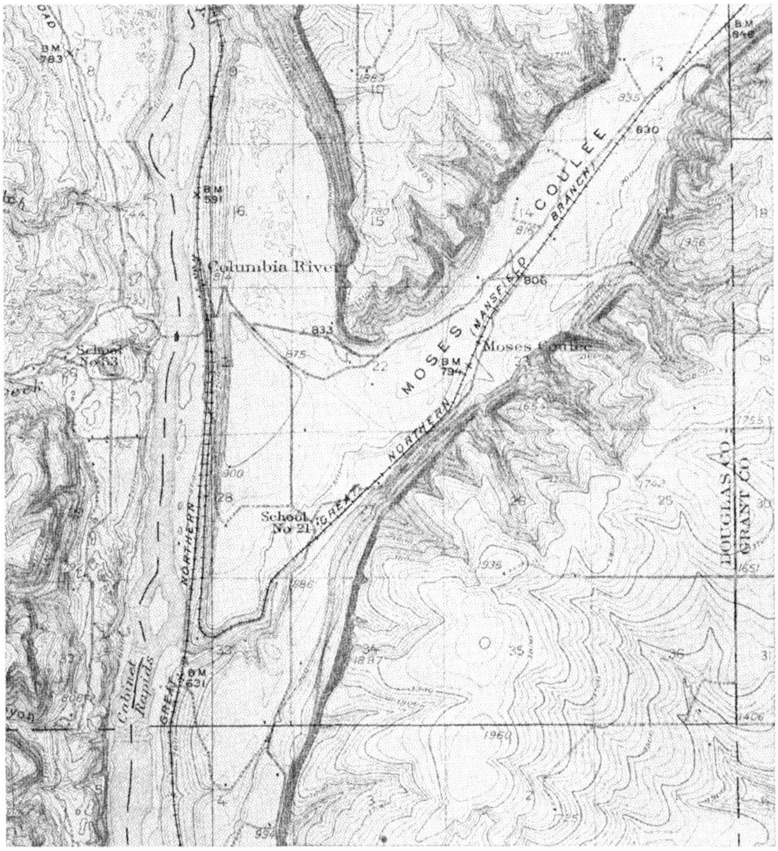

FIG. 9.—The lower part of Moses Coulee. Part of Malaga, Washington, topographic map. Note truncated spurs, hanging ravines, alluvial fans, and the great terrace in both Moses Coulee and Columbia Valley.

In Lower Moses Coulee (Fig. 9), the mouths of preglacial tributary ravines hang approximately 400 feet above the rock floor. Some of this discordance of grade is due to widening of the preglacial coulee, which here was a canyon, by the glacial stream but probably most of it is due to deepening during the glacial epoch.

In addition to the deepening of valleys already existing, the glacial flood actually made, *de novo*, drainage ways of greater width and depth than any previously developed on the plateau. This happened where divides were crossed and unusually high gradients down the farther slopes were found. Five such places are especially noteworthy: Devils Canyon, Palouse Canyon below Hooper, Drumheller Channels, Othello Channels, and Grand Coulee.

Preglacial Palouse River joined Snake River at Pasco, its subparallelism with the larger stream for 150 miles being structurally determined. The glacial flood from the north entered it in mid-length at several places between Winona and Washtucna. The volume of this flood was more than the valley could carry away.

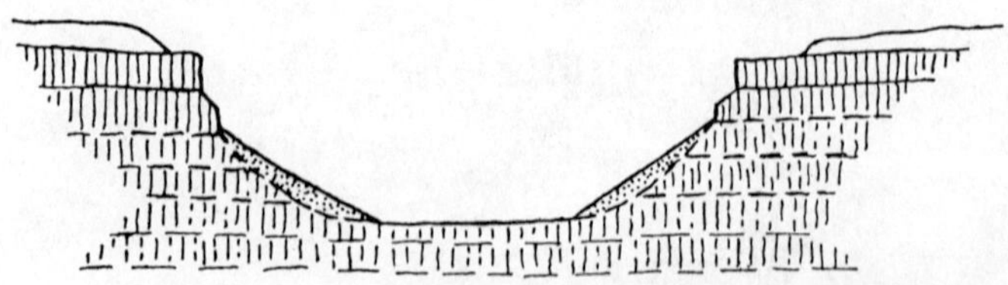

FIG. 10.—Devils Canyon, cross-section a mile and a half south of Kahlotus, showing (1) steepened loessial bluffs (33°), (2) narrow scabland above brink of basalt cliffs (3) canyon 450 feet deep and less than one-fourth mile wide, and (4) post-Spokane talus, three-fourths the height of the cliffs. Horizontal and vertical scale the same.

Two leaks across the divide to the Snake developed, one near Kahlotus, and one near Hooper, and in both very great gradients were encountered.

The Devils Canyon distributary, south of Kahlotus, cut 50 feet or so through the loess and then by recession of waterfalls over ledges of basalt in the north slope of Snake River Valley, it eroded a canyon 5 miles long, a quarter of a mile wide and 500 feet deep (Fig. 10). Every fall but the lowest retreated completely through the divide. The remaining ledge, like a dam separating the two canyons, is less than 100 feet above the floor of Washtucna Coulee and not half a mile wide.[1] An abandoned half of a double cascade stands in mid-length of Devils Canyon (Fig. 4).

[1] The Spokane, Portland and Seattle Railroad tunnels through this ledge. In 1916, Washtucna Lake was so high for a time that it overflowed through this tunnel into Devils Canyon.

A much larger volume of water spilled across the preglacial divide south of Hooper. Before the great canyoned channel now transecting this divide had been eroded, the stream was 10 to 15 miles wide. Several channels of canyon proportion were initiated, but one outran the others in its deepening and finally secured all the discharge. Many of the shallower canyons enter the southern part of the main one over abandoned waterfalls. These channels could have carried water only when the wide scabland of the divide had no deep canyon completely across it. Their falls could have developed only after a deep main canyon existed *below* their junction. It follows, therefore, that Palouse Canyon was cut by retreating waterfalls, though these were destroyed later in the epoch. One only survived, now notched considerably by the post-glacial work of the Palouse River. The falls today are 198 feet high. Palouse Canyon is another Devils Canyon in all save its greater width and the fact that the preglacial divide was cut entirely in two.

Drumheller Channels and Othello Channels are two remarkable cases where the glacial flood crossed anticlinal ranges.[1] In both, the anticline is asymmetrical and the waters flowed down the gentle slope. In both, the flood at first was wide but became concentrated later in the deepening canyons. The maximum depth of erosion in Drumheller Channels was 400 feet, about 100 feet of which was in a weak sedimentary formation (probably the Ellensburg), and 300 feet in basalt (Fig. 11). The width of the Drumheller denuded tract is about 10 miles. This particular scabland area has a more striking and more complicated development of channels and rock basins than any other on the plateau. It is the only area of this type now topographically mapped. Below the Channels, along the northern flank of Saddle Mountains, the ancient river eroded 300 feet into the broader scabland.

But Drumheller Channels is not wholly the product of the glacial flood whose history we have been following. It and the main canyon of Grand Coulee carried drainage from a later ice sheet, and it has carried Crab Creek since the later glacial epoch. The amount of deepening during each Pleistocene epoch is diffi-

[1] Drumheller Channels crosses the eastern nose of Frenchman Hills anticline, Othello Channels crosses the eastern nose of Saddle Mountains anticline.

cult to determine. Othello Channels carried less water than Drumheller Channels, consequently is a smaller tract. Furthermore, it received water during only the earlier epoch. The degree of development of its canyons and rock basins is comparable to that of the larger tract and makes it probable that most of Drumheller Channels were formed during the earlier epoch.

The features of Grand Coulee are of such magnitude and its history so complicated by local conditions that an entire paper might well be devoted to it. It affords the greatest example of canyon-cutting by glacial streams, not alone for the Columbia Plateau, but for the world. The field evidence indicates that no preglacial drainage route ever existed here. Scabland with shallow channels margins the upper part of the Coulee, though 1,000 feet higher than the adjacent coulee floor, and there are no tributaries in the mature topography such as are possessed by Lower Moses Coulee and Washtucna Coulee. A glacial river, 3 miles in minimum width, spilled southward here over the divide and down a steep monoclinal slope. Judging by present grades and altitudes of this structural slope, the stream descended nearly 1,000 feet on a grade of approximately 10°, a few miles north of Coulee City. Such a situation is unparalleled, even in this region of huge, suddenly initi-

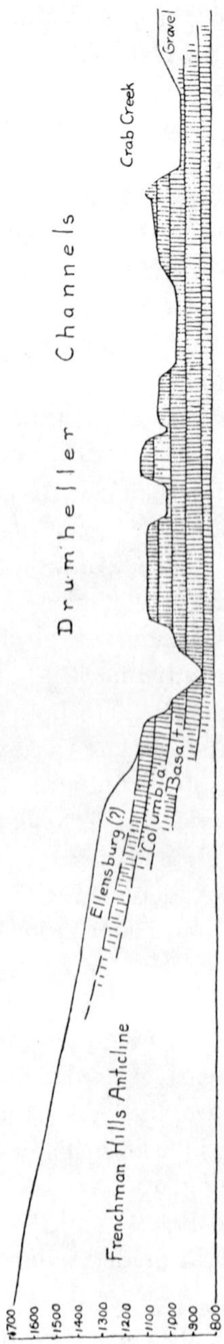

FIG. 11.—Cross-section of the head of Drumheller Channels, showing (1) structure at eastern nose of Frenchman Hills anticline, (2) the pre-scabland topography, and (3) the scabland channels. Probably no preglacial drainage crossed the anticline at this place. The lower 50 feet of the deepest channel is a rock basin.

ated, high-gradient rivers. Across this monocline between Columbia River and Coulee City, the canyon is 30 miles long and averages all of 2 miles in width and 800 to 900 feet in depth. In the making, at least 10 cubic miles of basalt were excavated and removed. Though a later flood of glacial waters used this route,[1] it did but little to deepen it.[2] By far the greater part of the erosion of Upper Grand Coulee was performed by the earlier glacial stream. It is very probable that this immense task was performed by the stoping of cascades and cataracts which retreated entirely through the monoclinal uplift to the deeper valley of Columbia River and thus left the great notch.[3]

It also seems probable that when the retreat of the ice sheet began, the plateau west of Grand Coulee was abandoned last. Earlier clearing of the Spokane and Columbia valleys to the east allowed all the drainage of the ice sheet to use the Grand Coulee route, which was then the lowest of all. Grand Coulee's greatest flood and probably its greatest erosion thus came after the other scabland routes had gone dry.

VOLUME OF THE GLACIAL STREAMS

If the channeled scablands of the Columbia Plateau are the erosive results of glacial waters, certain statements as to the volume of the streams can be made. Measurements are possible if remnants of the preglacial floor of the main valley exist in places where the valley brimmed over with the glacial flood to produce distributary courses. Should it appear that the amount of canyon-

[1] During the Wisconsin glaciation.

[2] Grand Falls, below Coulee City, consists of Dry Falls, Deep Lake Falls and a smaller unnamed falls a mile east of Deep Lake. The lip of the smaller falls is 125 feet higher than the floor of the channel leading to Dry Falls. Yet all were made by the same glacial stream and only Dry Falls and Deep Lake Falls were used and modified by the later discharge.

Furthermore, the Wisconsin ice did not cross Spokane River or Columbia River east of Grand Coulee and its waters were free to use the lowest of the ten earlier routes. Only Grand Coulee was so used, showing that it had been eroded by the earlier discharge to a depth not far short of that which it now has.

[3] This inference has no physiographic evidence in Grand Coulee to substantiate it, but is based on the known procedure of the glacial streams in similar situations, e.g., Lower Palouse Canyon, Devils Canyon, Frenchman Springs, and The Potholes.

cutting by these glacial streams has been overestimated, the depth of the stream to flood across the divide must be correspondingly increased. This view promptly runs into an absurdity, for the less the canyon-cutting is held to have been, the deeper and therefore more competent to erode the stream must have been.

One of the best cases for measurement is Washtucna Coulee at the head of Devils Canyon. Though there are but two small rock knobs out in the coulee floor, the summit of neither indicating the original valley bottom, there are good rock terraces to record it (Figs. 12 and 13). Near Kahlotus they lie 1,000 to 1,200 feet above tide. The glacial stream here, at the beginning of its history, overflowed through the loessial hills to Snake River, at an altitude of at least 1,350 feet. It was, therefore, from 150 to 350 feet deep. Its width averaged at least a mile.

And this was no ponded condition, for Washtucna Coulee opened widely into Esquatzel Coulee, an even more capacious valley, and Esquatzel in turn into Columbia and Snake valleys, and the glacial waters cut deeply into the bottom of both coulees. Figuring the preglacial floor as 1,000 feet A.T. at Kahlotus and as 675 feet A.T. at Eltopia, 34 miles farther down the valley, the great glacial stream had a gradient of about 10 feet to the mile.

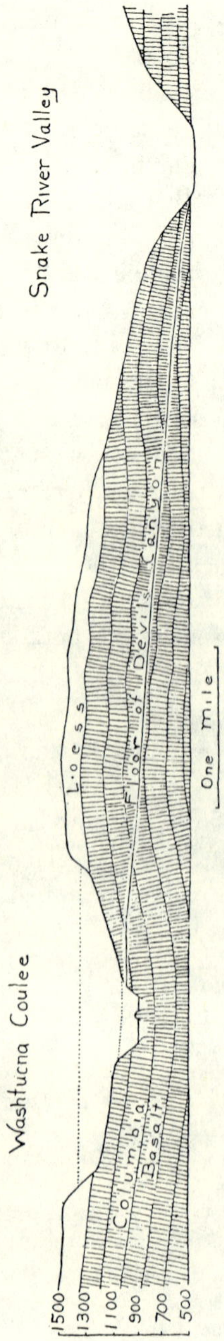

FIG. 12.—Cross-section of Washtucna Coulee and Snake River Valley, with longitudinal profile of Devils Canyon, showing (1) structure of the basalt, (2) its loessial cover, (3) rock terraces flanking Washtucna (Kahlotus) Lake, remnants of the preglacial valley floor, (4) the inner canyon eroded by the glacial stream, (5) steepened slopes of the loess, facing the coulee, and (6) approximate upper limit of the flood which spilled across to Snake River.

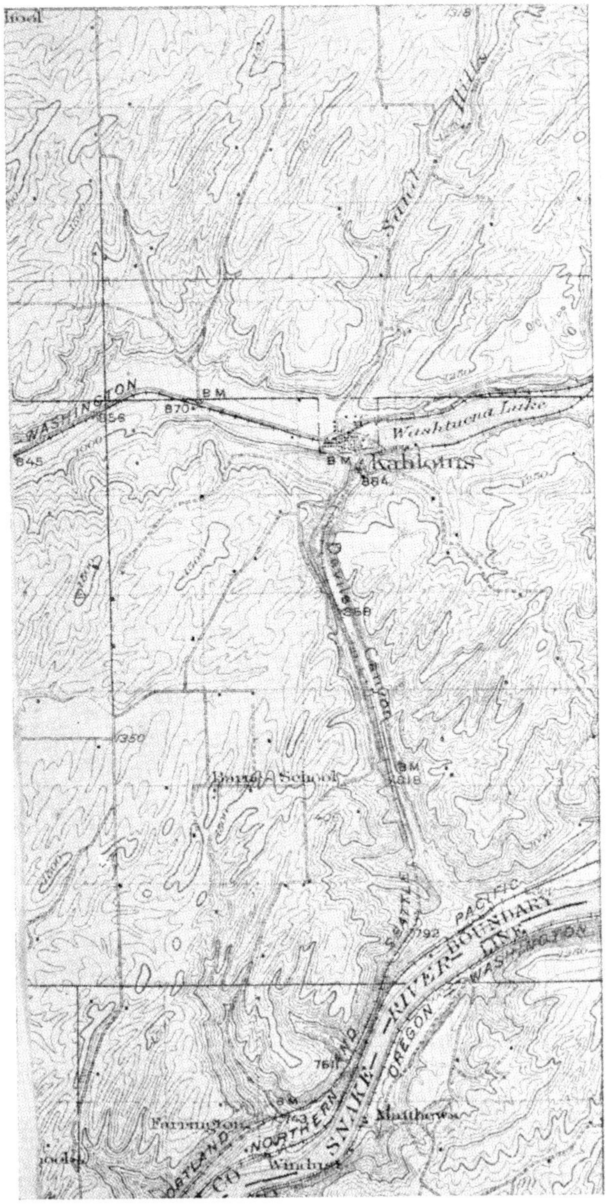

FIG. 13.—Devils Canyon. Part of Connell, Washington, topographic map.

Further evidence that glacial waters so filled Washtucna Coulee that the former valley became simply a channel is found in the upper limit of glacial stream gravels and of scablands. At the head of Devils Canyon, the highest scabland is 1,250 feet A.T., 250 feet above the brink of the cliffs of the canyoned channel. At Estes, a gravel bar deposited by the glacial stream lies 250 feet above the Coulee floor and about 125 feet above the rock terrace which marks the old valley floor. Near Sulphur, the highest scabland surface, at the base of the steep loessial bluffs, is between 1,100 and 1,150 feet A.T., 100 to 150 above the rock terrace. Northwest of Connell a terrace of sand and fine gravel lies at 1,000 feet A.T. It marks the upper margin of the scabland here and probably is a deposit of the glacial stream. It is 100 feet above the broad rock terrace. The canyoned channel here is cut 150 feet below the rock terrace.

Crab Creek Valley, below Odessa, received more water than it could carry away, at least before its central canyon had been eroded. It overflowed southwestward, by way of Black Rock Coulee and its associated scabland, to the Quincy Basin which Crab Creek itself entered farther north. Measurements here are only approximate but they indicate the order of magnitude of this glacial stream. Scabland and glacial stream gravel along the southern edge of Crab Creek Valley lie 300 feet or more above the present stream and extend a mile and a half back on the upland from the margin of the preglacial valley. This valley had been canyoned more than 100 feet by the glacial stream which, on the basis of these figures, was 200 feet deep at its inception.

The Telford scabland tract, 13 miles wide and 20 miles long, has been swept almost completely bare of the loessial deposit. The relief in a cross-section of the basalt surface now exposed, aside from the minor canyons, is about 50 feet. To have been so denuded, this tract must have had a sheet of running water of this depth completely over it.[1]

[1] That the ice sheet did not advance over the Telford denuded tract is shown by the presence of a few isolated loessial hills with characteristically steepened marginal slopes. One such group lies 7 or 8 miles north of Telford. It has a maturely eroded topography and a dark loessial soil without rock fragments of any kind. But it is cut by channels of glacial waters which eroded to the basalt. The fact that these waters went through the group, though the surface immediately north of them drops off into the deep canyons of Hawk Creek, a tributary of Spokane River, proves that glacial ice must have crowded up against the northern side of the group.

Further evidence of the depth of the glacial streams will be presented under the next subject.[1]

DEPOSITS MADE BY THE GLACIAL STREAMS

The record of Pleistocene glacial streams almost everywhere is one of aggradation. Glacialists commonly think of the subject only in such terms, textbooks discuss it only in that light; it is the orthodox conception. But in the Columbia Plateau exceptional factors controlled. The preglacial valleys in general were small and of relatively high gradient, the volume of the glacial streams was very great, the amount of detritus from the ice was very small, and the rock crossed was either loess or closely and vertically jointed basalt, both of which yielded rapidly to the torrents. The result of these conditions was great deepening of the valleys, and deposits made by the streams are of minor importance. Their character, however, adds to the weight of evidence already presented for the origin of channeled scabland by glacial stream erosion.

The deposits are almost wholly of gravel. Sand is a minor constituent and clayey material is lacking. The gravel and sand are almost wholly of basalt, though all deposits contain fragments of rock foreign to the plateau. The basaltic gravel is not well rounded though most of it is sorted and stratified and indubitably of stream origin. Foreset bedding is common, the direction of dip according with the slope of the scabland tract. In some places, the deposits are composed of very coarse material, with abundant, sub-rounded, basaltic bowlders 3 and 4 feet in diameter. These were originally bowlders of decomposition and were derived from flows with particularly large columns, underlying or in the immediate proximity of the deposit. Where a few erratic bowlders are associated[2] the deposit itself might be misinterpreted as a bowldery till.

The gravel deposits rest on irregularly eroded basalt, essentially a buried scabland surface. Nowhere do they lie on or beneath the loess. Neither the gravels nor the underlying basalt are

[1] If these enormous streams all came to the Columbia eventually, should not the great volume be recorded farther down the master stream? The writer has seen enough to convince him that it is so recorded, and hopes to publish on this subject in the future.

[2] As west of Lantz, for example.

decayed.[1] Cementation with calcium carbonate has begun but has not advanced far.

Most isolated loessial hills in a scabland tract have a deposit of gravel depending from their down-gradient end. Many knobs and buttes of basalt have similarly situated gravel deposits.

In many cases, the gravel deposits constitute discontinuous terraces on the margins of the scabland tracts, suggesting remnants of former valley fills.[2] But the evidence seems conclusive that all gravel deposits of the scablands are bars, built in favorable situations in the great streams which eroded the channels.

The rounded profiles and ground plans of many gravel deposits in the scablands are in accord with this interpretation. The unfilled canyons and rock basins, in intimate association with the discontinuous gravel deposits, indicate clearly that both are products of the same episode. The only alternative hypothesis is that channeled scabland was formed, then buried in gravel, then in large part re-excavated by streams little short of the magnitude of those which eroded the scablands. This has no other field evidence to support it and requires a much more complicated history. Furthermore, such deep canyons were cut when the scablands were made, and such noteworthy divide crossings were made that a reoccupation of all the scablands by glacial drainage from a second ice sheet would be impossible. And the hypothesis of dissection by the postglacial streams of the scablands is quite inadequate. Lakes and pools still stand in the rock basins on the channel floors, almost as they were left by the glacial flood.

Gravel deposits in the deeply canyoned scablands occur on the broad upper scabland surfaces, on the roughened slopes of the preglacial valleys and down in the canyons. The interpretation of these deposits as bars requires no change in general conditions, as does the alternative hypothesis; it simply requires that gravel be deposited locally as conditions might favor, all through the

[1] Local exceptions, as 1 mile southwest of Lamont, where ground water has been especially active.

[2] So the terraces in Pine Creek channel were interpreted in the earlier paper by the writer. That view is here abandoned for one much more consistent with all other features of the scablands.

epoch of erosion of the basalt. Deposits on the highest scablands[1] were made before the deepening canyons drew the waters into more restricted routes. Deposits in the deeper portions[2] were made during the latest stages of the episode. No gravel deposits were ever as thick as the rival hypothesis would require them to be.[3] That view demands that the canyons be eroded, then filled completely *with their own débris*, then re-excavated in large part.

However, there are two places on the plateau where the history of deposition by glacial streams has been somewhat different. One of these is Hartline gravel flat, the other is Quincy Basin. Both are structural depressions, not completely filled before the glacial floods arrived. The Hartline structural valley became filled with débris from the cutting of Upper Grand Coulee before Lower Grand Coulee had been eroded. The trenching of the lower coulee, and particularly the development and retreat of Grand Falls, incised the southern rim of the valley so that, by the close of the episode, the gravel fill, once the floor of the glacial stream, had been removed in its western part, and the remnant left 200 feet or so higher than the brink of Grand Falls.[4] The total fill in the Hartline structural valley is about 250 feet. It is composed of bowlders, cobbles, and gravel near Grand Coulee, and of sand 5 or 6 miles east, back from the main drainage line. An old channel from Grand Coulee crosses its northern and eastern part and leads into Deadmans Gulch, a distributary which spilled across the southern rim into Spring Coulee before Lower Grand Coulee had developed into the dominant notch that drained and dissected the flat. The terrace form of Hartline flat is the result of erosion of a once complete gravel fill. Its scarp is not constructional, as are those of bars. But the fill and the subsequent erosion occurred because of special local conditions, not because general conditions changed.

[1] As about Gloyd along the Black Rock distributary from Crab Creek Valley.

[2] As at the mouths of Duck Creek and Wilson Creek in Crab Creek Valley and at the junction of Crab and Coal creeks.

[3] For example, gravel deposits north of the town of Washtucna lie on the slopes of the coulee, 350 feet above the present floor. They antedate the deeper canyoned portions of the coulee.

[4] Estimated from the surviving eastern member of that complex waterfall, the only part which escaped modification by the Wisconsin glacial stream.

Quincy Basin, like the Hartline Valley, does not have a scabland floor. Both lay too low to be eroded. But both belong to the glacial drainage plexus. Quincy Basin probably contains more gravel than all the scablands of the plateau together. It was an enormous settling basin for the glacial rivers from Grand Coulee and Crab Creek. The flood of waters entering it was so great that at first three discharge ways were simultaneously in operation.[1] The southern and larger one obtained all the discharge later, and by deep notching of the basin's rim, caused the glacial waters which traversed the fill to incise their deposits. Two great channels and one smaller one were thus formed. The two large channels are each about 3 miles wide. Each was eroded about 100 feet deep during the later part of the episode. The one which contains Rocky Ford Creek also carried the later Wisconsin discharge and was further modified then.[2]

Erratic bowlders, some of them striated, are widely distributed at all altitudes on the basalt plain and the scablands. They also occur in valleys of the mature topography which open northward on to the basalt plain, and in some which open on to scabland tracts. The size, angularity, and striated surfaces indicate that these erratic bowlders were not rolled to their positions by running water. In the scablands, they must have been carried by berg ice on the great rivers. In their peculiar and limited distribution in the valleys in loess is evidence of small glacial lakes, in which the drift-bearing bergs floated.[3]

[1] Frenchman Springs, The Potholes, and Drumheller Channels.

[2] This interpretation is a modification of that published earlier by the writer. Further study of Grand Coulee, Quincy Basin, and Drumheller Channels has led to a magnification of the work of the earlier flood, and a minimizing of the results accomplished during the Wisconsin epoch. Grand Falls is now considered to be a pre-Wisconsin affair, none of the distributary canyons of Grand Coulee, except Dry Coulee, are thought to have functioned during the second flooding, the Adrian terrace is considered to be a part of the original fill and not of Wisconsin age, and all the deep canyons of Drumheller Channels are thought to date back to the earlier episode.

[3] Below an altitude of about 1,250, erratic bowlders occur on every formation and type of topography on the plateau. But these are a younger deposit (see *Journal of Geology*, Vol. XXVII [1919], pp. 489–506) and do not much overlap the scablands. Where overlap does occur, however, it is impossible to distinguish bowlders of the two categories by any difference in the amount of decay.

DEPTHS OF SNAKE AND COLUMBIA VALLEYS DURING THE EPOCH

Evidence on this question may be obtained at the debouchure of glacial drainage routes into these master valleys. Five of the nine such debouchures will be examined.

The rock floor of Moses Coulee is fully as low as that of Columbia Valley at the junction of the two. Both contain a great gravel fill here. Columbia River has cut through it, a depth of more than 300 feet. There is no such trenching in Lower Moses Coulee, but a well at Appledale penetrates 300 feet of this gravel without encountering bedrock.

The two cataracts of The Potholes and Frenchman Springs, which operated in the early part of the epoch, descended nearly the full height of the present Columbia Valley walls there. At The Potholes the glacial cataract can be traced down to less than 200 feet above the present Columbia (Fig. 7).

Koontz Coulee, 20 miles north of Pasco, is cut in the weak Ringold formation. It is 250 feet deep and a mile wide. It is floored with basaltic stream gravel from the scablands farther upstream. Though the Ringold silts extend down to the level of the Columbia at this place, the mouth of the glacial river channel hangs 200 feet above. No cataract could have been maintained here, as was done at The Potholes and Frenchman Springs, and the level of the Columbia of this epoch at this place is thus clearly recorded.

At the mouth of Palouse River, there are two parallel canyons in the scabland, one containing the river, the other dry. A basaltic butte separates them. It stands nearly in the center of the valley and its summit is between 350 and 400 feet above Snake River. It is a part of the original north wall of Snake River Valley, over which the gigantic cascade tumbled when the glacial flood broke across the preglacial divide from the north. This "Goat Island" testifies to the existence of a Snake River Valley at this place as deep then as now.

THE GLACIATION

Because the record of the ice sheet, from which came the streams that made the scablands, is best preserved on the basalt plain about the city of Spokane and along the south side of Spokane River, this has been named the Spokane glaciation.[1] It assuredly is not an early phase of the Wisconsin glaciation. That is recorded by pronounced moraines on the plateau west of Grand Coulee, in Columbia Valley north of the mouth of Spokane River, and in Colville Valley north of Spokane River. The Spokane ice left no terminal moraine and very little ground moraine. The reverse relation exists regarding the glacial waters of the two glaciations, for the Spokane glacial waters were of prodigious quantity and the Wisconsin waters of little consequence. Furthermore, a long time elapsed between the two glaciations as shown by the relative volume of talus accumulations in tracts swept by glacial streams of the two epochs. Post-Spokane talus in almost all places stands three-fourths to four-fifths of the total height of the basalt cliffs, post-Wisconsin talus stands about halfway up on the cliffs of Grand Coulee.[2]

For the absence of a terminal moraine along the southern edge of the area reached by the Spokane ice sheet, the writer has as yet no satisfactory explanation. It seems clear, however, that a moraine never was deposited, rather than that it was once built and subsequently removed. The functioning of some scabland tracts absolutely required glacial ice against the north slopes of the unglaciated hills at their heads. Floated granite erratics among some of these hills also require blocking of valleys by glacial ice. Yet there is no evidence of lateral drainage along the ice front in contact with the unglaciated hills, to which might be ascribed the removal of a terminal moraine.

The Spokane glaciation cannot be dated very far back in the Pleistocene, else the scablands should have a soil mantle of eolian sand and dust and disintegrated basalt, and the hundreds of lakes in the old channels should have been destroyed.

Leverett has suggested that a pre-Spokane till beneath loess at Cheney is of Kansan age.[3] If it is, and if the post-Spokane glaciation is correctly ascribed to the Wisconsin epoch, the Spokane

[1] J H. Bretz, "Glacial Drainage on the Columbia Plateau," *Bulletin of the Geological Society of America*, Vol. XXXIV (1923), pp. 573–608.

[2] This question of differentiating Pleistocene epochs by talus accumulation will be discussed more fully in a separate paper.

[3] Frank Leverett, *Bulletin of the Geological Society of America*, Vol. XXVIII (1917), p. 143.

glaciation should be either Iowan or Illinoisan in age. Farther than this, the writer does not care to go. Ordinary criteria in use east of the Rocky Mountains for differentiation of drift sheets cannot safely be used for the correlation of these glaciations in Washington. The only one relied on here is the moraine-building habit of the Wisconsin ice sheet, a character which seems to have been world-wide.[1]

There were no channeled scablands on the Columbia Plateau before the Spokane glaciation. A mantle of loess, with a mature topography, completely covered it. The evidence for this conclusion is found in the great and remarkably persistent width of the Cheney-Hooper scabland tract throughout a length of 70 miles, and the various distributary courses out of it, some of which never were eroded to the basalt. These features never could have been formed, had spillways like those of the present existed. But with early escape southward retarded by the loessial hills and their small drainage ways, a wide spreading among them necessarily occurred, and some distributaries were able to cross to Crab Creek drainage.

A puzzling situation regarding glacial drainage exists in the vicinity of the small Spangle lobe. There is no adequate drainage route around it for glacial waters which came from Idaho and western Montana and entered the Cheney-Hooper scabland tract. Two spillways exist north of Mica, between Lake Spokane[2] east of this lobe and Pine Creek channel. Both have erratics in them, the highest at 2,550 feet A.T., but neither carried much water. There is no error involved in mapping this lobe because an ice dam at Spangle is required for the operation of the Pine Creek channel. This channel carried far more water than the Mica spillways, water derived directly from the Spangle lobe. Yet it also is inadequate for the drainage in question. And much more water went down the Cheney-Hooper scabland tract, in proportion to the immediately tributary ice edge, than passed through any other

[1] No till or other evidence of pre-Spokane glaciations has been found beneath the loess (save only the Cheney deposit). No till has been found in the scablands. The writer is unable to agree with J. T. Pardee who states (*Science*, Vol. LVI [December 15, 1922], pp. 680–87) that till occurs at "scores" of places on the plateau south of the limit of Spokane glaciation, as mapped in Plate III.

[2] Thomas Large, *Science*, Vol. LVI (September 22, 1922), pp. 335–36.

scabland tract except Grand Coulee in its later stages. It seems necessary, therefore, to assume a prominent *subglacial* drainage line, across the area covered by the Spangle lobe. This is best located along the preglacial valley of Lake (Marshall) Creek and the rock basin of Farrington (Fish) Lake. Out of this rock basin, just beneath the edge of the ice, the waters from the east emerged and joined those coming directly from the ice.

If the battle between the diluvialists and the glacialists, out of which has emerged our conception of Pleistocene continental glaciation, had been staged in the Pacific Northwest instead of the Atlantic Northeast, it seems likely that the surrender of the idea of a debacle might have been delayed a decade or so. Fully 3,000 square miles of the Columbia plateau were swept by the glacial flood, and the loess and silt cover removed.[1] More than 2,000 square miles of this area were left as bare, eroded, rock-cut channel floors, now the scablands, and nearly 1,000 square miles carry gravel deposits derived from the eroded basalt. It *was* a debacle which swept the Columbia Plateau.

[1] Except in the Hartline and Quincy structural depressions.

3

Reprinted from pages 97–99 and 257–259 of *Jour. Geology* **33**:97–115 and 236–259 (1925)

THE SPOKANE FLOOD BEYOND THE CHANNELED SCABLANDS

J HARLEN BRETZ
The University of Chicago

ABSTRACT

The channeled scablands of the Columbia Plateau in Washington are extensive elongate denuded tracts of basalt, deeply scored by huge high-gradient glacier-born rivers. As physiographic features they are unique. All the rivers which produced them converged to Snake and Columbia valleys, and the record of the tremendous flood thus engendered has been found all the way from the plateau scablands to Portland. Columbia River Valley below the confluence of Snake River carried a flood which at Wallula Gateway was 2 miles wide and reached 750 feet above present river level, at Arlington was 7 miles wide and 450 feet above present river level, at Lyle was 3 miles wide and 430 feet above present river level, and at Portland was nearly 20 miles wide and 350 feet above present river level.

The field evidence for the flood consists of (1) noteworthy, even spectacular, denudation and erosion of slopes below the upper surface of the flood, (2) spreading of the river in favorable places back among preglacial hills, (3) deposition of great gravel bars, some of them nearly as thick as the flood was deep, (4) damming of Snake River above the entrance of the glacial rivers by a delta built upstream against the course of the Snake, and (5) deposition of the Portland Delta, 350 feet thick, 200 square miles in area, and with bars 100 feet high on its surface.

THE PHYSIOGRAPHIC SETTING FOR THE FLOOD

Channeled scabland is limited to that part of the Columbia Plateau which lies between, and above the confluence of, the Snake and Columbia River canyons (Fig. 1). The slope of the plateau in general is southwestward, away from Columbia River on the north and toward Snake River and the lower part of the bounding Columbia on the south and southwest. Almost all the drainage of the plateau is consequent on this slope.

The Spokane ice sheet crossed the Columbia Valley north of the plateau and advanced a few miles out on it along almost the entire northern margin. Water from the melting of this ice flowed in the preglacial consequent drainage system, entering at or near the heads of these valleys and traversing their entire lengths. The ablation of the Spokane ice sheet must have been extraordinarily rapid, for the volume of water was very great, almost incredibly great, and in spite of high gradients to draw it off, the pre-existing valleys first entered were inadequate to carry it all, and the flood spread widely in a complicated group of anastomosing routes. This pattern is unique in that all the routes were greatly eroded, many of them being developed into canyons of notable proportions by the glacial streams, and some of them crossing prominent preglacial divides on the plateau.[1]

The pre-Spokane drainage of the plateau which entered the Snake was conveyed to one main route, that of Esquatzel Coulee, near Pasco. During the Spokane epoch, two other discharge ways were formed, one through Devils Canyon, near Kahlotus, and one through Palouse Canyon, between Hooper and Perry. The Palouse Canyon route now carries most of Snake River's share of the plateau run-off, Devils Canyon carries none, and Esquatzel Coulee carries but a small portion. This change from preglacial conditions was due to the profound erosion by glacial waters in Palouse Canyon. Greater volume and higher gradient at this place determined the change.

Columbia River had two considerable pre-Spokane tributaries from the plateau, Moses Coulee and Crab Creek. These proved inadequate to carry the Spokane flood, and six others were opened: "The Potholes," Frenchman Springs, and the Koontz Coulee group of channels, the first two being simply cataracts over the cliffs of Columbia Valley. All of these discharge ways leave the west side of that part of the plateau concerned in this study. None of them now carry plateau drainage.

[1] Set forth in some detail in the writer's paper on "The Channeled Scablands of the Columbia Plateau," *Jour. Geol.*, Vol. XXXI (1923), pp. 617–49.

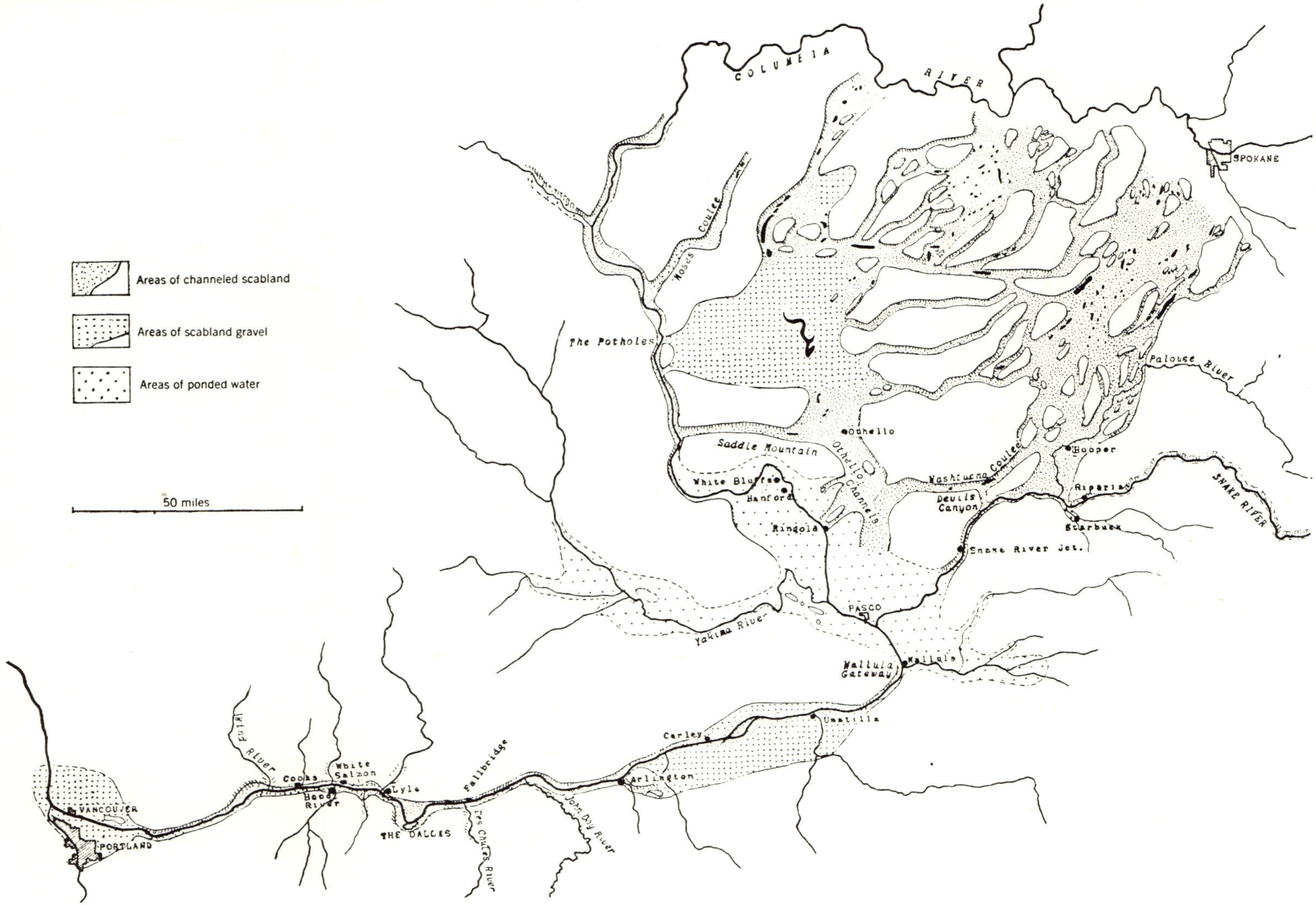

FIG. 1.—Sketch map of the region affected by the Spokane flood

[*Editor's Note:* Material has been omitted at this point.]

THE VOLUME OF THE FLOOD

If the volume of the Spokane flood can be measured anywhere, it can be measured at Wallula Gateway. Certain assumptions must be made, however, as to the capacity of that notch at the inception of the flood. The platform on which "The Sentinels" stand appears to be a remnant of the pre-Spokane floor at about 500 feet A.T. Spring Gulch is a hanging tributary valley which enters the Columbia just below the Gateway. The abrupt steepening of gradient in it occurs at close to the same altitude. The position of "The Sentinels" indicates that the pre-Spokane width of the notch did not exceed $\frac{3}{4}$ mile.

The waters rose to about 1,025 feet A.T. in mid-length of the Gateway, and had a surface gradient through the Gateway of about 15 feet to the mile. A cross-sectional area of the flood at its inception, to include that which spread back on the summits of the cliffs, amounts to approximately 3,485,000 square feet. The surface of the flood descended to 775 feet A.T. at the mouth of Juniper Canyon, about $8\frac{1}{2}$ miles below mid-length of the Gateway, and its cross-sectional area there appears to have been about 2,006,000 square feet.

These figures have been used by Mr. D. F. Higgins, of the University of Chicago, as a basis for computing the volume of the initial flood before any noteworthy deepening or widening had lowered its surface. Mr. Higgins used Chezy's formula $v=c\sqrt{rs}$, in which v is velocity, r the hydraulic radius (area of section divided by wetted perimeter of same), s the slope of the surface (expressed as tangent of angle of slope), and c is a constant. The value of c depends on the character of the channel, and was chosen to indicate a maximum of impedance from irregularities of sides and from eddies and cross-currents generated by turns in the course of the valley. This formula was devised for and is applied to small-capacity drainage lines, such as sewers, ditches, etc., and its use here involves the assumption, which cannot be true, that conditions in this enormous river were similar. However, it is the nearest approach to any formula which can be used. The actual volume probably was greater than the computed one secured by its use.

On the basis of surface slopes, the velocities indicated are 20.6 feet per second for the current in the Gateway itself and 30.4 feet per second for the current just below Juniper Canyon. On the basis of cross-sectional areas, the velocities appear as 17.5 feet per second in the Gateway and 35.7 feet per second at Juniper Canyon. Using mean velocities and computing for discharge (area × velocity), the enormous quantity of 66,132,000 second-feet, or 38.9 cubic miles per day, is secured. As above noted, this figure is probably below the actual amount. As it stands, it represents the melting of about 42 cubic miles of ice daily. The maximum measured flood of the Columbia at The Dalles, 100 miles downstream, is 1,170,000 second-feet. The Spokane flood thus was much more than 50 times the volume of present-day floods in the Columbia.

Another quantity of importance in interpreting the Spokane flood is the area of ice which, by surface ablation, could yield this prodigious quantity of water. Using the average maximum summer solar radiation of Lincoln, Nebraska, and Madison, Wisconsin (about 500 calories per square centimeter), Mr. Higgins finds that a maximum depth of summer melting by direct insolation today would be 2.66 inches. At this rate of melting, approximately 1,000,000 square miles of ice would be required to yield the water for the Spokane flood. But if the drainage of the Cordilleran ice sheet was controlled by the existing major divides and river valleys, not

100,000 square miles could have contributed to the discharge which crossed the Columbia Plateau. Obviously, the rate of melting must have been far greater than any direct insolation could have produced.

Computations such as these strongly incline one to doubt the actual occurrence of the flood. The writer has repeatedly been driven to this position of doubt, only to be forced by reconsideration of the field evidence to use again the conception of enormous volume. It is the only adequate explanation of the phenomena. These remarkable records of running water on the Columbia Plateau and in the valleys of Snake and Columbia rivers cannot be interpreted in terms of ordinary river action and ordinary valley development. The factors of declivity, of valley shapes, of rock, and of time cannot be adjusted to explain them. Enormous volume, existing for a very short time, alone will account for their existence. The field evidence all the way from Spokane to Portland, a distance of 300 miles, knits together in a consistent whole to support this explanation. Though there are a few features of the record whose origin is not yet understood, none are known which weaken the hypothesis.

FINAL STATEMENT

That the Spokane flood occurred is clear. That it was short-lived and waned very promptly after diversion around the northern margin of the plateau seems probable. The exceptional conditions which caused it are still obscure. All phases of glacial behavior considered, the writer at present can conceive of only two possible explanations for this flood. One is a very rapid and short-lived climatic amelioration in which warm winds and warm rains produced great surface ablation of a much-fissured ice sheet before the front of the ice retreated from the northern margin of the plateau. Apparently no such climatic change is recorded elsewhere, and the rapidity demanded seems impossible of realization.

The other possible explanation is that the Spokane flood was a gigantic "jökulloup," such as have occurred in Iceland repeatedly when volcanic activity has broken out beneath the cover of an ice cap. But nothing has been found in the literature to suggest Pleistocene vulcanism in the area which was drained across the Columbia Plateau during this episode.

Part II

THE SPOKANE FLOOD DEBATE

Editor's Comments on Papers 4, 5, and 6

In 1927 the Geological Society of Washington invited Bretz to give a lecture "Channeled Scabland and the Spokane Flood." The invitation was purposeful: a veritable phalanx of doubters had been assembled to debate the flood hypothesis. The principal flood antagonists were all scientists in the U.S. Geological Survey: W. C. Alden, James Gilluly, E. T. McKnight, G. R. Mansfield, and O. E. Meinzer.

THE WASHINGTON MEETING

Paper 4 (Bretz 1927a) presents the published record of the Washington meeting. Note that the first discussant, W. C. Alden, warned cautiously of the difficulties with the hypothesis. Lacking personal field experience in the region, he suggested that the rock basins might be collapsed lava caves, but he realized that the major features indicated stream erosion.

O. E. Meinzer argued that the glacially swollen Columbia could easily have cut the dry falls and deposited the great gravel fan of the northern Quincy Basin. He described the Quincy Basin as containing an extensive series of terraces. Moreover, the high-level channels were explained by progressive abandonment as the glacial Columbia progressively cut down to lower levels.

One difficulty that Meinzer appreciated from his field work in the Quincy Basin (Schwennessen and Meinzer 1918) was the fact that four great spillways led out of the region where water had obviously been ponded. Bretz (1923) had shown that the upper limits of the torrents that poured through these spillways occurred at the same altitudes. Rather than ascribing this coincidence to contemporaneous operation, Meinzer actually suggested that the spillways had been cut one at a time, and subsequent minor earth movements had later brought them to an equivalent altitude. Bretz (personal communication, 1978) recalls that Meinzer first formulated this unusual hypothesis in response to a query from Bretz at Meinzer's Ph.D. final examination at the University of Chicago.

James Gilluly was another of those at the Washington meeting who was upset with Bretz's hypothesis. Although he had not studied the Channeled Scabland in the field, he presented an imaginative and persuasive argument for the creation of the unusual landforms by the long-continued erosion of present-sized streams. He took exception to a minor point concerning the use of talus heights as time indicators and then attacked the major weak point in the flood hypothesis: the source of the water.

Ironically, Gilluly was a student at Franklin High School in Seattle when Bretz was a science teacher there. Nevertheless, the two men had few interactions either before or after the 1927 meeting. Their mutual friend Aaron Waters (1969, pp. A22–A23) records one revealing story about the relationship:

> Many years later during a field excursion across the eastern Washington plateau, Gilluly was greatly surprised by the overall scale of the Scabland features: the breadth and depth of the water-torn coulees, the fills of coarse outwash gravel topped by giant ripple marks, the huge abandoned cataracts, dry waterfalls, and deep plunge-pool basins left by the Pleistocene floods. As he gazed upward to the 200-m wide brink of Palouse Falls—over which the present diminutive Palouse River tumbles in a narrow 2-m wide ribbon, making only a delicate tinkle, as it dimples the placid surface of the huge plunge-pool below—Gilluly's expression changed from a look of astonishment to a broad grin. Clearly his mind had reconstructed the wild surge of a thrice-swollen Columbia River leaping from the brink with the roar of ten thousand Niagaras. Then he spoke loudly—perhaps competing with the noise of that vanished waterfall—"How could *anyone* have been so wrong?"

Among the spectators at the 1927 Washington lecture was J. T. Pardee. A year prior to Bretz's first scabland work, W. C. Alden, who was chief of Pleistocene geology, U.S. Geological Survey, had sent Pardee to study the scablands. Pardee published a brief article (Paradee 1922) proposing that the Cheney-Palouse scabland tract had been created by glaciation of rather unusual character. Bretz visited Pardee's field locations a season or two later and found that Pardee's "glacial" deposits were actually flood bars (Bretz 1974). Correspondence in the 1920s between Alden, Bretz, and Pardee suggests that Pardee was really considering a hypothesis that the scablands might be related to drainage from a large Pleistocene lake that he had studied in the western part of Montana (Pardee 1910). Apparently Alden dissuaded him from that idea (Bretz 1974). In his memorandum of September 25, 1922 to David White, chief geologist of the U.S. Geological Survey, Alden noted of Pardee's work: " . . . very significant phenomena were discovered in the region southwest of Spokane. . . . The results so far . . . require caution in their interpretation. The conditions warn against premature publication." David White later showed this memorandum to Bretz and asked if he knew what Alden's middle name was. When Bretz replied in the negative, White said, "It's Cautious, Bretz, Cautious."

It seems clear that the source of the great scabland flooding was known even as Bretz was struggling to defend his hypothesis at the Washington meeting. One story has it that during the discussion Pardee leaned over to his friend Kirk Bryan and said, "I know where Bretz's flood came from."

Bretz finally solved the source problem for the Spokane Flood in 1928. Although Harding (1929) without acknowledgment made the first announcement of Bretz's idea, Bretz (1930a) published the concept that scabland flooding resulted from an abrupt failure of the ice dam that retained glacial Lake Missoula (Pardee 1910, pp. 376–386). Bretz (1932a) clearly illustrated the relationship of Lake Missoula to the Channeled Scabland.

DEFENDING THE HYPOTHESIS

After the Washington meeting Bretz continued to answer various criticisms of his flood hypothesis (Bretz 1927b, 1928a). He also established some new lines of inquiry into the problem. His detailed paper on the scabland bars (Bretz 1928b) is reprinted in part as Paper 5. The description of the region at the mouth of the Palouse River is especially revealing. Only Bretz's theory can explain these field relationships. Bretz ended the paper with a plea that his idea be tested, not by emotion or intuition, but by the established principles of the scientific method.

In Paper 6 (Bretz 1928c) Bretz presents an especially eloquent statement of the field evidence for the Spokane Flood. This paper perhaps best illustrates Bretz's famous writing style. His writing has a unique forcefulness that conveys the emotion of scientific investigation. The original paper was also noteworthy for its colored plate (not reproduced here). Bretz (personal communication, 1978) recalls that the American Geographic Society shocked him with a demand for $200 to help defray the printing costs. Although he could not afford this fee, Bretz knew the illustration was essential, for it made clear the detailed anastomosing pattern of channelways that he had mapped in the field. Bretz eventually raised the money by solicitation among friends in Spokane.

The last of Bretz's field studies mainly added details to the story presented thus far. Bretz (1929) showed that each of the valleys entering the eastern margin of the scabland spillways contained flood deposits emplaced by phenomenally deep water flowing *up* the tributaries away from the scabland channels. Along the Snake River he traced these deposits to beyond Lewiston, Idaho, more than eighty-five miles upstream from the nearest scabland channel. The conclusion again defied conventional wisdom (Bretz 1929, p. 509):

> Up-valley currents of great depth and great vigor are essential. . . . No descending gradient of the valley floor can be held responsible. The gradient must have existed in the *surface* of that flood. The writer, forced by the field evidence to this hypothesis, though warned times without number that he will not be believed, must call for an unparalleled rapidity in the rise of the scabland rivers.

By the early 1930s Bretz's scabland studies were essentially complete. Bretz (1930b, 1932a) had published the last of his field results. This period was marked by planning for XVI International Geological Congress, to be held in 1933. Bretz campaigned to include at least one field excursion to the Channeled Scabland. He even wrote a guidebook (Bretz 1932b). Although the trip was approved, Bretz was selected to participate in a major expedition to Greenland that prevented his participation in the geology congress. The trip was led instead by Ira S. Allison, professor of geology at Oregon State University. Allison was familiar with the region from his own studies (Allison 1933), and he proved to be an able substitute.

REFERENCES

Allison, I. S. 1933. New version of the Spokane Flood. *Geol. Soc. America Bull.* **44**:675–722.

Bretz, J H. 1923. Glacial drainage on the Columbia Plateau. *Geol. Soc. America Bull.* **34**:573–608.

Bretz, J H. 1927a. Channeled Scabland and the Spokane Flood. *Washington Acad. Sci. Jour.* **17**:200–211.

Bretz, J H. 1927b. The Spokane Flood; a reply. *Jour. Geology* **35**:461–468.

Bretz, J H. 1928a. Alternative hypotheses for Channeled Scabland. *Jour. Geology* **36**:193–223, 312–341.

Bretz, J H. 1928b. Bars of the Channeled Scabland. *Geol. Soc. America Bull.* **39**:643–702.

Bretz, J H. 1928c. The Channeled Scabland of eastern Washington. *Geog. Rev.* **18**:446–477.

Bretz, J H. 1929. Valley deposits immediately east of the Channeled Scabland of Washington. *Jour. Geology* **37**:393–427, 505–541.

Bretz, J H. 1930a. Lake Missoula and the Spokane Flood. *Geol. Soc. America Bull.* **41**:92–93.

Bretz, J H. 1930b. Valley deposits immediately west of the Channeled Scabland. *Jour. Geology* **38**:385–422.

Bretz, J H. 1932a. The Grand Coulee. *Am. Geog. Soc. Spec. Pub. 15*, 89 pp.

Bretz, J H. 1932b. The Channeled Scabland. 16th Internat. Geol. Congress, U.S. 1933, Guidebook 22, Excursion C-2, 16 pp.

Bretz, J H. 1974. *Memories*, pt. III. Unpub. manuscript distributed by the University of Chicago, Department of Geophysical Sciences and J Harlen Bretz, 136 pp.

Harding, H. T. 1929. Possible supply of water for the Channeled Scablands. *Science* **69**:188–190.

Pardee, J. T. 1910. The glacial Lake Missoula, Montana. *Jour. Geology* **18**: 376–386.

Pardee, J. T. 1922. Glaciation in the Cordilleran region. *Science* **56**: 686–687.

Schwennessen, A. T., and Meinzer, O. E. 1918. Ground water in Quincy Valley, Washington. *U.S. Geol. Survey Water-Supply Paper* **425**: 131–161.

Waters, A. C. 1969. James Gilluly, pioneer of modern geological ideas. *Earth-Sci. Rev./Atlas* **5**:A19–A27.

4

Reprinted from *Washington Acad. Sci. Jour.* **17**:200–211 (1927)

CHANNELED SCABLAND AND THE SPOKANE FLOOD

J H. Bretz

The 423d meeting was held at the Cosmos Club January 12, 1927, President BUTTS presiding. The Secretary announced the election to active membership of W. N. WHITE and A. M. PIPER both of the U. S. Geological Survey, and also the resignation of FRANK TWEEDY and S. W. BEYER.

Program: PROFESSOR J. HARLAN BRETZ, University of Chicago: *Channeled scabland and the Spokane flood.* That part of the Columbia Plateau which lies north of Snake River in Washington bears a remarkable system of erosional and depositional land forms. They are extraordinary in their magnitude, in their extent and distribution, and unique in their relationships. Running water is generally conceded to have caused them but the unparalleled results indicate unparalleled conditions under which it acted.

The channeled scablands constitute the erosional part of the record. They cover almost 2000 square miles, about $\frac{1}{6}$ of the area of this part of the plateau. They are elongate tracts, oriented with the gentle dip slope of the underlying basalt flows, mostly bare rock or with a thin cover of coarse basaltic rubble, commonly with canyons in them, and are bounded by steep slopes of the deep loessial soil of the plateau. They constitute a curious anastomosing pattern, the down-dip convergences inherited from an earlier normal drainage pattern and the divergencies, equally numerous, produced by crossing of divides of this older pattern. There are hundreds of tracts of the higher loess-covered areas in the scablands, from a fraction of a square mile to many townships in area, all discontinuous and bounded by the scabland areas. The steep marginal slopes in loess are in striking contrast to the gentle slopes of the older drainage pattern surviving within each isolated loessial tract. Canyons in the scablands are multiple and anastomosing, amazingly so in some tracts; deep canyons and shallow ones uniting and dividing in a labyrinthine fashion about bare rock knobs and buttes unlike any other land surfaces on the earth. Certainly but few of these canyons are inherited from the older pattern.

The scabland drainage was discharged from the northern glaciated portion of the plateau through ten openings into the loess-covered area and led thence by nearly one hundred different routes of varying lengths to nine discharge-ways into Snake and Columbia rivers on the south and west. The canyon plexus is the most striking feature of the scablands and probably is most significant of conditions of origin. These canyons are interpreted as channels, not valleys, hence the term "channeled scabland."

The depositional land forms associated with channeled scabland are chiefly great mounded masses of little worn basaltic gravel. They occur on the down-gradient side of eminences and in other protected places in the scablands, and in the Snake and Columbia valleys below the entrance of the scabland drainage routes. They are not eroded forms, they possess aggradational slopes and they inclose depressions or by their position aid in inclosing depressions between themselves and adjacent rock walls. All attempts to interpret them as dissected remnants of terraces or originally continuous gravel deposits have failed. They are gravel bars of huge size.

A brief summary concerning the more significant features and relationships follows.

Canyons of the scablands. Largely channels of huge rivers, eroded during the Spokane epoch.

A—Rock basins in the canyons. Thousands of them. Commonly elongate with the canyons, generally constituting the canyon floor. Lengths as great as eight miles, depths as great as 200 feet. Some canyon floors essentially a series of rock basins. Formed by large vigorous streams plucking the columnar basalt. In no other way can most of these basins be explained. Some are potholes at the foot of extinct waterfalls.

B—Plexus grouping of canyons. Occur on crossings of divides of the older drainage pattern, the four largest groups ranging from 6 to 10 miles wide. Developed subfluvially like the high-water anastomosing channels of the present Columbia at the Dalles. Alternative explanation demands a remarkably braided pattern of an eroding stream, with narrower strands in many cases cutting deeper than broader ones.

C—Cataracts. Hundreds of extinct waterfalls, many of which during recession became wider, several two to three miles wide. Unless the record of very large streams, they should show the "horseshoe" concentration from any initially great width.

Areas surrounded by the channeled scabland tracts. Residuals of a once continuous loessial cover, with maturely eroded drainage ways. 100 to 200 feet of loess removed over large areas.

A—Aligned scarps of loess facing the scablands. Slopes 30° to 35°. The bluffs left by undercutting of streams whose width was that of the adjacent scabland, from half a mile to 15 miles. They truncate minor valleys of the older drainage pattern.

B—Small isolated loessial hills on the scabland. Slopes as above indicated, with "prows" pointing up the scabland gradient. Some are miles from any other loess. Others, in groups, record abrupt introduction of a large volume of water which simultaneously entered several of the pre-Spokane drainage ways and eroded them to bedrock, leaving these remnants of the former divides.

Trenched divides. Several remarkable cases where a canyon plexus has three or four closely spaced rock-basined gashes 200 to 400 feet deep across a divide, yet only one case where one of them cut deeply enough to divert subsequent drainage. Water must have been 100 to 300 feet deep above preglacial valley bottoms on the north to have crossed. No piracy nor headward erosion nor local drainage has been responsible. Good evidence that no post-Spokane uplift has occurred in these places.

Deposits on the scabland and in Snake and Columbia valleys. Discontinuous originally. Their features clearly record actual building of each individual deposit. Any explanation must start with this.

A—Gravel chiefly. Pebbles little worn, 90 to 99 per cent basalt, unweathered.

B—High deposits, above brink of canyons 400 feet deep and at foot of loessial scarps, yet identical with other deposits down in the canyons.

C—Bar forms, undissected, foreset bedding conforming to slopes where required by this hypothesis. Associated depressions as much as 50 feet deep where vigorous eddies existed. Some bars, 20 to 100 feet high, blocked subsequent drainage.

D—Deltaic bar, 5 miles long and 200 feet thick in Snake and Tucannon valleys, with foresets dipping up these valleys from point of entrance of scabland stream. In striking contrast with Snake River gravels immediately upstream, which are in 60-foot terraces, dissected and with large alluvial fans built out on them, and are composed of 90 to 95 per cent nonbasalt, well-rounded gravel.

E—Quincy structural basin. More than 15 townships covered with basaltic gravel in terrace-like and mesa-like forms. Channeled canyons tributary to and distributary from this settling basin. The forms, however, are bars, as field study has amply demonstrated. No interpretation as terraces will account for many significant relationships.

Anastomosis of entire scabland tract on the plateau. Contemporaneous occupation of all scabland routes seems indicated. No evidence on glaciated tract of marginal drainage to supply, in turn, any one or two of ten entrances to the scabland during any conceivable shifting of ice edge. All channels seem to have headed on margin of the glaciated tract. Anastomosis due to the huge volume of glacial water and the abrupt introduction, thus flooding a multitude of minor drainage ways of the plateau and crossing a multitude of minor divides. Insufficient time for eroson of a few adequately capacious spillways. Debouchure into Snake and Columbia valleys at very different levels, indicating varying depth of different channelways and a lowering water-level in these valleys during the discharge.

Wallula Gateway high-level scabland. A short narrow canyon south of junction of Snake and Columbia Rivers, 20 miles from nearest plateau scabland. Yet with same features of subparallel lateral canyons, rock basins, knobs and buttes as high on canyon walls as in the Snake and Columbia upstream. All scabland drainage passed through this canyon and the flood reached 900 feet above present river bottom, perhaps 650 feet above canyon bottom at beginning of Spokane episode. Constriction here caused the ponding recorded in lower scabland tracts on plateau and made possible the plexus crossings of divides. Erosion of Gateway canyon was rapid enough to lower the ponded waters while the scabland rivers were still running. No other conceivable cause of ponding is indicated elsewhere in the Columbia valley below the plateau.

Columbia Valley below Wallula Gateway. For 150 miles a descending series of scabland tracts and gravel deposits in Columbia valley.

A—Bars in mouths of tributaries. Basalt gravel, ranging in height up to 600 feet above the Columbia. Delta foresets which prevailingly dip back into tributary mouths.

B—Portland delta, area 200 square miles, foreset-bedded throughout, basalt, gravel channels and great bars, remarkable eddy depression on upstream side of a rock island in the delta.

There are many apparently possible alternative explanations for the remarkable features of the preceding list. Virtually every one of these, when applied, involves exceptional combinations of factors and no one of them will explain more than one or two of the fifteen listed phenomena. Most of them have been tested in the field and rejected. These extraordinary features must be treated as a *genetic system.* Their assemblage on, and limitation to, this little corner of the globe cannot be coincidence, as required by alternative hypotheses. The only genetic interpretation yet proposed which is inherently harmonious and which fits all known facts is that of a great flood of water abruptly issuing from the Spokane icesheet. The unfilled rock basins with gravel bars perched on their walls indicate abrupt cessation of this flood.

The cause of this Spokane flood is unknown. It may have been a "Jökullaup" or glacier flood produced by subglacial vulcanism but this hypothesis must stand or fall on field data not yet secured. (*Author's Abstract.*)

Discussion: W. C. ALDEN: Professor Bretz frankly points out the difficulties met in applying his explanation of the origin of the remarkable features of the Columbia plateau. It is not easy for one, like myself, who has never examined this plateau to supply offhand an alternative explanation of the phenomena. I have read Professor Bretz's papers on the subject with great interest but I am left with the feeling that some things essential to the true explanation of the phenonena have not yet been found. The "channels" appear to be due to stream erosion. The main difficulties seem to be: (1) The idea that all the channels must have been developed simultaneously in a very short time; and (2) The tremendous amount of water that he postulates as coming from the melting of the ice sheet in so short a time to do the work. It seems to me impossible that such part of the great ice fields as would have drained across the Columbia plateau could, under any probable conditions, have yielded so much water as is called for in so short a time. It also seems as though the estimated capacity of the Wallula Gateway, when fully opened, is too great for this gorge to have served as a bottle neck to hold above it a flood of such dimension to the level called for in the explanation offered. It appears that ice sheets of three distinct stages of glaciation invaded the borders of this region and may have afforded conditions of repeated floodings of much smaller volume. It would seem that a more extended study of the glacial phenonena is required about the heads of the scabland "channels" to determine, if possible, more exactly just what sort of glacio-fluvial discharge actually occurred. Perhaps the explanation of the phenomena does not actually necessitate contemporaneous development of all the "channels," nor in so short a time. The problem would be easier if less water was required and if longer time and repeated floods could be allotted to do the work. The conditions of repeated glaciation in the basins of Clarks Fork and the Columbia are not yet well enough understood to afford very sure bases for postulations as to stream flow therefrom. It is important and highly desirable that means may be provided for a more extended study of the Pleistocene phenomena of these basins and of the adjacent mountains.

JAMES GILLULY: The question of the existence of a Spokane flood rests on the interpretation of many highly abnormal field facts. The evidence presented by Professor Bretz is assuredly convincing as to (1) the anomalous, indeed unique, drainage features of the Columbia Plateau, (2) their direct dependence upon glacial waters, and (3) the necessarily large volume of many of these streams. However, certain criteria used to determine the actual quantities of water involved appear somewhat questionable. Both Russell and Jenkins have recognized a ponded condition of the Snake River at Wallula Gateway, but Russell has attributed it to monoclinal deformation rather than to a flood, and, while Jenkin's views are not clear, he presumably agrees with Russell. Presumably then, without definite evidence to negative this theory, scoured basalt at high levels here is not conclusive of a sudden deluge of Spokane waters. The overflow may well be much older than Spokane glaciation. The evidence presented by Professor Bretz to fix the Wallula cutting as Spokane and not earlier is the height of talus. In an article published in 1925 he pointed out that "most of the talus (at Wallula Gateway) is higher than three-quarters though some is typical." Thus the date of the overflow is not fixed accurately as the date of the Spokane glaciation, even if we grant that the three-quarter talus criterion of age elsewhere applied by him is valid.

This talus criterion, however, is very doubtful as an accurate time gauge on several counts: (1) The range in rainfall from point to point on the plateau is from 5 to 20 inches. If three-quarters talus is found at both places it appears then that the height is probably a relatively stable stage in topographic development, rather than a measure of total elapsed time since the cliffs were formed. (2) The validity of the criterion depends on the assumption that all fragments falling from the cliffs remain in the talus heaps. This assumption is assuredly unwarranted, for in the formation of talus rock is disrupted into finer fragments, which waste by weathering and are subject to removal by wind and by even the most ephemeral streams. The Colorado Plateau offers numerous examples of cliffs which have retreated scores of miles yet still have vertical faces. Similar tendencies must prevail here, although basalt is decidedly more resistant than those sandstones. (3) The well known differences in rate of weathering of canyon walls dependent on their directional trends also give one pause in accepting uniform talus heights as more than a very rough measure of the age of cliffs. If the talus is of uniform height in both east-west and north-south canyons it strengthens still farther the suggestion that the three-quarters stage of talus is relatively stable, rather than a good time measure. (4) Even disregarding these three points which appear to me of great cumulative importance as tending to throw doubt on the validity of the talus age criterion and accepting Professor Bretz's analysis, elapsed time since a cliff was formed varies as the square of the proportional talus heights. Seven-eights talus then means 49/36 as great age (or 134 per cent) as three-fourths talus. Wallula Gateway, a narrow gorge for a great river, is likely to have had its talus sapped from time to time in its history, so that even granting that three-quarters talus is ordinarily valid as a time measure (which for the reasons stated above is very doubtful), the flooding of the top walls of the Gateway is probably pre-Spokane, and the deepening of the canyon a much longer process than granted by Bretz. This idea is strengthened by considering the quantitative factor. According to Bretz's old measurements the discharge at Wallula Gateway was 38.9 cubic miles per day, or over 50 times the present flood volumes of the Columbia and, proportionally to drainage basin, over 200 times as large as the greatest recorded floods of the Mississippi. Now he believes the water reached even greater height and hence must have produced a still greater flood. To explain this great flood without retreat of the ice front, he has suggested—(1) a very sudden climatic amelioration (only to rule it out as very improbable) and (2) that subglacial volcanism resulted in sudden melting of large quantities of ice which formed the flood. This mechanism is wholly inadequate as, even allowing most generous thermal properties for basalt and perfect efficiency of transfer of this heat for melting the ice, even a 10-day flood would require 17 cubic miles of basalt or a layer about 9 feet thick over 10,000 square miles. Further computations show that the rate of cooling of basalt is so slow that if we assume the subglacial surface replaced by molten basalt it would require over 9,700 square miles of basalt exposure beneath the ice to produce only 350 cubic miles of water which would only maintain the flood postulated by Bretz for ten days. But the areal geology of the Columbia basin is sufficiently well known to completely eliminate any possibility of late Pleistocene volcanism of such magnitude. Even this preposterously low estimate of flood duration is eliminated by the physical factors (such as resistance of basalt, depth of channels above Wallula referred to that constriction and others) involved. How much less competent then, must we admit is the jokullaup hypothesis to furnish the tremendous flood volume postulated by Bretz. The incompetence of the postulated mechanism to furnish the flood volume required by Bretz's interpretations of the field evidence seems to call for a reinterpretation of that evidence. That, as suggested by G. O. Smith, Meinzer, and Ferguson, such a reinterpretation is apt to be considerably more complex than the suggested flood hypothesis, seems exceedingly probable. That the actual floods involved at any given time were of the order of magnitude of the present Columbia's, or at most a few times as large, seems by no means excluded by any evidence as yet presented.

E. T. McKnight: Three dry coulees cut southwestward from the "Othello channels" at the east end of the Saddle Mountains in central Washington across the aggraded surface of the Ringold formation and debouche at varying elevations as hanging valleys along the White Bluffs of the Columbia. All are markedly U-shaped in cross section. The master channel of the three is Koontz Coulee which debouches at Ringold at an elevation of 150 feet above the river. It is $1\frac{1}{2}$ miles wide in its lower and better defined portion and its bed lies from 200 to 300 feet below the flat undissected Ringold plain which borders it abruptly on the northwest. In its upper half it has cut through the soft Ringold sediments and has exposed the underlying basalt in the form of scablands. The two remaining dry coulees are of comparable size, averaging less than half a mile in width and from 75 to 150 feet in depth, much smaller than Koontz Coulee. They head at essentially the same point in the west rim of Koontz Coulee, 10 miles above its mouth, at an elevation of 250 feet above its bed, and after following divergent courses reach the bluff of the Columbia at points 2 and 15 miles, respectively, above the mouth of Koontz Coulee and at elevations of 425 and 300 feet, respectively, above the Columbia. Both in turn show remnants of still shallower braided channels at higher levels, appearing as short shunts off of the primary channels. The southernmost dry coulee maintains its U-shape practically up to the point where its profile breaks off abruptly to the level of the Columbia; the enclosing walls of the northern coulee, on the contrary, begin to recede diagonally and to flatten at a point 5 miles back from the river so that the incised character of the coulee is soon lost, though the old drainage line can still be traced through to the bluff overlooking the town of White Bluffs.

That all three of the dry coulees above described were produced under climatic conditions widely different from those prevailing at the present time is amply proved by the sage-covered depressions in the two smaller coulees, and by the scattered depressions in the upper regions of Koontz Coulee. Mr. Bretz has interpreted these dry coulees as the product of huge glacial torrents that emptied into the Columbia when its flood waters stood at or slightly above the levels of the present debouchures of the coulees. That these are at different levels is believed to be due to the different stages in which they were abandoned during the lowering of the flood of the Columbia, Koontz Coulee being the last one abandoned. The bed of the river at the time of this flood is believed to have been at essentially its present level. The writer believes that the three dry coulees were formed by glacial waters at a time when the Columbia lay several miles west of its present channel and have been converted into hanging valleys by the lateral shift of the river to the east against the soft sediments of the White Bluffs. That this lateral shift is a reality is proved by the vigorous manner in which the river is at present undercutting the bluff, producing numerous land slides, and by the fact that the Ringold formation of pre-glacial Pleistocene age, has been largely removed by lateral planation from the region west of the river although it formerly filled the basin as far west and southwest as the Yakima and Rattlesnake ranges, 15 to 20 miles southwest of the White Bluffs. The shift of the river is down the slope of the pre-Ringold basalt floor. Restoration of the profile of Koontz Coulee indicates that the Columbia lay 3 miles west of its present position when the coulee was formed. The two smaller coulees at the higher level are interpreted as distributaries formed when Koontz. Coulee lay at that level and abandoned after it had gained the mastery.

It is believed that the explanation of Mr. Bretz is inadequate to explain (1) the fact that the two smaller dry coulees previously described, although heading at the same place, reach the White Bluffs above the Columbia at different elevations; (2) the fact that a V-shaped valley along the southern base of the Saddle Mountains, which was produced entirely by drainage off of the southern slope of the mountains, also forms a hanging valley 275 feet above the Columbia, although in this case, as with the northernmost glacial coulee, the south slope of the valley breaks back several miles from the river so that the lower course of the old drainage is almost obliterated. Mr. Bretz interprets the coarse stream gravels and cobbles that cover extensive areas west of the river at White Bluffs and Hanford as the debris of the Spokane flood. The writer believes them to be the normal channel deposits of the Columbia during its eastward shift over the area in pre-glacial, glacial, and post-glacial times. The features described in this summary appear on the Scootenay Lake, Hanford and Coyote Rapids quadrangles.

G. R. Mansfield: Mr. Bretz cites the occurrence of numerous anastomosing channels, with associated rock basins, some of which are 8 miles or more long, carved in basalt to depths of 200 feet or more, and locally on some divides as deep as 400 feet; he notes the occurrence of hundreds of abandoned waterfalls with "potholes" at their bases, also in basalt, and speaks of the recession of some of these falls; and yet he ascribes all these phenomena to a single flood of relatively brief duration. Basalt is a hard rock and very resistant to corrasion, but it possesses a well known columnar jointing, which supposedly renders it susceptible to undermining and to plucking, and Mr. Bretz relies on this property to account for the unprecedented rapidity of erosion of the basalt which his hypothesis requires. I am not convinced that so much work could be done on basalt in so short a time, even by such a flood as is postulated. The Dalles of the Columbia, which Mr. Bretz says are typical scabland channels, and the various falls in Snake River, can furnish pertinent and definite evidence with regard to rapidity of erosion and of the recession of falls in basalt if systematic observations are carried on for a few years. It does not seem to me necessary to assume that all the scabland channels, or even that all parts of the same channels were occupied by water at the same time. Mr. Bretz notes that deep and shallow canyons unite and divide in labyrinthine fashion. Possibly some of the shallower channels were formed earlier than some of the others and now hang on the sides of more favorably located channels. Mr. Bretz based much of his argument for contemporaniety of all the channel phenomena upon the relative heights of talus piles beneath basaltic cliffs along the canyon sides. Probably few, if any, direct observations are available regarding the rate at which talus piles in basalt may form. From such observations as I have made in basaltic country in the past fifteen years I should say that the rate is so slow that considerable time intervals would fail to register significant differences. Such a measuring stick should not be applied too rigidly. Again contemporaneity of erosion can not well be assumed from the identity of materials composing high level and low level gravel bars. Since all the material, high or low, may be presumed to have come from much the same general sources, identity is to be expected. Some differences in state of weathering might be looked for if the time interval between high level and low level deposits was great, but even here other factors such as texture and mineral composition of the rock fragments would enter in and mere differences in weathering would not be conclusive. The general nature of the phenomena suggests conditions similar to those attending the ice front in New York State, where temporary channels and falls now abandoned were developed. Although these were temporary, geologically speaking, some of them appear to have persisted for long periods of years. The scablands seem to me better explained as the effects of persistent ponding and overflow of marginal glacial waters, which changed their position or their places of outlet from time to time through a somewhat protracted period. Not enough is known in detail of the glacial geology and physiography of the region to furnish an adequate basis for any connected story of events here. The hypothesis of a single tremendous flood should not be accepted without further detailed regional study.

O. E. Meinzer: I have seen only the part of the region under discussion, that including Quincy Valley, Grand Coulee, and Moses Coulee. As Doctor Bretz has stated, the erosion features of the region are so large and bizarre that they defy description. However, the Columbia River is a very large stream, especially in its flood stages, and it was doubtless still larger in the Pleistocene epoch. Its erosive work in the Grand Coulee and Quincy Valley, impressive though it is, appears to me about what would be expected from a stream of such size when diverted from its valley and poured for a long time over a surface of considerable relief that was wholly unadjusted to it. The dry falls in the Grand Coulee resemble Niagara Falls and are evidently the product of normal stream work. The deep gorge of the coulee below the dry falls was apparently excavated by the same orderly and long-continued process of head-end erosion as the gorge below Niagara Falls, and it could hardly have been produced in a short time by a flood of whatever magnitude. Quincy Valley, into which the waters of the Grand Coulee discharged, evidently became the scene of a lake in which sediments were deposited and which at first discharged westward into the valley of Columbia River, forming several cataracts that retreated some distance in normal fashion before they were abandoned. Later all the water was discharged through the present outlet, the lake was drained, and the stream cut into the sedimentary deposits in Quincy Valley to a depth of about 150 feet, forming a broad stream valley with a series of extensive terraces. All these were orderly and long-continued processes of erosion and sedimentation. The features of erosion and deposition are indeed very impressive, but they are, I believe, of a kind and size that would be expected from so large a river as the Columbia must have been in the Pleistocene epoch. The rock-cut terraces of the Columbia River Valley also indicate a succession of long periods of stream erosion. The short gorges below the cataracts that discharged the overflow of the lake in Quincy Valley open upon a stream terrace that stands 900 to 1,000 feet above sea level and 400 to 500 feet above the present river level. The floor of Moses Coulee is continuous with a terrace 800 to 900 feet above sea level and about 300 feet above the river. Though the evidence is perhaps not conclusive, it seems to me that the upper of these two terraces represents the floor of the

Columbia River Valley before the river was diverted to form the Grand Coulee and the lake in Quincy Valley, the lower terrace represents the floor of the Columbia River Valley at a later time when the Moses Coulee was cut, and the gorge below the lower terrace represents the later erosion work of the Columbia.

Having seen only this part of the region, in which I believe the existing features can be explained by assuming normal stream work of the ancient Columbia River, I am naturally loath to accept a theory of an abnormal flood for the scablands farther east. Before a theory that requires a seemingly impossible quantity of water is fully accepted, every effort should be made to account for the existing features without employing so violent an assumption. I suggest that full weight be given to the following considerations, all of which have, of course, already received careful study by Dr. Bretz: 1. The Pleistocene Columbia was necessarily a very large stream, especially in times of flood. 2. Its waters were diverted over country of considerable relief that was wholly unadjusted to it. 3. It is probably not necessary to assume that all of the region was channelled simultaneously. Is it not more probable that the water flowed successively over different tracts as the ice front changed? 4. Unless there is conclusive proof, it should not be assumed that along any drainage line the erosion work at high and low levels was done simultaneously. It would seem more probable that the work of erosion proceeded during a long time and that the high-level channels were abandoned as the stream cut down to lower levels. 5. Tilting and folding of the rocks have in this region occurred in recent geologic time, probably during and since the cutting of the Pleistocene channels. For example, there seems to be evidence that the upper terrace of the Columbia River Valley has been deformed. This recent deformation may account to some extent for channels cut through ridges that can not otherwise be well explained except by assuming excessive depths of flood water.

J. T. Pardee: Other things being equal, the amount of talus beneath a cliff would vary considerably according to whether the exposure faced toward or away from the sun. Disruption of the rock due to temperature changes and alternate freezing and thawing of water would be most effective on southern exposures.

Dr. Bretz: *Reply to Mr. Gilluly:* There are old weathered gravels down in Snake River canyon both above and below the entrance of the scabland rivers, and much lower than the upper limit reached by the glacial waters in this canyon. The canyon therefore is older than the Spokane episode or any gravel deposits in the scablands, and a canyon at Wallula Gateway must have been there in pre-scabland time. I believe that the field evidence for this conclusion would convince any of my critics. Scabland and gravel up on the walls of the Gateway do not date back to the initiation of the canyon, and some sort of ponding must be provided. Without a ponding episode, the upper scabland must be older than the tributary canyons at the Gateway. But even the map shows how the glacial waters cut up the shoulders between the small tributary canyons already in existence. A pre-scabland floor of the Gateway about 200 feet above present river level is suggested by the hanging condition of Spring Gulch. Russell's speculations on ponding in the Columbia Valley dealt with the berg-carried erratics, not with scabland, and called for a valley glacier down in the gorge at or near The Dalles, or for a subsidence of the entire region. I have believed that this second suggestion is correct. There is a possibility that final solution of the scabland problem will tie this berg-drift into the scabland story. Jenkins did not suggest any disagreement with my bottle-neck hypothesis, though his article was avowedly a discussion of the ponding. There are certain to be differences in talus height because of local variations such as I have indicated in one of my papers. But where these differences are definitely traceable to local factors, such as variations in basaltic structures or spill of local runoff over the cliff or an active stream in the valley below or other conditions, it seems permissible to so treat them and yet to draw generalizations. The blocky material on the surface of talus in the scablands indicates a rate of growth in excess of the rate of disintegration into fine material. Bryan has indicated this in one of his papers on Pedestal Rocks. Comparison with sandstone cliffs is hardly trustworthy. It is much more likely that talus accumulations in the scablands have been added to, rather than diminished, by wind action. And in the empty channels I do not believe that removal of talus material by any process has occurred in appreciable amount in post-scabland times. Furthermore, the unfilled basins of the channeled scablands, existing even close to the upper limits of glacial waters in Wallula Gateway, have lingered overlong for a region whose talus has reached a profile of equilibrium. I have not considered "equilibrium" attained until the talus covers the whole face of a cliff. A soil from disintegration of the basalt should cover a stable talus, and this is essentially lacking in the scablands or at the Gateway. The "jökullaup" sub-hypothesis I never have defended and never shall, until adequate field evidence from the required volcanic tract is forthcoming. But must it be dismissed because of Mr. Gilluly's computations? Are they conclusive? After indicating the untrustworthiness of figures, because of possible overlooked factors not represented in the computations, Gilluly moves into the very position I am scolded for taking. And the accounts of Icelandic jökullaup—are they merely early experiments in sensational journalism? Granting, however, that we must never again look toward subglacial vulcanism for a Spokane flood, must this close our minds to the flood hypothesis? I believe that my interpretation of channeled scabland should stand or fall on the scabland phenomena themselves. Perhaps, however, my attitude of dogmatic finality is proving contagious. I am glad to have convinced Mr. Gilluly that the scabland high up on the Gateway walls is stream work. I anticipated a skeptical attitude on even this point, for it is an extraordinary place for scabland, by any hypothesis for the great system on the plateau north of Snake River. The only known records for pre-Spokane ice on the plateau are about Spokane and Cheney where there is a weathered till beneath the Palouse loess. The whole record of glacial waters across the plateau and through the Gateway is post-loess! I do not yet share Gilluly's belief that pre-Spokane glacial waters are recorded in the Gateway, nor his confidence that the history can be diagnosed readily at long distance. Even a bed-side practitioner may err, I understand.

Reply to Mr. McKnight: Undoubtedly the Columbia has undercut White Bluffs and thus shortened Koontz Coulee. But its eastward shift is not down the slope of a pre-Ringold basalt floor. The basalt is at least 100 feet below the river surface at Ringold, at the foot of White Bluffs. And Gable Mountain, a rugged eminence five miles long and at least 700 feet above the river, stands in the eastern half of the tract which the river, by this conception, has made by lateral planation. Jenkins thinks that the Ringold silt underlies the gravel of this wide part of the Columbia Valley. The 200-foot range in altitude of the gravel deposits about Gable Butte and Gable Mountain appears to be wholly in material of the same age. If the gravels of this part of Columbia Valley are normal affairs of valley or channel deepening, the older and higher portions should lie successively farther west from White Bluffs. But no succession of terraces of different ages has been found; instead, the relief appears to be a matter of great bars and abandoned channels with "holes" in them as low as present river level, all disposed with reference to the two basaltic interruptions as they should be if the entire valley at one time were a river bottom. While the main floor of Koontz Coulee hangs only 150 feet above the Columbia, three other glacial spillways, all within a mile and a half of the mouth of Koontz Coulee, hang 250 to 400 feet above, and all have approximately the same gradient. If their profiles are projected, they will reach at least twice as far out into Columbia Valley. By Mr. McKnight's explanation, there must have been as much lateral planation by the Columbia while the glacial waters were using Othello Channels as in all subsequent time. This projection of profiles assumes a Columbia River bottom, at about 500 feet A. T. at that time. How Othello Channels and its distributaries (Koontz, etc.) could have discharged across the Ringold flat at an upper limit of 1150+ while the Crab Creek syncline was undoubtedly an open route (for it also carried the glacial waters) to the capacious Columbia Valley is a puzzle for which I have found no answer save great volume and ponding at Wallula Gateway. If my information were limited to this district alone, I probably would not have arrived at present interpretations. It is in the remarkable interrelationships of the channeled scabland *ensemble* that the conception of a Spokane flood finds support.

Reply to Mr. Mansfield: I have had no success in fitting the field evidence to the idea of shifting dischargeways across the scablands. I cannot get the glacial streams to cross at Palouse Canyon, Devils Canyon, etc., without a ponding farther down the Columbia, nor to cut the canyons or canyon groups at these separated places without maintaining all dischargeways while the ponded condition is being removed. A labyrinthine group admittedly records a succession of events in that the glacial river was drawn down from original wide spreading as the canyons were eroded. But such cutting apparently must be done while the Wallula ponding was being lowered, or else that ponding must be repeated for each successive epoch of occupation. Only the Moses Coulee and Grand Coulee-Drumheller Channels plexus groups would have developed without the backing up postulated. Mr. Mansfield's idea of obtaining the rate of erosion of the basalt at The Dalles is excellent, but probably can be applied only to the features exposed at low water, and altered only during high water. The deeper and inaccessible parts of such a channel group should be the most rapidly changed. I hope that Mr. Mansfield and others of the United States Geological Survey will be able some time to study the channeled scablands in detail. These features should take their place in the literature as a group of land forms without parallel, and a genesis that can be agreed upon should be established.

Reply to Mr. Meinzer: The erosion of Grand Coulee could not have taken a longer time than the sum of whatever glacial stream occupation it has had. The margin of the ice sheet must have remained within 35 miles of its maximum advance in order to produce the Grand Coulee diversion. We may multiply whatever time interval this represents by different glaciations to get a longer period but it does not seem possible to consider this canyon a normal product of long-continued stream erosion. The scabland plexus east of Grand Coulee, from Coulee City south to the Quincy basin, is certainly not a normal affair, for three pre-existing valleys were entered and much eroded by glacial waters from the north before deepening of the lower coulee occurred sufficiently to contain the discharge. The lake sediments of Quincy Valley lie beneath thick basaltic gravel deposits which are disposed, as in so many other places in the channeled scablands, in great mounds, several miles long, with gentle back slopes and side slopes that are constructional profiles. These are separated by three wide channels of contemporaneous origin, leading from Crab Creek and Grand Coulee across the basin to Drumheller Channels. My earlier interpretations were that these channels (Moses Lake lies in one of them) were dissected out of a once-continuous gravel fill, but many features could not be harmonized with this, and I have been trying the Spokane Flood idea since. Several years ago Mr. Meinzer suggested to me that tilting or some other adequate cause had diverted discharge from The Potholes and Frenchman Springs cataract to Drumheller Channels, and I have tried since to apply this in the field. But I cannot find any evidence to support it. The rock terraces of Columbia Valley to which Mr. Meinzer refers were there in present development when the Flood occurred. The prominent one at the Potholes alcove rises northward along the Columbia about 300 feet in four miles. The surface of the southern part of it, below about 1250 A. T., is typical scabland and has a sharply defined cliff back of it. The northern part, above the level of the flood as indicated by upper limits in all three spillways out of Quincy basin, carries no such record and has no bare rock in the cliff back of it. Slope of the cliff and talus, which is soil covered to the very top, indicate a much older feature. As Professor Davis would say, the cliff in its southern part has been "refresht." This northward rise of the rock terrace is due to the warping in Babcock Ridge, a low anticlinal into which the Columbia has cut its valley. That rock terrace was here, in its present warped condition, I think, before the glacial discharge occurred. The close approximation of upper flood limit on this terrace, in the two cataracts named and in Drumheller Channels seems to indicate that no warping comparable to what the terrace shows has occurred since the glacial flood. This is also the upper limit of scabland in Wallula Gateway. The upper limits in such widely spaced features as Othello Channels, Devils Canyon and Snake River Canyon do not vary more than 100 feet from 1250. There is, of course, no precise marker in the topography for upper limits, for most of these features are river-bottom forms, but I do not think my figures miss it by more than 100 feet. Such coincidence of altitudes as I have indicated is remarkable if different episodes are involved or if the region has been warped since Spokane time. Moses Coulee below Three Devils abandoned cataract has 200 feet or more of debris, mostly gravel, on the rock floor which therefore is as low as in Columbia Valley at the junction of the two valleys.

W. P. Woodring, W. W. Rubey, *Secretaries.*

5

Reprinted from pages 643–644, 655–662, and 700–701 of *Geol. Soc. America Bull.* **39**:643–701 (1928)

BARS OF CHANNELED SCABLAND[1]

BY J HARLEN BRETZ

(*Read before the Geological Society December 29, 1927*)

CONTENTS

INTRODUCTION

Gravel deposits in which the combination of topography, composition, structure, and magnitude is unlike anything described in geologic litera-

[1] Manuscript received by the Secretary of the Society December 28, 1927.

ture are widely distributed over the channeled scablands of the Columbia Plateau in Washington and thence along the Snake and Columbia rivers as far down as Portland, Oregon. They record extraordinary conditions of origin. No hypothesis for channeled scabland which does not account for them can be worthy of consideration. The writer's interpretation of channeled scabland [2] is so great a departure from prevailing conceptions of rivers and river work that numerous suggestions have been made by other geologists looking toward more acceptable, more conventional explanations. In none of them, however, is there an adequate appreciation of the character of these gravel deposits, perhaps because no adequate descriptions have been made by the writer. The endeavor in this paper is to show clearly just what these deposits are and to show the limitations of various hypotheses for them.

[*Editor's Note:* Material, including figure 5, has been omitted at this point.]

VICINITY OF THE MOUTH OF PALOUSE RIVER

One of the most remarkable tracts of channeled scablands is the preglacial Snake-Palouse divide, now trenched by Palouse Canyon. Before the inception of the glacial discharge four small tributary valleys, eroded into the basalt, drained southward into the Snake, but nothing worthy of the name of valley drained northward. The crest of the divide was much closer to the old Palouse Valley. Of the four Snake River tributaries on this 10-mile stretch, three are wholly included in the scabland, one of them having its head stoped back completely through the divide to constitute the present Palouse Canyon trench. The fourth headed east of the ravaged portion of the divide and only its lower two miles are involved in the scabland. Here conditions were especially favorable for deposition of flood gravel and here lies a deposit of singularly significant character.

The terminus of the minor divide north of this tributary, and nearly at right angles to the course of the flood, was overrun by the glacial waters and for nearly two miles the northern wall of the tributary valley

[2] J H. Bretz: The channeled scablands of the Columbia Plateau. Jour Geol., vol. 31, 1923, pp. 617-649.

J H. Bretz: The Spokane flood beyond the channeled scablands. Jour. Geol., vol. 33, 1925, pp. 97-115 and 236-259.

is completely buried from top to bottom in basalt gravel (figure 5). The south wall is good rugged scabland with no gravel on it. Near the western end of this spur the gravel bar deposited on it is 245 feet thick. Its structure is foreset, the beds dipping 21 degrees down the north wall into the tributary. This gravel certainly came over the divide from the north. As shown by the foresets, it did not come from the tributary valley above the scabland portion and it was not eddied back from the

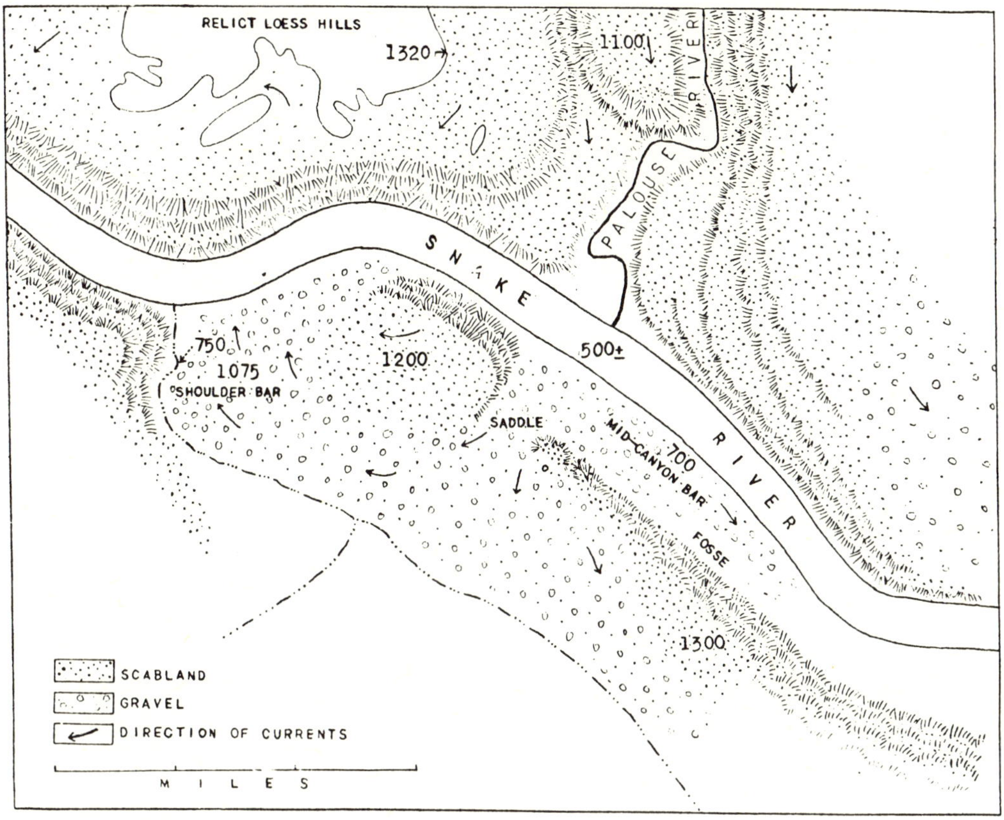

FIGURE 6.—*Sketch Map of Region at Mouth of Palouse River, Washington*

mouth. The valley contains the deposit and is therefore pre-scabland. The top of the bar is about 1,050 feet above tide and its base, at about 800, is 200 feet above Palouse River, half a mile away. Scabland surfaces, with rolled basalt cobbles and a few erratics, extend up to 1,325 feet above tide south of the east end of the bar.

If the combination of features shown by this deposit can be explained in conventional terms, the writer lacks the ingenuity required to do it.

An explanation logically should not be attempted without consideration of the more massive and more peculiarly placed gravel deposits in Snake River canyon at the mouthing of the great glacial stream.

One of these deposits is an elongated mid-canyon hill 200 feet above the Snake. It extends up the canyon two miles above the mouth of Palouse River (see figure 6). Snake River flows north of this hill, with basalt cliffs on one side and a 200-foot scarp of gravel on the other (figure 7). Basalt cliffs constitute the southern wall of the canyon also, and between them and the mid-canyon hill there is a depression, a fosse, 160 feet below the hilltop. The deepest part of this was a closed basin originally, about 50 feet below the lowest place in its rim. The hill slope toward the south wall is relatively gentle. At the west end the hill is wider and is built up against the south wall, which is as typically scabland as anything on the plateau to the north.

The material of this gravel hill is more than 90 per cent basalt. The fragments are very little worn. In contrast, the pebbles of nonbasaltic rock are conspicuously round and are unlike foreign pebbles in the scabland gravels of the plateau. Comparison with Snake River gravel, in present bars and in older terraces for 100 miles farther up the river, makes it as clear that this 5 per cent nonbasaltic material came down the Snake as it is clear that the angular basaltic débris came from the Palouse drainage. The comparison also makes it clear that the mid-canyon hill, the 200-foot trench of the Snake on one side, the 160-foot empty fosse on the other, and the scabland basalt cliffs to the south are wholly unlike anything farther up the Snake. East of the entrance of the scabland glacial flood, gravel deposits in the Snake Canyon are terraces, not bars. They are 75 per cent or more of nonbasalt pebbles. Nowhere in the first 50 miles up the Snake from the Palouse River junction are these terraces known to be 75 feet above the river. They are topographically older than the bar deposits, for considerable alluvial fans have grown out on them, while the fosse depressions of the bars are still unfilled. The rock walls adjacent to the terraces are much less vigorously expressed with ledges and cliffs than are those facing the great deposits of basaltic gravel in the scablands and at the mouth of the Palouse River.

But this mid-channel hill of gravel is not alone, nor is it the largest of its kind in Snake River Canyon. The south side of the canyon, opposite the mouthing of the glacial river from the north, possesses features so extraordinary that the writer almost fears that his description will be doubted. Only a very sketchy map (figure 6) can be presented here, although it shows relations correctly. The sketch shows a short tributary

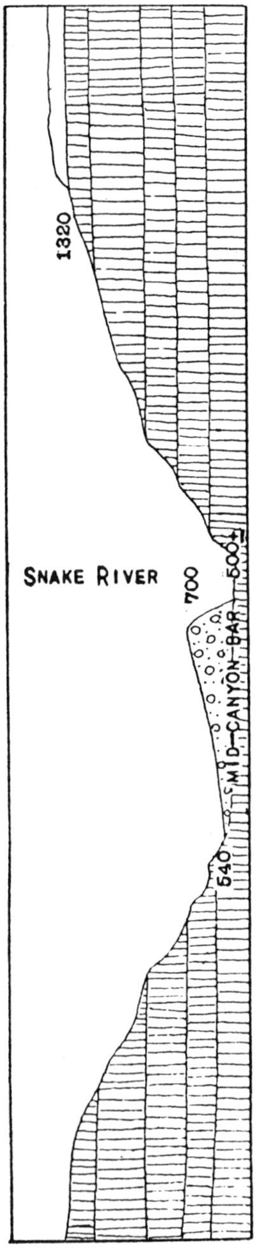

FIGURE 7.—*Cross-section of Snake River Canyon through the Mid-canyon Bar*

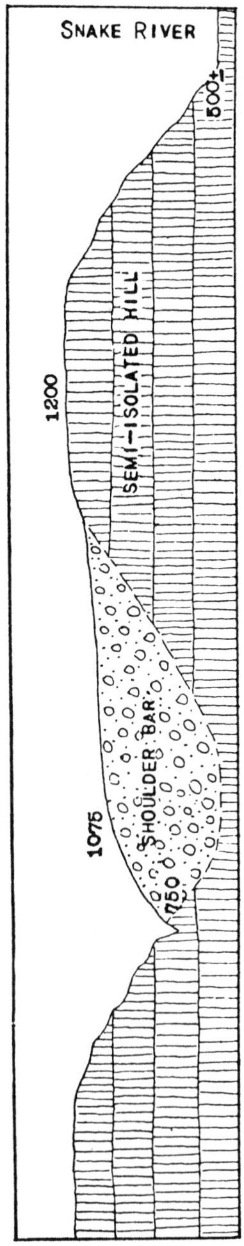

FIGURE 8.—*Cross-section Semi-parallel with Snake River Canyon through the Shoulder Bar*

View is near the mouth of Palouse River.

of the Snake flowing westward for a few miles parallel with it and entering it a little below the mouth of Palouse River. The divide between this tributary and the Snake has a saddle in it directly opposite the entrance of the Palouse, making a semi-isolated hill of the western terminus of the divide. The hill summit is about 1,200 feet above tide, 700 feet above Snake River. Both the saddle and the hill to the west are strikingly eroded into scabland forms, but on the south side, over into the tributary valley, there is an extensive gravel fill most of whose surface descends southward for half a mile with a 4-degree slope. Its placement is similar to that of the gravel in the minor tributary on the Snake-Palouse divide. There is no other gravel in the tributary valley. Another tributary draining northward and entering the one subparallel with the Snake is blocked by the extreme southern margin of this gravel deposit. The margin is a steep frontal slope descending 40 feet to the silt-covered flat which floors the valley immediately to the south.

The down-valley terminus of the divide is the most interesting place here on the south side of the Snake River. The semi-isolated hill, very greatly scoured and plucked on the north side and summit, is heavily mantled with gravel on the south side, almost from the summit down to the broad fill with the 4-degree slope. This heavy mantle is yet thicker on the southwest, and on the west side the gravel constitutes a great shoulder, projecting into and nearly closing the mouth of the tributary valley (figure 7). The top of this great shoulder is about 1,075 above tide, 550 feet above the Snake and hardly 100 feet below the rocky semi-isolated summit. The gravel in it is at least 325 feet deep. It is foreset-bedded, dipping 26 degrees northward *toward* the Snake, paralleling the 26-degree northward slope of the shoulder.

The gravel is 12 per cent nonbasalt. The nonbasaltic pebbles are all very well rolled and many are polished. Almost all are more than an inch in diameter. The basalt pebbles are subangular and for the most part small, so that much of the gravel is fine. In this sorted, fine gravel are no recognizable foreign pebbles.

The 325-foot shoulder of gravel extending westward from the semi-isolated scabland summit crosses the axial line of the preglacial tributary valley. The present discharge from this tributary swings around the tip of the shoulder and, after passing it, straightway returns to the axial line, thus in effect flowing eastward for a short distance *down* its own valley wall. In this short stretch it has cut a very narrow cleft in the basalt 100 feet in maximum depth—a canyon that is completely out of harmony with all other features of the tributary valley, and, let it be repeated, is essentially down the original slope of the valley wall (see figure 2, plate 19).

Though there are many other remarkable features of the district where the glacial flood entered Snake River Canyon, they will not be outlined here. Enough has been presented to show the nature of the problem. Conclusions that seem unavoidable are as follows:

1. The basaltic gravel south of Snake Canyon was derived from the scablands of Palouse River drainage, not from the walls of Snake River Canyon farther east, nor from the deeply soil-covered slopes of the tributary.

2. The scabland buttes and cliffs of the walls of Snake River Canyon are associated with, and only with, the deposits of basalt gravel in the canyon. This is a genetic association, not fortuitous.

3. The profiles of the hills, mounds, and mantles of this gravel are constructional, not erosional. Witness the fosse on the south side of the mid-canyon hill, the long foresets of the 325-foot shoulder which dip parallel with the surface, and the partial blocking of the tributary mouth.

4. If constructional, then these slopes are the original profiles of the deposits.

5. If original forms, they must be considered as gigantic bar deposits.

6 (*a*). If they are bars, then the mid-canyon bar, which is bulkiest at the west end and tails out up the Snake away from the mouth of the Palouse, was built by a great current setting up the Snake from the entrance of the huge scabland river.

(*b*). If they are bars, then the sheet of gravel extending south from the saddle was built by water flowing vigorously across that saddle southward out of the Snake and into the minor tributary valley.

(*c*). If they are bars, then the 325-foot shoulder of gravel which depends westward from the semi-isolated hill, which partially blocks the mouth of the tributary, and which has long foresets dipping into Snake Canyon, is a bar built largely by water returning to the Snake after crossing the saddle.

7. The base of the great shoulder bar is as high as the summit of the mid-canyon bar; yet, from all field evidence of degree of weathering and erosion and of continuous aggradational profiles across the saddle between the two bars, they are contemporaneous forms.

8. The easily identifiable pebbles of Snake River gravel in the great shoulder bar are more than 450 feet higher (above the Snake) than in any known terraces for 50 miles east of the entrance of the scabland flood.

[*Editor's Note:* Figure 2 has not been reproduced here.]

9. The upper limit of the well-expressed scabland just west of the mouth of the Palouse is marked by a pronounced and ungullied scarp in loess whose base is 1,320 feet above tide, or 800 feet above Snake River. The summit of the rock hill, semi-isolated by the saddle through the divide, is about 1,200 and not three miles distant. The summit of the great shoulder bar, therefore, was under 150 to 250 feet of flowing water.

If the description alone was not sufficiently challenging to raise doubts as to the actual existence of these phenomena, the above almost unescapable conclusions may succeed. The writer's answer to such doubts can only be, "Come and see!" The physiographic features of channeled scabland are spectacular enough to repay the visitor. And they are unique. They can not be classified under textbook analyses of land forms.

According to the Spokane Flood hypothesis, the glacial water, which crossed the Palouse-Snake divide in a stream nearly 10 miles wide, here entered the Snake in such volume and with such current that much of the flood crossed the canyon,[6] poured through the saddle and returned in a broad curve down the lower two miles of the tributary to reenter the canyon from the *south* side and to build the great shoulder bar with foresets dipping northward. Pre-Spokane terrace deposits in the Snake (like those now existing above the junction of Palouse River) were destroyed by the entering flood, and their characteristic pebbles, in a trip not to exceed three miles, were dragged 450 feet higher up the subfluvial slope and were thoroughly intermingled with the angular basalt gravel in the shoulder bar. Another diverging current flowed up the Snake Canyon, though no *outlet* exists in that direction, and built a bar 200 feet thick beneath its surface. Data to be submitted in another paper indicate a vigorous current, full of floating ice, moving *up* the Snake at least 25 miles above the mouth of the Palouse and probably much farther. The preglacial depth of Snake River Canyon here was essentially its present depth. This is attested by the accordant junction of many unviolated tributaries of the preglacial drainage system. The minor valleys of this system in many places on the plateau and along the Snake contain heavy deposits of gravel, sand, and silt, thrown back into them by the flood water. There is no escape from these conclusions, therefore, by holding that Snake Canyon has been deepened here since glacial water first crossed the divide from the north.

[6] Some of the water took the normal course, directly down Snake Canyon from the mouth of the Palouse, but some flowed vigorously up Snake Canyon to make a great pond whose record will be described in another paper.

The mid-channel bar was built in a strong current more than 200 feet deep. The shoulder bar, if built during the same episode, attests a depth of 550 feet. The scabby summit of the semi-isolated hill indicates that the depth was at least 675 feet, and the loess scarp, with scabland basalt ledges and scabland gravel deposits at its base, 1,320 above tide, requires a total depth of 800 feet of flowing water at Snake Canyon at this place. These are shocking conclusions.[7] But this brief account of some bars of channeled scabland is only well started. For more data let us return to the plateau.

[*Editor's Note:* Material has been omitted at this point.]

[7] Their validity depends on (1) the pre-scabland existence of the Snake and Palouse valleys; (2) the actuality of the divide crossing by glacial waters, and (3) the bar genesis of the gravel deposits. Evidence is submitted in many places in this and other papers to establish the first point. There should be no need to repeat it and there is no possibility of challenging it except by denying the existence or the described character and setting of the features noted. A reader unconvinced will remain so until he sees the region himself or until some investigator he will trust has reported on it. The divide crossing (item 2) exists and its very youthful features postdate the divide and the valleys on either side. The gravel deposits (item 3) exist and if correctly described clearly must be bars. Since no one bar at the mouth of the Palouse extends from the bottom of the canyon up to the level of the loessial scarps, one may demur at the conclusion that they all postdate the canyon's present depth; but to call for two or more episodes of rivers great enough to build such bars (one or more to predate the present canyon depths) does violence to the older terrace gravels farther up the Snake, to the character of the episode (for such bar-building obviously was unique in Pleistocene history), and to the existence of other bars whose uneroded forms do extend almost all the way from valley bottom to the scarp-marked upper limit of glacial waters.

The writer appreciates that hesitancy to accept the flood hypothesis, even definite denial of his interpretations of the land forms themselves, will exist until other workers enter the region. He has come to realize that probably no amount of printed description will dispel a belief that selection of data has been made to support a favored hypothesis rather than to find the true explanation; but if the field evidence for the flood exists as described, it is obviously useless to attempt an explanation by the hypothesis of a long and complicated series of Pleistocene drainage changes. (See the writer's article, "Alternative hypothesis for channeled scabland," Journal of Geology, vol. 36, 1928, numbers 3 and 4.)

[*Editor's Note:* Discussions of various other bars have been omitted at this point. The same features are described in Paper 11.]

Conclusion

As stated in the introduction, the purpose of this paper is to show that the gravel deposits of channeled scabland are not interpretable as ordinary valley fill; and that, if they be interpreted as bars, very great streams must be dealt with and only one episode can be granted. The local origin of the débris has been pointed out repeatedly. The incorporation of decayed material into otherwise fresh material has been indicated. The presence of soft, almost incoherent, rock fragments in the deposits has been mentioned. The existence of clearly defined spillways out of preglacial valleys has been described. The meaning of these facts, when considered with the unique topographic forms, is obvious.

This paper covers only the more significant bar deposits of the region. There are slack-water deposits yet to be described; there are inter-relations among the channels to be set forth. There is an extraordinary relation between loessial scarps and the preglacial valley depths, only touched upon here, and an extraordinary limitation of these scarps to a definite *niveau*. There are new data concerning the glaciated plain north of the plateau scablands. There are significant items concerning the mechanics of erosion of basalt, and an adequate interpretation of the *ensemble* is still to be drawn up. The writer has not yet adjusted all items to a sequential series of events and has no satisfactory explanation for the cause of the hypothecated flood, but the argument for the hypothesis is not yet closed.

Ideas without precedent are generally looked on with disfavor and men are shocked if their conceptions of an orderly world are challenged. A hypothesis earnestly defended begets emotional reaction which may cloud the protagonist's view, but if such hypotheses outrage prevailing modes of thought the view of antagonists may also become fogged.

On the other hand, geology is plagued with extravagant ideas which spring from faulty observation and misinterpretation. They are worse than "outrageous hypotheses," for they lead nowhere. The writer's Spokane Flood hypothesis may belong to the latter class, but it can not be placed there unless errors of observation and direct inference are demonstrated. The writer insists that until then it should not be judged by the principles applicable to valley formation, for the scabland phenomena are the product of river channel mechanics. If this is in error, inherent disharmonies should establish the fact, and without adequate acquaintance with the region, this is the logical field for critics.

6

Reprinted from *Geog. Rev.* **18**:446–477 (1928)

THE CHANNELED SCABLAND OF EASTERN WASHINGTON

J Harlen Bretz
University of Chicago

[With separate map, Pl. VI, facing p. 476]

NO one with an eye for land forms can cross eastern Washington in daylight without encountering and being impressed by the "scabland." Like great scars marring the otherwise fair face of the plateau are these elongated tracts of bare, or nearly bare, black rock carved into mazes of buttes and canyons. Everybody on the plateau knows scabland. It interrupts the wheat lands, parceling them out into hill tracts less than 40 acres to more than 40 square miles in extent. One can neither reach them nor depart from them without crossing some part of the ramifying scabland. Aside from affording a scanty pasturage, scabland is almost without value. The popular name is an expressive metaphor. The scablands are wounds only partially healed—great wounds in the epidermis of soil with which Nature protects the underlying rock.

With eyes only a few feet above the ground the observer today must travel back and forth repeatedly and must record his observations mentally, photographically, by sketch and by map before he can form anything approaching a complete picture. Yet long before the paper bearing these words has yellowed, the average observer, looking down from the air as he crosses the region, will see almost at a glance the picture here drawn by piecing together the ground-level observations of months of work. The region is unique: let the observer take the wings of the morning to the uttermost parts of the earth: he will nowhere find its likeness.

The Thesis

Conceive of a roughly rectangular area of about 12,000 square miles, which has been tilted up along its northern side and eastern end to produce a regional slope approximately 20 feet to the mile. Consider this slope as the warped surface of a thick, resistant formation, over which lies a cover of unconsolidated materials a few feet to 250 feet thick. A slightly irregular dendritic drainage pattern in maturity has been developed in the weaker materials, but only the major streamways have been eroded into the resistant underlying bed rock. Deep canyons bound the rectangle on the north, west, and south, the two master streams which occupy them converging and joining near the southwestern corner where the downwarping of the region is greatest.

Conceive now that this drainage system of the gently tilted region is entered by glacial waters along more than a hundred miles of its northern high border. The volume of the invading water much exceeds the capacity of the existing streamways. The valleys entered become river channels, they brim over into neighboring ones, and minor divides within the system are crossed in hundreds of places. Many of these divides are trenched to the level of the preëxisting valley floors, others have the weaker superjacent formations entirely swept off for many miles. All told, 2800 square miles of the region are scoured clean onto the basalt bed rock, and 900 square miles are buried in the débris deposited by these great rivers. The topographic features produced during this episode are wholly river-bottom forms or are compounded of river-bottom modifications of the invaded and overswept drainage network of hills and valleys. Hundreds of cataract ledges, of basins and canyons eroded into bed rock, of isolated buttes of the bed rock, of gravel bars piled high above valley floors, and of island hills of the weaker overlying formations are left at the cessation of this episode. No fluviatile plains are formed, no lacustrine flats are deposited, almost no débris is brought into the region with the invading waters. Everywhere the record is of extraordinarily vigorous subfluvial action. The physiographic expression of the region is without parallel; it is unique, this channeled scabland of the Columbia Plateau.[1]

General Features of the Region

In but few regions of comparable area does a single formation have so great a thickness, possess so nearly intact the original surface, and appear so consistently in the relief features as does the Columbia basalt in eastern and central Washington. The plateau is high where the basalt is upwarped, it is low where the basalt is downflexed, it is rough where the basalt is dissected, and relatively smooth where dissection has not occurred. Its main drainage lines are largely consequent on the structure of the basalt, as are its more prominent divides.

The flows of this great formation were originally horizontal. Now, however, there are but few tracts whose underlying flows are not

[1] The author's previous writings on this subject include: Glacial Drainage on the Columbia Plateau, *Bull. Geol. Soc. of America*, Vol. 34, 1923, pp. 573–608; The Channeled Scablands of the Columbia Plateau, *Journ. of Geol.*, Vol. 31, 1923, pp. 617–649; The Age of the Spokane Glaciation, *Amer. Journ. of Sci.*, Ser. 5, Vol. 8, 1924, pp. 336–342; The Spokane Flood Beyond the Channeled Scabland, *Journ. of Geol.*, Vol. 33, 1925, pp. 97–115 and 236–259; Channeled Scabland and the Spokane Flood, *Journ. Washington Acad. of Sci.*, Vol. 17, 1927, pp. 200–211 (an abstract and a number of criticisms); What Caused the Spokane Flood?, *Bull. Geol. Soc. of America*, Vol. 38, 1927, p. 107 (abstract presenting the idea that subglacial vulcanism may have been responsible); The Spokane Flood: A Reply, *Journ. of Geol.*, Vol. 35, 1927, pp. 461–468 (answering E. T. McKnight: The Spokane Flood: A Discussion, *ibid.*, pp. 453–460); Alternative Hypotheses for Channeled Scabland, *Journ. of Geol.*, Vol. 36, 1928, pp. 193–223; 312–341.

gently tilted or markedly folded. In general, the dip is westward in the eastern part and southward in the western part. Several sharply localized folds occur in the western part, and one in the southeast; elsewhere the warping is so gentle that the terms used for folded structures are not applicable. The drainage pattern faithfully reflects the structure, though it is obvious that not all streams can be consequent.

FIG. 1—Mature topography of the loess. (Photograph by Donald Bell.)

Though the west-flowing Spokane and Columbia rivers skirt the northern edge of the plateau for more than a hundred miles, they receive in this distance the drainage of only about 1300 square miles, so close to their route is the high edge of the gently tilted region. Structure determines the convergence of two-thirds of the entire region's drainage to the Snake and Columbia along its southwestern margin.

Rocks both older and younger than the basalt are elements in the topography of the plateau. The older rocks constitute local elevations, "steptoes,"[2] in the extreme eastern and northern parts of the plateau above the plateau surface. The younger rocks, all sedimentary, are a wide-spread mantle which originally completely covered the basalt. They are composed of piedmont waste, stream gravel, volcanic ash, lacustrine silt, and loess. The loess is much more extensive than all the other deposits. It is deepest in the eastern portion, but in all nearly level parts of the plateau it is deep enough to have been an important item in recording the episode of scabland-making. The essential fact regarding these sediments is that they are all much less resistant than basalt to erosive attack, especially by agents which erode too rapidly for weathering to keep pace and to transform the resistant basalt into incoherent soil. An essential item concerning erosion of the basalt is that, despite its hardness, it is

[2] I. C. Russell: Geology and Water Resources of Nez Perce County, Idaho, *U. S. Geol. Survey Water Supply Paper 53*, 1901, p. 22.

prevailingly cut with the close-set joints of columnar structure. Incipient decay and physical parting extend down these joints far below the depths to which the cores of the columns are decayed.

The rainfall of the plateau today is inadequate to maintain a single permanent and continuous stream from its own run-off. Yet the region is strongly marked almost everywhere by running water.

FIG. 2—Eastern wall of Moses Coulee at its mouth. Minor tributaries of the preglacial valley left hanging. Cliff summit 850 feet above coulee floor. (Photograph by L. C. Robinson.)

Aside from the scablands and the more strikingly folded tracts, this stream-determined topography expresses maturity of dissection (Fig. 1). In the eastern part of the region, where the loessial mantle is from 100 to 300 feet thick, the smaller valleys are entirely in loess. The larger ones, more deeply eroded, have their lower portions in basalt and are more youthful in appearance. Their canyon-like characters are due unquestionably in part to the greater resistance of the basalt, but evidence is not lacking to suggest that much of the mature aspect of the plateau may antedate uplift.

In addition to the mature slopes of the loess mantle and the canyon-like larger valleys on and bounding the plateau, the marks of running water include the scablands. These scabland tracts are exceedingly youthful features and are distributed along almost every drainage line from the northern high edge of the plateau sufficiently important to bear a local name. Yet in large part the scablands are not valleys. Some scabland tracts are completely lacking in valleys; some include several valleys, the separating divides themselves definitely scabland; some scabland tracts which for miles follow valley courses have margining scabland above the valley walls as wide as, or even wider than, the valley itself.

With but one exception, all valleys in the scabland have some tributaries from adjacent maturely dissected loess-covered country.

Except at and near their entrance to the greatly eroded scablands, these tributaries lack all trace of either the erosional or depositional modifications that produced the scablands. Their almost unvarying accordant junction with the valleys in the scabland indicate that those severely scoured valleys existed in present dimensions before the ravaging glacial waters entered.

All of the scabland drainage ways are united in a most remarkable anastomosing pattern. The convergences in this pattern clearly were inherited from the preëxisting drainage system. But the divergences, equally numerous, are the consequence of a flooding of the preglacial valleys so great that the minor divides, even some of the major divides, of the plateau were crossed, and adjacent routes were entered. Some of the routes thus entered were carrying glacial water from the ice edge. Others did not head far enough north and, without this flooding, never could have shared in the episode. Pronounced divergences took origin more than 50 miles in a direct line from the edge of the ice. It is nothing short of amazing to find that some of these divergences required rivers 200, 300, and even 400 feet deep: not ponded conditions of that depth, but vigorously flowing rivers.

Four Major Tracts of Scabland

In terms of separate origin on the northern high part of the plateau, the youthful scabland areas are resolvable into four different portions. The greatest of these is the easternmost. It heads in the general region of Cheney, 20 miles southwest of Spokane, and enters Snake River Canyon with the drainage of the Palouse River system. It may therefore be termed the Cheney-Palouse River tract. The next spillway group to the west, the Telford-Crab Creek tract, heads in the vicinity of Telford station, about 40 miles from Cheney. The third is Grand Coulee, a single great canyon across the divide, 40 miles west of Telford. The fourth and westernmost is Moses Coulee, also a separate canyoned route but heading on the southern slope of the plateau instead of crossing the divide.

The Cheney-Palouse River tract is 75 miles long and about 20 miles wide. Not one-fifth of it is occupied by loessial islands, though it was all covered with loess before the glacial discharge occurred. Several streams enter it from loess-covered country to the east, and one stream, Cow Creek, has its entire drainage area within the scabland tract. The wide spreading of the glacial waters is strikingly shown by this extensive scabland and its seventy-odd loessial islands. Yet more striking is the existence of eight well-marked subsidiary spillways which diverged from the west side of the Cheney-Palouse River tract at distances from the ice edge ranging from 3 to 60 miles. All these required operation of the major dischargeway for their own functioning; all were unquestionably contemporaneous at their initiation; no succession of operation could possibly have occurred by change in position of the glacial margin.

The Telford-Crab Creek scabland tract, next to the west, is as wide as the Cheney-Palouse River tract, but it loses its unity about 15 miles south of the plateau divide because of the number and capacity of divergent channels. However, convergence of all these (and of all but one of the Cheney-Palouse River divergences) occurs in a wide shallow structural sag (Quincy Basin) in the western part of the plateau. In this basin is the largest single area of gravel left by the glacial waters. It covers at least 24 townships. That all the channels of this Telford-Crab Creek spillway were in operation at one time cannot be gainsaid. There is continuous scabland northward from all channel heads of this tract across the plateau divide to the limits reached by the ice sheet. Some 40 to 50 square miles of this drain northward to the Columbia at present and did so in preglacial time; yet they clearly carried a wide, south-flowing sheet of glacial water to supply all the divergent channels.

Both Grand Coulee and Moses Coulee are canyons with maximum depths of 900 to 1000 feet (Fig. 2). The canyon character of these tracts is due to greater structural gradients of this part of the plateau. In both, there were preëxistent stream canyons (except in the northernmost 20 miles of Grand Coulee), but in both there was great deepening and widening by the glacial waters. Grand Coulee discharged a broad divergent sheet of water southeastward from its southern portion, but Moses Coulee had no divergences.

Causal Conditions Obtaining Over the Area

The anastomosing habit of scabland tracts over the entire area demands causal conditions that likewise existed over the whole region. These conditions were essentially:

1. A volume of glacial water too great for the existing valleys to carry.
2. A marked contrast in erodibility of the loess and basalt.
3. A high gradient for streams of the volume indicated by the abandoned channels.

Nowhere does the field evidence suggest any other cause, nowhere is it inconsistent with this explanation, and almost nowhere does it fail to show several mutually corroborative phenomena in support of this explanation.

More than 50 well-defined cases of divergence across loess-covered divides are known and at least three cases where such divides, parallel with (instead of transverse to) the direction of flow, were destroyed for 15 to 45 miles between the preglacial valley courses. In eight

FIG. 3

FIG. 4

FIG. 5

FIG. 3—Palouse Canyon between the two falls across the former Palouse-Snake divide. Cut about 400 feet deep. Looking north into the preglacial valley.

FIG. 4—Upper Palouse Falls. Close to the divergence from the preglacial valley.

FIG. 5—Palouse Canyon below the lower falls. Portion modified from a short preglacial tributary of Snake River. Rounded relict hill of loess on sky line.

places the divide crossings in the anastomosis were notably eroded in basalt. The most striking case is the diversion of Palouse River from its preglacial valley (Washtucna Coulee) to a very narrow canyon (Palouse Canyon) and the consequent shortening of the river by about 40 miles. Palouse Canyon (the lower ten miles of the river) is 400 to 800 feet deep in basalt (Figs. 3, 4, 5). Part of it is modified from a short preglacial tributary to the Snake, but the very narrow straight portion across the old divide clearly is a product of the Cheney-Palouse River discharge. A waterfall 196 feet high still survives in this canyon (Fig. 6).

Palouse Canyon is not the only record of the breaching of the divide. The whole summit for ten miles from east to west (transverse to the canyon) is marked scabland. It reaches 350 feet above the floor of the abandoned valley and 800 feet above the bottom of Snake River canyon to the south. Yet both valleys are preglacial. And in both the water flowed vigorously westward and southwestward toward the Columbia.

All scabland areas descend toward Snake River or Columbia River and enter these large valleys. Here the effect of the glacial waters might presumably be indistinguishable from the work of these two rivers before and since the scabland-making episode. Such a presumption would be based on the conception of these river valleys as valleys also for the glacial streams. The river-channel characters, the key to scabland interpretation, would give place to normal valley characters. It should then follow that the extraordinary anastomosis on the plateau was the consequence of shallow streams (for their widths cannot be disputed) or that there was sequence, not synchrony, in their operation. If the great widths and great depths actually existed contemporaneously in all channelways, both the Snake and the Columbia should similarly have become channels. They should have developed some cases of divergence in consequence of the flooded condition, and their tributary valleys which did not head in the scabland should have been notably ponded by the great volume. This is a crucial test of the writer's thesis.

FIG. 6—Lower Palouse Falls, 196 feet high. (Photograph by L. C. Robinson.)

The valley system of southeastern Washington is older than any scabland. This is amply attested by the graded slopes, the development of tributary systems, the relations of loess to scabland, and the presence of old stained gravels in valley bottoms. In many places on the plateau it is clear that the scabland features are superposed on this older drainage pattern in loess and basalt. Palouse Valley, for example, was as deeply entrenched in basalt east of the scabland as it is today; and, if a dam 100 feet high were built at the head of the canyon through the divide, the abandoned route (Washtucna Coulee) would be resumed. This abandoned preglacial course, now without drainage, is more than 40 miles long and 200 to 350 feet deep in basalt (Fig. 7). Minor tributaries from a large loess-covered tract to the north join it at accordant grade, yet belong to the pre-scabland mature drainage system.

There were eight different entrances of the scabland rivers to the Snake and Columbia valleys, one of them multiple. Three of them, entering the Columbia, are over precipices 200 to 400 feet high (Fig. 8), and two of them, entering the Snake, are very narrow canyons across the preglacial Snake-Palouse divide (Fig. 9). Judged by the criteria of an accordant junction and a system of pre-scabland tribu-

FIG. 7—Washtucna Coulee at Washtucna Lake. Scabland walls well shown. Compare this abandoned valley of the Palouse River with the canyon it now occupies. (Photograph by Robert Landon.)

taries, the only scabland routes entering the master valleys that were inherited from the older drainage system are Moses Coulee, Crab Creek, and Palouse River (via Washtucna Coulee).

Deposits in the channeled scabland system constitute another remarkable feature. There are two kinds; the channel deposits, prevailingly of pebbles, cobbles, and boulders, and the pond deposits back in valleys entering the scabland from unglaciated areas. This classification is not a rigid one, however, for in some places a ponded condition succeeded current flow.

The gravel deposits in the channels depart as much from conventional valley fills as the scablands do from valley forms. They are so situated with reference to the canyoned channels and the rock basins that they cannot be remnants of valley trains or similar deposits unless two episodes of enormous discharge be postulated. Their shapes are such that by no possibility can they be erosional in origin. They are local deposits, built beneath the surfaces of the great rivers. As channel deposits, they are contemporaneous with the erosional forms in basalt and in loess. Many of them are hills a hundred feet in height, some of them are 200, 300, and even 400 feet above their lower base. There is no term in geological literature for them except that of "bars" (Fig. 10). Their locations are all in conformity with the genesis implied by that term, as also are their structures, their shapes, and their composition. But if they are bars, the glacial streams in which they grew far exceeded any other glacial streams in depth. Were this conception of great depth based alone on these so-called bars, it could command but little respect. Rivers of this character must have left other evidence. Some of that evidence has already been briefly reviewed. The great widths shown on the map are undeniable, the physiographically recent great scouring of the basalt over hundreds of square miles is almost startling, the scarped islands of loess still carrying the mature topography of pre-scabland time are indisputable. And such divide crossings as Palouse Canyon, formed while the pre-glacial valley a mile wide and 350 feet deep was carrying a great stream away from the threatened divide, offer strong corroborative evidence. But there is still more evidence of the flooded condition of the valleys of eastern Washington. It lies in the other type of deposit noted, the silt and sand of the ponded tracts.

A generalization to which no exceptions have yet been found is that every non-scabland valley entering the scabland contains débris carried back into it from the scabland. This débris may consist of scattered erratic boulders, cobbles, and pebbles; or of stratified silt and sand; or it may be a gravel deposit partially blocking the entrance. All three types are found in many of these valleys. The erratic character of some of the débris is not limited to the larger fragments, it is to be recognized even in the sand deposits. And some remarkable lenses of berg-laid till in the stratified sand and silt have been found.

These features are common in valleys that receive and always have received their run-off entirely from loess and basalt. In each valley the records above listed occur only up to an altitude corresponding to the upper limit of the glacial waters where that valley opens into the scabland complex (Fig. 11). The two most notable cases of ponding yet found are in Snake[3] and Walla Walla valleys. In Snake Valley, the upper limit of erratic cobbles and boulders and of stratified silt and sand is 1350 feet A. T., and the altitude reached by the glacial flood across the Palouse-Snake divide is 1325 to 1350. At the entrance of the adjacent Walla Walla Valley, 40 miles farther down the Snake, the upper limit of the glacial waters was about 1100 feet A. T., and the deposits under discussion are known up to 1080. Maximum depths of the two ponds thus recorded are about 800 feet each.

These records of ponded waters are not topographically expressed except at the very entrance of the tributary valleys into the scabland. Gravel deposits here commonly have the form of embankments which, when built, partially or even completely blocked the tributary valleys. Elsewhere the materials generally constitute a mantle on all slopes below the upper limit, thicker on lower slopes but easily overlooked if one passes through by train or automobile. In contrast, the great mounded channel deposits are almost as spectacular as the associated basalt scabland.

---

[3] East of the entrance of Palouse River.

Fig. 8

Fig. 9

Fig. 10

Fig. 8—Frenchman Springs Cataract. A western spillway from Quincy Basin. The total height of the irregular cliff is 400 feet.

Fig. 9—Devils Canyon. A narrow gash across a preglacial divide. No stream other than local run-off has used it since it was formed.

Fig. 10—Bar in Crab Creek Valley, south of Downs. Except at the left, a partially enclosed depression lies between the bar and the cliff.

The junction of the two master valleys, the Snake and the Columbia, in the southwestern part of the river-scarred plateau is 120 miles by direct scabland routes from the head of the Cheney-Palouse River spillway and 140 miles from the Telford-Crab Creek spillway head. All glacial waters of the anastomosis converged here and escaped thence to the Pacific along the lower Columbia. The junction occurs in a wide structural valley where there is very little scabland but a large area carrying huge gravel mounds. Nineteen townships are covered deeply with this débris. The criteria by which the mounded gravel deposits in the scabland are adjudged glacial river bars lead to the same conclusion here. The relations between these bar deposits and their basin differ somewhat from the relations in the Quincy Basin up on the plateau. These 100-foot to 400-foot gravel mounds, though clearly having constructional profiles, have steeper slopes and better-marked river channels among them. They more nearly approximate river bars in appearance, as a scrutiny of the generalized contour lines on them (Pl. VI) will show. And they almost nowhere approach the upper limits or the outer limits of the glacial waters. In Quincy Basin, along the Telford-Crab Creek spillway group, they closely approach or actually reach both. This basin had but one outlet; the Quincy Basin had four. Scabland on the walls of this outlet (Wallula Gateway) is perfectly definite up to 1150 feet A. T. and probably extends to 1200 (Fig. 12). This gap is cut across an asymmetrical fold 12 miles south of the Snake-Columbia junction, a fold which constitutes the south wall of the structural valley. If Wallula Gateway was approximately as deep in preglacial times as at present[4] the waters flowing through it during the scabland-making episode were nearly 900 feet deep.

FIG. 11

FIG. 12

FIG. 13

FIG. 11—Terrace in Touchet River valley, tributary of the Walla Walla River. (Photograph by Robert Landon.)

FIG. 12—Scabland on east wall of Wallula Gateway. Smaller butte 300 feet, larger butte 450 feet above river.

FIG. 13—Near Celilo, Ore. Scarp eroded in ancient gravel formation; scabland at base a rock terrace 500 feet above the Columbia.

Obviously, another explanation is at hand

[4] New data seem to show this. Earlier interpretations called for considerable deepening by the glacial waters.

for the 900-foot vertical range of the marked scabland of the Gateway—that the upper scabland is older; so much older that it antedates the present depth of the canyon across the anticline. The evidence for a full development of the valley system draining the plateau before the glacial flood occurred has been outlined. The alternative hypothesis apparently cannot be harmonized with it.

FIG. 14—Steamboat Rock, in the middle of upper Grand Coulee. The largest mid-channel butte of the scabland. (Photograph by C. L. Beane.)

But before the matter be considered closed, the valley of the Columbia beyond the Gateway should be examined. Its character must have an important bearing on the problem.

Channeled Scabland Beyond the Plateau

From Wallula Gateway to the Willamette River, west of the Cascade Range and 200 miles beyond the Gateway, the Columbia basalt is the physiographically dominant formation. The plateau is generally considered to extend westward to the Cascade Range, but it is considerably more warped and folded than in eastern Washington. In a general analysis three portions may be recognized along the river: two downwarped areas separated by an area of upwarp. The river valley in the downwarps is wide and without notable walls; across the upwarp it is a canyon with maximum depth of a thousand feet. Across the folds of the Cascade Range farther west it is a great canyon or "gorge" 4000 feet in maximum depth; and still farther west, about Portland, there is a great widening which probably is largely structural.

The "wides" and "narrows" along 200 miles of Columbia River below the junction of the Snake have almost the same modifications that are found in similar situations above that junction. The wides carry great gravel deposits with bar forms almost intact. The stratification in these bars dips prevailingly away from the Columbia. The material is commonly but little worn, is poorly sorted, and is almost all local in origin. Striking contrast exists between present Columbia River gravel and this bar material. The bars are not all spread out in the structural wides, many are built in reëntrants in the walls of the canyoned portions. Some are built high above the river, some extend almost from river level up to heights as great as 400 feet. Mouths of tributary valleys are favorite places for them;

FIG. 15—Head of Spring Coulee, a divergent canyon leading southeastward out of lower Grand Coulee. (Photograph by L. C. Robinson.)

their form, composition, and structure indicating clearly that they were built by the greatly swollen Columbia, not by the tributary streams. Not one of these deposits can be explained as old stream gravels left during the deepening of Columbia Valley, or as recording former valley trains, without doing violence to the field evidence. No other river valley in the world possesses such features, so far as the writer has been able to learn. Like those on the plateau above the junction of Snake River they are channel forms, not valley forms.

As with the depositional part of the record, so with the erosional part. There are channels, rock basins, and buttes of very youthful character, eroded in basalt, at high levels above the Columbia. They are river bottom features. They assuredly do not date back to the original cutting of the Columbia Valley. Their occurrence as actual divergences from the Columbia across divides to tributary valleys, and the building at such places in the tributary valleys of great bar dams or partial dams, demand the same extraordinary relations of volume to capacity as in the scabland system on the plateau.

The erosional features are most pronounced in the narrows. Where the contact of weaker formations on basalt was reached by the immense glacial river the overlying material was swept off to produce a broad basalt terrace, etched with the characteristic vigorous strokes and margined by a youthful scarp in the superjacent formation (Fig. 13). To build these great bars in the lower Columbia Valley, to erode the high-level scabland and its margining scarps, and to allow the divergences into the lower courses of tributaries demand depths but little less than those at Wallula Gateway and in Snake River Canyon. A remarkable combination of all these features near the mouth of John Day River records 750 feet of flowing water in the Columbia.

The record of the great glacial river has been found as far down the Columbia as Portland, Ore. Here it appears to end in a deltaic gravel deposit which covers about 150 square miles. The surface of the deposit ranges from river level (which is nearly sea level) up to 360 feet A. T. Most of it lies above 200 feet. It is not a plane surface, and its irregularities are not due to erosion since the deposit was made. Though the structure everywhere is foreset, the direction of dip is so diverse in some places that, taken with the absence of a plane surface, it does not conform to the conception of a delta. A specific explanation for the ensemble would be difficult if one had no acquaintance with channeled scabland and its bars. But for the requirements of that explanation, the Portland Delta is ideal. It is a great assemblage of bars and channels, correctly disposed and constructed for the aggraded floor of a huge river here losing its current at the head of a broad drowned valley. The detailed relations to certain rock hills which are surrounded by the deposit are incapable of any other explanation than that of subfluvial origin of the whole deposit.

Erosional Forms in Basalt

The realization that a remarkable episode of glacial history was recorded in eastern and central Washington first came from an attempt to map and explain the bizarre erosional forms on the plateau surface. Running water, not of local origin, was clearly indicated as the map grew. Surviving areas of the earlier topography, a normal drainage pattern, were available everywhere for comparison. Certain unusual forms in the scabland and certain relationships in its pattern early led to the conception of prodigious streams. This conception demanded the existence of other features and relationships which had not yet been recognized, and it afforded a clue to the origin of some features which in themselves did not so strikingly tell of great volume. The development of the hypothesis has continually led to the finding of new field relations and new criteria, all of them consistent with the explanation, none of them pointing to any other interpretation. Furthermore, essentially every kind of evidence required by the hypothesis has now been found.

In the remainder of this paper, a genetic classification of the physiographic forms of channeled scabland will be presented.

Canyons: Grand Coulee

The empty canyons and dry cataracts of the plateau are easily the most conspicuous scabland forms. Among the canyons Grand Coulee takes precedence, followed by Moses Coulee. Grand Coulee has many times been explained as a former course of the Columbia across the plateau.[5] For a time it was considered to have been initiated by a great fissure,[6] and this idea was built into the first interpretation of the coulee as a glacially diverted Columbia.[7] Grand Coulee is nearly 50 miles long. It has an upper canyoned part about 28 miles long, a short stretch in mid-length across a structural sag (Hartline Basin), and a lower canyon about 20 miles long. At least four definite upfolds in the basalt are recognizable across the coulee, their existence conditioning the variations in character and the divergences.

Upper Grand Coulee is a simple canyon whose average depth is perhaps 900 feet. The width of its floor is nowhere less than a mile and a half and in one place is nearly five miles. In the middle of this wide place stands Steamboat Rock, a butte of basalt a square mile in summit area and 800 to 900 feet high (Fig. 14). Its summit is a part of the original plateau surface, now trenched by this glacial river canyon. Like the margining cliff summits, its top is marked by scabland buttes and channelings in basalt.

Lower Grand Coulee begins with a very irregular abandoned waterfall, 400 feet high, over nearly horizontal basalt flows. Most of the lower coulee, however, is eroded in tilted flows, along a monoclinal fold whose eastward dip is responsible for lower cliffs on that side. At least five pronounced divergent canyons lead southward and eastward out of the lower coulee and the mid-length structural basin (Fig. 15). All cross the minor upfolds of the basalt here. The head of the lowest of these is about 200 feet above the adjacent floor of Grand Coulee.

Grand Coulee is a part of the Crab Creek drainage system, but today it is a closed basin. Its lowest place, Soap Lake at the south end of the lower coulee, is shut in on the south by extensive gravel deposits. The lower coulee is made over from a preglacial drainage line out of the structural sag. Tributary furrows on the monoclinal slope to the west, expressing the general maturity of the plateau, hang hundreds of feet above the surface of the lakes on the coulee floor. The preglacial eastern wall, however, has been almost completely overrun by glacial waters and is striking scabland.

The upper coulee apparently is a glacial river canyon *de novo.*

---

[5] The earliest published suggestion of this appears attributable to Samuel Parker, who crossed the Grand Coulee in 1835. In his "Journal of an Exploring Tour Beyond the Rocky Mountains," Ithaca, N. Y., 1838, he speaks of the "Grand Coulé, which was undubitably the former channel of the river [Columbia]," p. 294.

[6] Probably first suggested by T. W. Symons in "The Upper Columbia River and the Great Plain of the Columbia," *47th Congr., 1st Sess., Senate Ex. Doc. No. 186*, 1882.

[7] I. C. Russell: A Geological Reconnaissance in Central Washington, *U. S. Geol. Survey Bull. 108*, 1893.

Unlike the lower coulee, its walls have no notches where preglacial drainage furrows hang above the bottom and therefore no rounded gable-like headlands in the upper profile. It is also unlike the other two scabland channel routes across the plateau divide (Cheney-Palouse River and Telford-Crab Creek) in that it is deeply entrenched. The others show scarified basalt over wide tracts, but the chief result of

FIG. 16—Part of "Dry Falls," the abandoned cataract in Grand Coulee. Summit of cliff is 400 feet above the plunge basin lake. This cataract was in use during a later episode of glacial diversion. Looking northeastward into Hartline Basin. Battleship Rock on right. (Photograph by C. L. Beane.)

glacial waters in them was to carry away the overlying loess. The upper limit of glacial water at the head of the Grand Coulee route is essentially the same as in the other two (2500 to 2550 feet A. T.), but the floor is nearly a thousand feet lower than in either of them. There are at least two reasons for this. The original spill across to the mid-coulee structural sag descended the monoclinal slope already noted. This afforded an initial descent of nearly a thousand feet in a very few miles, and the great river probably sloped back rapidly to cut the canyon. Furthermore, the coulee was in use in a later glacial episode than the one to which the rest of the scabland is ascribed. The ice of this later episode did not cross the Spokane and Columbia valleys east of the head of Grand Coulee, though to the west it advanced farther than the earlier ice sheet. Apparently Grand Coulee had already been deepened considerably in excess of the other routes and thus functioned alone across the plateau divide during this later episode. Some deepening probably occurred at this time.

The upper part of Grand Coulee, therefore, shows the following characteristic features of the canyons eroded entirely by glacial rivers on the plateau.

1. It crosses a preglacial divide. Though not an essential character, this is the prevailing relation to divides.

2. It possesses scabland on the "rim rock," above and behind the brink of the cliffs. Every canyon shows this, though not necessarily along its entire length.

3. It has steepened slopes in the loess or other sediment overlying the basalt. These are present wherever scabland occurs above the brink of the canyon walls.

FIG. 17—Head of lower Grand Coulee looking south from the brink of the old cataract. Pothole Lake at right. Battleship Rock in middle left of the view. The monocline which constitutes the west wall of the lower coulee determines the sky line. (Photograph by E. J. Evans.)

4. It has islands of loess on this upper scabland. They are found on all such scabland.

5. It has island buttes of basalt in mid-canyon.

Lower Grand Coulee lacks only the location on a preglacial divide. It shows some additional features that surely were not inherited from the preglacial drainage route.

1. Rock basins on the floor are numerous. All glacial river trenches in the scabland possess these features, except perhaps upper Grand Coulee.[8]

2. A large trans-coulee cataract exists. Most of these glacial river canyon trenches have more than one, though large cataracts also exist outside of the canyons.

3. There are several striking canyons diverging from the main canyon. These are really minor divide crossings in themselves.

Moses Coulee best illustrates one feature of these canyons, the remarkable deepening and widening locally produced in the preglacial routes. Douglas Creek, an affluent which enters the coulee 15 miles above the Columbia, may well be considered the major streamway, not the tributary, of the preglacial pattern; but its headwaters were not

[8] The floor of the upper coulee is covered with gravel and silt; and, if there are rock basins here, they are buried.

entered at the time the coulee carried a glacial river, and as a consequence the deepening and widening of the coulee left Douglas Creek Valley hanging 350 feet above the gravel fill in the coulee bottom. Wells show the fill to be more than 200 feet deep, and the amount of "hang" is correspondingly increased. Widening of Moses Coulee is responsible for perhaps a hundred feet of this total of 550 feet. Douglas Creek, though a permanent stream, has gashed the lip of the hanging valley for less than a mile back during all post-trenching time. Since the later ice sheet discharged but little water through Moses Coulee, the slow rate of normal valley deepening, compared with the rate of glacial stream canyoning, is obvious.

Cataracts and Cascades

Their name is legion. Scarcely a scabland area is without ledges transverse to its length. In some places, whole series of them are in sight from well-chosen viewpoints. Most of them have rock basins at their base, commonly empty basins. Those less than 20 feet high are almost negligible in developing the argument, so common are those of larger dimensions. Sheer descent of their full height is rare, though Palouse Canyon has one 196 feet high (now carrying Palouse River), Grand Coulee's abandoned falls (Fig. 16) are 400 feet high, and Quincy Basin has three such spillways out of the western part, with vertical transverse cliffs 200, 400, and 400 feet in maximum height. In most cases the glacial rivers descended in gigantic cascades, Moses Coulee having a complex one a mile wide and a mile long, in which the descent was 400 feet.[9]

Recession of these cataracts and cascades left canyons that may be mere niches or recesses in the basalt wall over which the water fell or that may be miles in length. Many of the smaller ones came into existence only through the sweeping off of the loess and the uncovering of buried basalt ledges of the pre-loessial topography. Staircase Rapids, near Washtucna, is an example.

Among these cataracts there were a few double falls each member of which receded at approximately the same rate, so that the island in mid-channel became very much elongated, like a great blade, as the falls receded and the canyons lengthened. Potholes Cataract, one of the three western spillways out of Quincy Basin, has the most striking mid-cataract blade. It is a mile and a half long, a quarter of a mile in maximum width at the base, and its cliffs are from 100 to 275 feet high (Fig. 18).

It seems entirely probable that upper Grand Coulee was cut through the divide by the stoping of cataracts or cascades down the steep monocline north of the structural Hartline Basin. Steamboat

[9] Not 600 feet, as previously reported by the writer.

Rock thus would be explained as a Goat Island around which both falls retreated to unite again as one and thus to isolate it. An actual case of this sort is Battleship Rock which stands just in front of the great cataract at the head of the lower coulee (Fig. 17). Another illustration of canyon-making by cascade-stoping is Devils Canyon, a short spillway from Washtucna Coulee to the Snake, 15 miles west of Palouse Canyon.

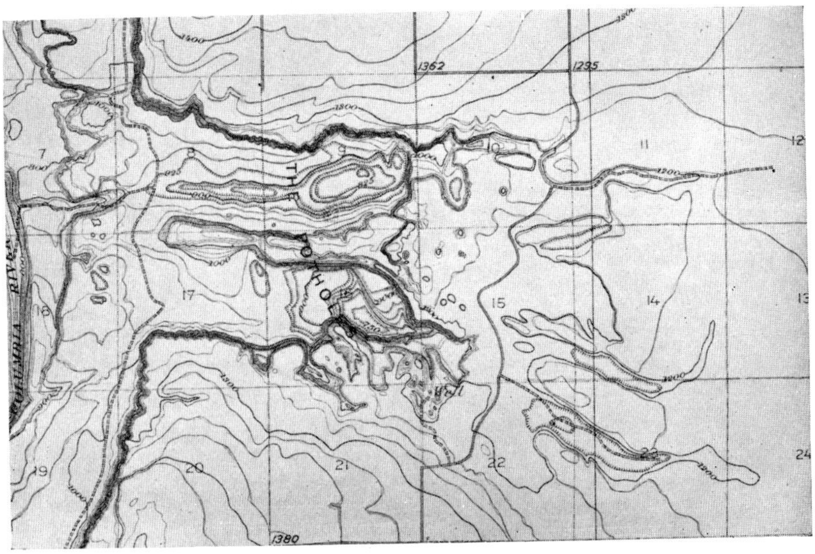

FIG. 18—Potholes Cataract, one of three cataract spillways from the west side of Quincy Basin. (From Quincy sheet, U. S. Geological Survey.)

Rock Basins and Mid-Channel Buttes

There is no feature of channeled scabland more challenging than the rock basins. There are thousands of them, from shallow saucers to the great canyoned basin of Rock Lake on the eastern margin of the Cheney-Palouse River spillway, more than seven miles long and having a reported depth of 250 feet near its northern end. The rock basins are not limited to the canyons and the cascade chutes or the foot of cataracts. They occur in great profusion on the broad uncanyoned heads of the two eastern scabland tracts. They occur on the torrent-swept summit of the Palouse-Snake divide, on the slopes of the preglacial valleys involved in the scabland discharge, and on the scabland benches above the rim rock of these preglacial valleys. Though not the only closed depressions left in the area after the scabland-making episode, they are the most common.

Most of the basins in the eastern and northern higher part of the plateau contain lakes and swamps today. Farther southwest, where the rainfall is less, many are empty most of the year. Some of these empty rock basins are 200 feet deep, though they have been receiving wash ever since the glacial flood subsided. Many are so steep-walled that they can be entered only at favorable places, and the bottoms of a few are accessible only by rope or ladder. The steepness of their walls is comparable to that of the associated canyon walls

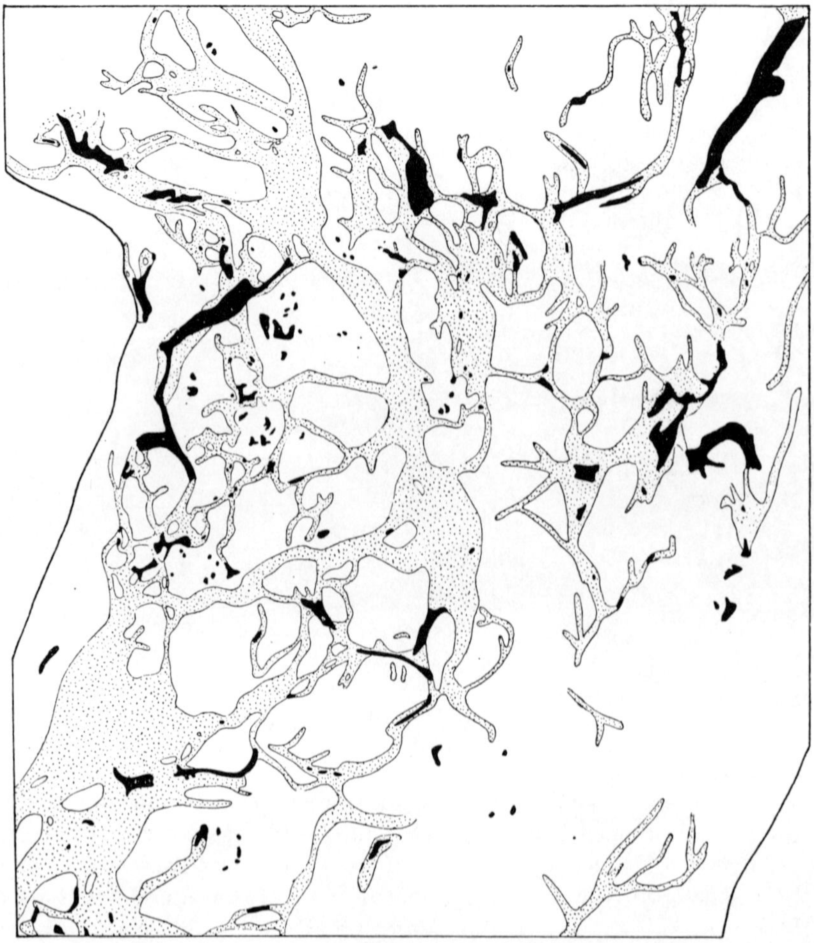

FIG. 19—Channels, rock basins, buttes, and general scabland of a part of the Drumheller Channels. Rock basins black, channels stippled, buttes and general scabland blank. The beginning and ending of many channels in the plexus is better shown here than on the contour map, Figure 21, as also is the extreme irregularity of the channel pattern. The larger rock basins are notably linear and are largely in channels. It must be understood, however, that the highest scabland is itself channel floor of the great spillway. Scale approximately 1:120,000, slightly more than half an inch to a mile.

and butte walls and has the same explanation. Both have their slopes determined by the character of the basalt. Where that has large, well-developed columnar structure (and where glacial rivers had adequate gradient) slopes are steepest: where this structure is absent or poorly developed, the rock basins have gentler slopes.

Buttes in mid-channel are far more numerous than abandoned cascades and cataracts, and most of them cannot have been formed by cataract recession. The association of isolated hills and hillocks of basalt with rock basins in the scabland channels is characteristic. Where the rock basins are deep, the buttes are prominent. Though one

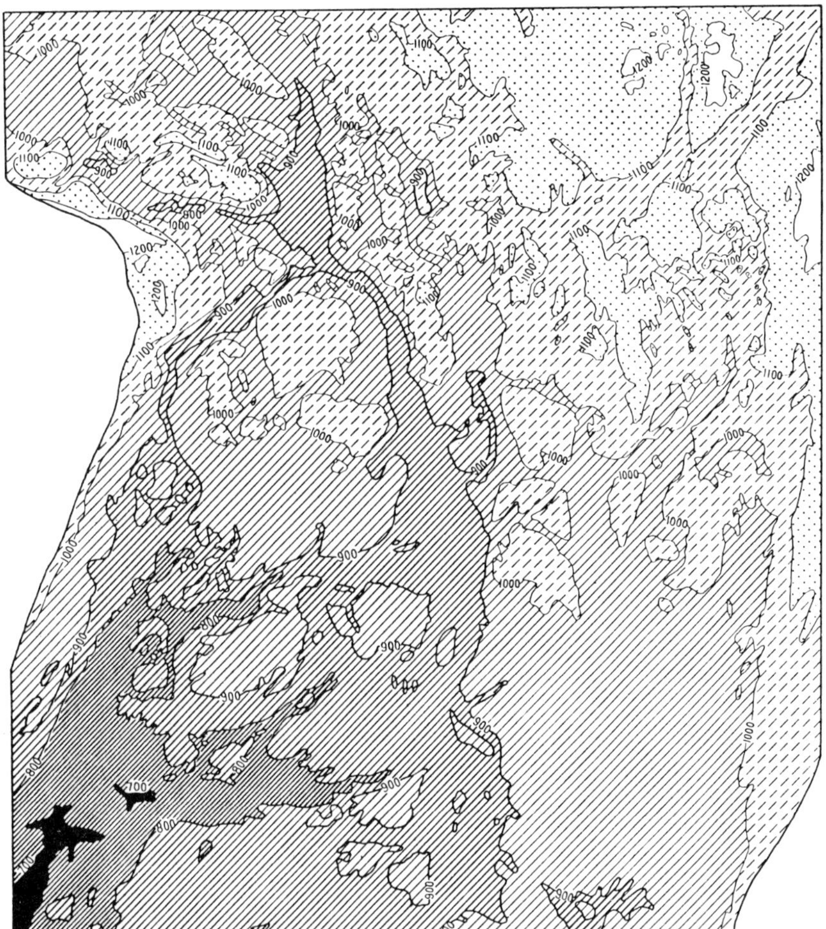

FIG. 20—Hypsometric map of a part of Drumheller Channels. Distribution of higher and lower tracts is more graphically shown than by the intricate and unusual contour pattern of the map, Figure 21. The contour interval is 100 feet. The highest areas (blank) are over 1200 feet; the lowest (solid black) less than 700 feet A.T. Comparison with Figure 19 will show the location of several of the largest (including the deepest) rock basins and many of the channels on the higher part of the scabland plexus.

is the reverse of the other, they are genetically associated forms. The conditions that gave rise to one produced the other. The maximum depth of erosion by glacial waters in any one place is the sum of butte height and basin depth. In such tracts the spacing of buttes and basins is very close.

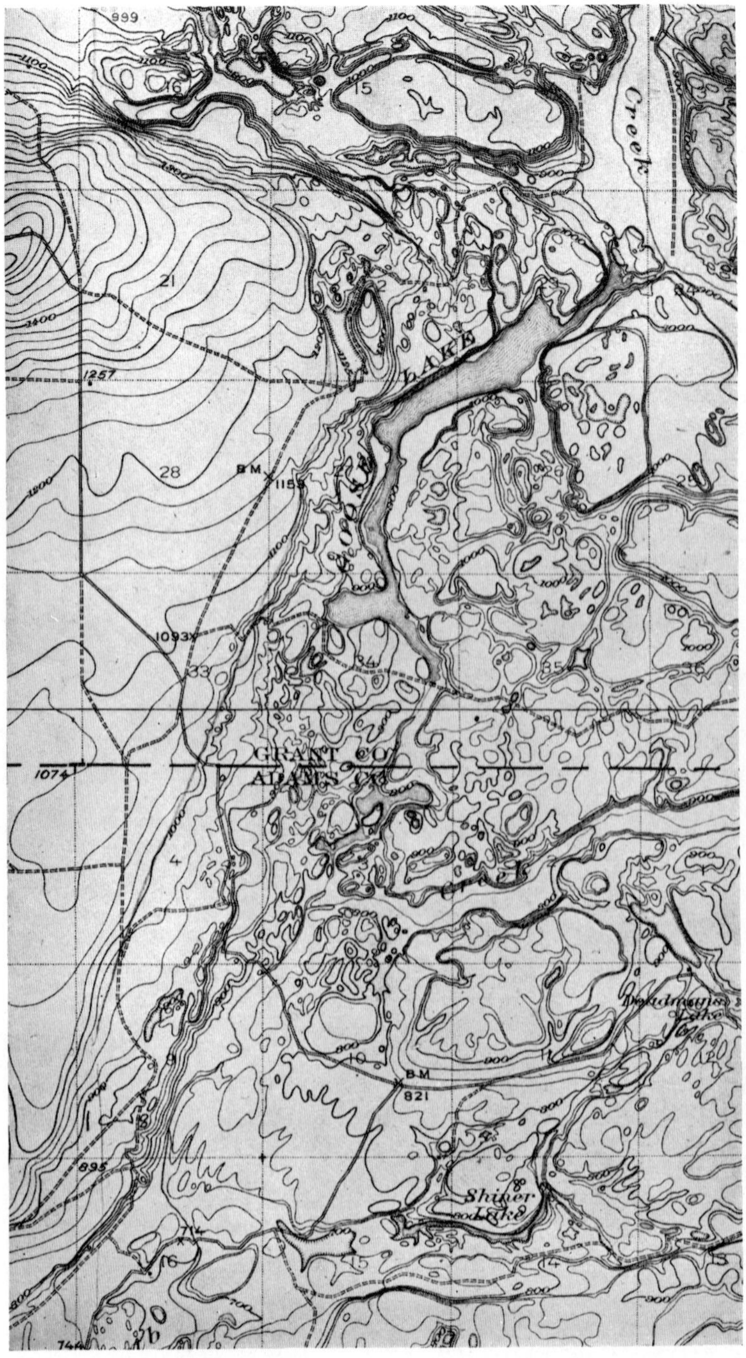

FIG. 21—A part of the Drumheller Channels, the southern spillway from the Quincy Basin. (Reproduced from the Corfu (above) and Othello (page facing) topographic sheets of the U. S. Geological Survey.)

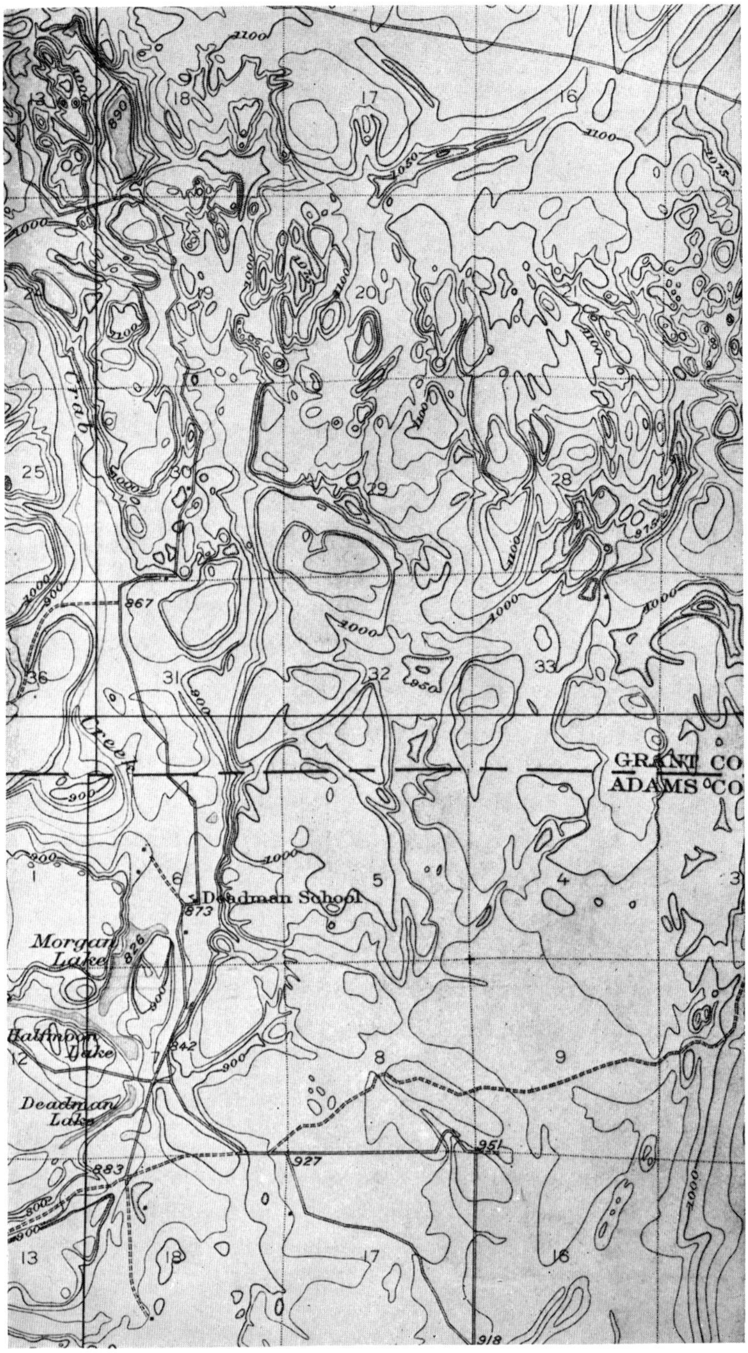

The area shown is slightly less than that of Figures 19 and 20 with which it is to be compared. The scale is one inch to a mile, nearly twice that of Figures 19 and 20. The contour interval is 25 feet.

Widely spaced buttes also occur on broad scabland surfaces which have no other marked relief (Fig. 22). The summits of some of these are higher than the loess-basalt contact in adjacent marginal scarps. They appear to be remnants of pre-loess elevations on the surface of the basalt and are not a record of great erosion by glacial water. But the closely associated buttes and basins have no adequate explanation other than the one proposed.

Plexus Groups: The Drumheller Channels

There are a few tracts of scabland where all the erosional features above described exist in a closely related group, in sharp contrast with the adjacent higher preglacial topography. The most marked of these is Drumheller Channels, the southern dischargeway from Quincy Basin (Figs. 19, 20, and 21). This spillway is a labyrinth of abandoned channels and rock basins among buttes, hills, and higher scabland surfaces of the basalt. The maze is eroded across the nose of a plunging anticline in the basalt. Its total width is nine miles, and in a single cross section the total vertical range of the scabland is 375 feet. It is difficult to count the channels because of the anastomosis and the indefiniteness of some of them, but nearly 150 definite channelways can be identified. They range from less than a mile to more than ten miles in length, and their floors are as much as 200 feet below adjacent butte tops. Almost every butte top has channels across it or heading on it, almost every large channel has buttes in it, and in many cases assignment of definite boundaries must be arbitrary. Most of the channels have continuous descending gradients, but there are many whose floors rise in places in what was undoubtedly the downstream direction.

Relations among these channels are instructive. The larger ones can be traced continuously through the maze, but many smaller ones and some large ones flare out downstream and lose their channel character in the generalized scabland. A larger number head in the scabland plexus 200 to 300 feet above the adjacent deeper and continuous channels. Junctions of small channels with the larger ones are but rarely accordant. The numerous small channels are on butte tops or on the higher tracts among the buttes. Their divergence from, or entrance to, the larger channels is almost invariably hanging.

Another remarkable feature of Drumheller Channels plexus is its rock basins. One hundred and eighty-two of these are shown by the 25-foot contour interval of the Corfu and Othello topographic sheets of the U. S. Geological Survey. The larger ones are in the more prominent channels and are elongated with the channel shapes. Many of the smaller ones are on butte tops, unassociated with definite channel forms.

FIG. 22

FIG. 23

FIG. 24

FIG. 22—A mid-channel butte. Absence of other basalt buttes or ledges at its level indicates that it records a hill of pre-loess topography. (Photograph by L. C. Robinson.)

FIG. 23—Trenched spur buttes in Crab Creek valley near Wilson Creek station.

FIG. 24—Scarp in loess three miles northwest of the mouth of Palouse River. Base of scarp 1,320 feet A.T., about 800 feet above Palouse. A valley in the loess left hanging by erosion shows on the right. (Photograph by Donald Bell.)

Drumheller Channels constitutes one of the most peculiar groups of topographic forms to be found anywhere. It suggests karst topography, which it is not. The discontinuity and irregularity slightly resemble those of a morainic region or a tract of dunes. But the field evidence is convincing; the topography is the result of running water. For its development the tract must be covered with a current deep enough and vigorous enough to erode these channels and rock basins where there previously had been nothing remotely resembling them. It is permissible to think that, as erosion progressed, the glacial waters were concentrated in the deeper continuous channels across the ravaged anticline. But the deepest rock basin of the plexus, lying in the east-central portion, is two miles and a half from the nearest of these continuous channels, its rim 150 feet above that channel floor, and its bottom not 25 feet above. The demand for an initial great volume is obvious.

Trenched Spur Buttes and Junction Buttes

The deeper preglacial valleyways entered by the glacial streams possess a type of scabland eminence that closely resembles the mid-channel butte but has a somewhat different genesis. This type, the trenched spur butte (Fig. 23), occurs on the inside of curves of the original valley. The great glacial rivers, filling these valleys to the brim and even overrunning them, could not tolerate the leisurely preglacial curves. The submerged slip-off slopes were the most exposed places in the extemporized channelways and were most vigorously attacked. A common mode of attack was the cutting of a gorge or trench across the slip-off spur. This provided a more direct route and left a scabland butte, the terminal part of the spur, isolated in the widened watercourse. Widening by a method peculiar to this region and this episode is recorded by the buttes of this type. They are not common forms, for they require a particular setting and a particular behavior of the glacial river. They are not limited to the plateau scablands. Snake River Valley possesses two remarkable trenched spur buttes, well shown on the Wallula topographic sheet of the U. S. Geological Survey, and there is an equally remarkable situation involving trenched spur buttes in the Columbia Valley 115 miles by river below the mouth of the Snake.

In a few places where large glacial streams, occupying preëxisting valleys in basalt, joined to become one, the tip of the separating divide was severely eroded and cut into buttes much like a trenched spur series except for situation. The erosion seems to have resulted because one of the confluent rivers had superior volume and a part of it poured across the divide tip into the other valley. The junction of Crab Creek and Coal Creek, a few miles east of Odessa, possesses four such buttes in line.

Fig. 25

Fig. 26

Fig. 27

Fig. 25—Relict hill of loess, 180 feet high, surrounded by scabland. Near Hooper, Wash.

Fig. 26—Cluster of relict hills, remnants of a minor divide. Near Hooper, Wash. The hill shown in Figure 25 is one of this group.

Fig. 27—Channel in loess, looking north into Washtucna Coulee, whose preglacial floor is 350 feet lower. Nunamaker School, southwest of Washtucna, Wash.

Erosional Forms in Overlying Sediments

The erosional forms of channeled scabland are not limited to the basalt. Bearing their own distinctive characters, they occur in loess and other sediments that overlie the basalt. They may be classified under the terms scarps, relict hills, and channels.

By far the most common form is the scarp, a steepened slope in the overlying formation (Fig. 24). It is almost invariably parallel to a glacial river course, very seldom transverse. It invariably is marginal and almost invariably rises above the highest limit reached by glacial water at that place. It is analogous to the basalt cliffs of the scabland but always occurs at a higher level, its base essentially at the altitude of the highest rugose basalt surfaces.

Scarps commonly truncate the slopes of the mature topography. Their relations to it are like those of sea cliffs to older erosional slopes. Their heights may be as great as 200 feet, their slopes as steep as 35°. Their alignment along the margins of scabland river routes, their consistent development whether the glacial river was a quarter of a mile wide or twenty miles, their formation in piedmont waste or gravel material, their orientation in various directions, their absence at the summit of loess-capped basalt cliffs in non-scabland valleys, and the occurrence of rock basins, cataract ledges, canyons, and gravel bars close to their bases—all these circumstances debar any other interpretation of them than as integral parts of the scabland system.

The "islands" of loess which make striking features on many broad scabland tracts are almost invariably elongated with the gradient of that tract and are almost invariably scarped on the sides and upgradient end (Fig. 25). Convergence of the lateral scarps to this upgradient end in many cases produces a definite prow, like the conventional pilot of a locomotive. Small "islands" may have their summits determined by the upward convergence of the lateral scarps. Larger ones commonly carry the mature topography that characterizes the loessial mantle east and south of the scabland system. Many such hills or hill groups are separated by narrow scabland channels. Some stand alone in the scabland, miles from any others of their kind. There are more than 350 of them that do not exceed five miles in length. They are commonest in elongated clusters close to the margin of a much larger tract of loess hills (Fig. 26). These clusters record one or more minor divides, the valleys among them having been completely destroyed and the divides themselves narrowed by marginal sapping until cross channels through them appeared. All gradations can be found between a group of channeled valleys in a loessial tract to a group of isolated relict hills with wide scabland separating them.

It is not difficult to see in these island hills the residuals or relicts of a once complete mantle of weaker material. The difficulty lies in explaining them by any other hypothesis. If the glacial waters diverted across the plateau had been in volume comparable to that which made the spillways from Labradorean or Keewatin ice in the northeastern and central states, only a few preglacial valleys would have been demanded to care for the discharge. The fact that two of the four source routes were stripped of their loess (except for a few relict hills) over a width of 20 miles is very significant. It is in harmony with the other phenomena of scabland, all of which call for an extraordinary volume and depth, despite the notable gradients that existed.

Glacial spillways without scabland floors exist in the anastomosis (Fig. 27). All are small features. The divides crossed by the water entering them stood too high and too far from the main routes to allow any great volume. In some of these there is good basalt scabland at the head where the divide was crossed. Most of them have definite margining scarps. Gravel deposits on their floors, typically containing a few foreign cobbles and boulders, are also common. In large part, they are still the preglacial minor valleys, though most of them are clearly set off from adjacent valleys of this category by the features above noted.

Groups of such channels, closely spaced and nearly parallel, cross some of the relict hill areas. They are taken to represent the initial violation of the whole of the vanished loess cover. Where the broad scabland areas now exist, this initial stage was succeeded by deepening and, especially, widening until lateral planation had destroyed many thousands of these hills of the preglacial dendritic pattern.

This conception is not especially difficult to accept if no capacious preglacial valleys existed among them. Such was undoubtedly the case in the northern uncanyoned scabland tracts. But the same groupings of closely spaced channels are found farther south where there clearly were preglacial valleys 300 feet to 400 feet deep in basalt. The same flooding across minor divides to make the minor channels in loess occurred here also, despite the presence of capacious valleys below the upland loessial topography. The need for great volume is nowhere set forth more strikingly.

Depositional Forms of Channeled Scabland

It has been pointed out that all the peculiar features of channeled scabland knit together into a harmonious scheme on the thesis of tremendous volume of the escaping water. Should one, however, argue that the scabland denudation can be explained by assuming smaller streams and allowing more time, he must logically test this interpretation by the deposits of these streams. They afford probably a more crucial test of any hypothesis than do the erosional forms. As befits this paper, only the deposits with definite topographic expression will be considered.

All scabland channels possess discontinuous mounds, hillocks, or hills of stream gravel. Essentially none possess gravel terraces. The mounded deposits may occur in rock basins or perched high on channel walls. In form they are characterized by a rounded crest or summit which is commonly elongated with the channel, an enclosed or partially enclosed fosse between the crest and the adjacent channel or basin wall, and the complete absence or only the incipient stages of gullying (see Fig. 10). They are constructional forms; as definitely so as kames, eskers, cones, or deltas, though differing from all of these.

In situation they commonly lie either on the downgradient side of mid-channel buttes, trenched spur buttes, relict hills, or other eminences in the scablands; or in alcoves and recesses of the channel courses. Not a single prominent non-scabland valley entering the complex is known to be without them at the place of entrance.

In distribution they occur all through the scabland complex, from the summit of the plateau divide near the head of the channel groups to the great gravel deposit at Portland west of the Cascade Range.

In structure they are prevailingly current-bedded. Many of them possess long foreset strata like those of delta fronts, paralleling a steep slope in the mound. Their steep faces, like delta fronts, were growing faces. Débris that was swept over them came to rest in quieter places in the great rivers.

In composition these hills of gravel are dominantly of local material. Many along the Snake and Columbia have 75 to 90 per cent local material, though the present river gravels in close juxtaposition may have this ratio of local and foreign material completely reversed. Some have abundant pebbles, cobbles, and even boulders of loess, of silt, and of soft calcareous material associated with the hard basalt. Many sections show a large percentage of angular decayed basalt, well mingled with fresh "bright" fragments of the same rock. Decay of the deposits as a whole has extended but a foot or two below the surfaces. No differences have been found in amount of decay of the higher and lower deposits.

In magnitude these hills range from low mounds covering an acre or so to huge piles measuring a hundred to four hundred feet from lowest base to crest and covering a square mile or more. A few of them extend through the entire vertical range of scabland in their neighborhood. Their magnitude, both in actual size and in relation to the valleys in which they occur, is the most arresting feature among all their surprising characteristics.

They are unlike any other detrital accumulations except the much smaller features of river channels commonly called bars. With these there is exact parallelism except for size. When considered in their setting in the scabland system, with all its other evidence for great volume and great erosion, they are seen to be an integral part. They should exist! And if they are bars, the great scoured channelways should exist! Again this assemblage of unique land forms in the Pacific Northwest is seen to be a genetic group. A lively imagination is required for the acceptance of the hypothesis, but a scientific imagination withal.

The Causal Episode

It is difficult to convey by words and pictures a visualization of the observable relationships of channeled scabland. Such terms as "canyon," "cataract," "scarp," "gravel deposit," calling up images of the usual topographic forms, suggest the usual explanation, which involves far less water and far more time than the writer's hypothesis demands. Put a glacial Columbia first in one channel, then in another, then in a third, instead of running all of them at the same time; use more than one glacial episode if necessary; upwarp the transected divides to give the present high altitudes above valley floors. Procedures such as described here are unheard-of. Other ice sheets have come and gone and left no such record. How could such quantities of water be yielded from so small a front and with so little retreat?

It may be that there are other significant facts yet to be discovered. But the writer is convinced that the relations outlined in this paper do exist and that no alternatives yet proposed by others or devised by himself can explain them.

That unique assemblage of remarkable physiographic forms on the Columbia Plateau in Washington, described here as the channeled scabland system or complex, records a unique episode in Pleistocene history. Special causes seem clearly indicated. But what these causes were is yet an unsolved problem.

Part III

THE REVISIONISTS

Editor's Comments on Papers 7 and 8

7 **ALLISON**
Excerpts from *New Version of the Spokane Flood*

8 **FLINT**
Excerpts from *Origin of the Cheney-Palouse Scabland Tract, Washington*

The 1933 International Geological Congress field trip was an appropriate beginning to a concentrated assault on Bretz's hypothesis. Ira S. Allison (personal communication, 1978) recalls that several famous geologists attended, including du Toit of South Africa and Bailey of Scotland. They were careful observers, looking for the evidence on their own, rather than waiting for the pronouncements of the trip leader. The field trip was an appropriate prelude to the type of scrutiny of his hypothesis that Bretz himself had strongly urged—new field studies.

THE ICE-JAM HYPOTHESIS

Allison published the first important revision of the flood hypothesis. Note that his view (Paper 7) was not a denial of the Spokane Flood, but rather a modification. Allison (1933) argued that floating berg ice was the critical factor in the flood. He presented detailed evidence for the ponding of floodwater all the way from the Columbia River gorge to the Wallula Gap. This ponding was produced ("in spite of the obvious difficulties of such an explanation") by a blockade of ice in the Columbia Gorge. The blockade grew gradually headward until it extended into eastern Washington. As water was dammed to higher levels, it spilled across secondary drainage divides creating the enigmatic hanging valleys, high-level gravels, and widely distributed erratics.

A key point in Allison's paper is contained in its last sentence:

"Perhaps this revision will make the idea of such a flood more generally acceptable" (Allison 1933, p. 722).

Allison (written communication to Bretz, 1978) subsequently abandoned his ice-damming hypothesis. Nevertheless, he still maintains that the role of ice carried by the flooding has not been fully appreciated by most investigators. Some recent studies have been reexamining this important question (Fecht and Tallman 1978).

THE GLACIAL HYPOTHESIS

E. T. Hodge, a professor at Oregon State College, Corvallis, was a persistent critic of Bretz's flood hypothesis. During the 1930s Hodge gave numerous seminars and oral papers attacking the Spokane Flood concept. His alternative (Hodge 1934) postulated a complicated alternation of ice advances and drainage changes. He argued that the basalt was quarried by glacial erosion and that channel complexes in the basalt were produced by the diversion of meltwater streams around blocks of stagnant glacier ice and jams of berg ice. The theory was presented only in outline form and was never adequately supported by published field evidence.

In 1935 Hodge founded the Geological Society of the Oregon Country. During the next three years he edited its *Geological Newsletter*, in which he published several additional abstracts and notes restating his glacial hypothesis. Bretz (personal communication, 1978) recalls one confrontation with Hodge at a scientific meeting in Pullman, Washington. Bretz had been invited to keynote the meeting with an address on the Spokane Flood. Hodge was so incensed that he wrote to the meeting chairman demanding that he be allowed to debate Bretz formally during the address. This request was refused, but, undaunted, Hodge then invented a subsidiary society with a concurrent meeting time. Bretz acceded to an informal debate on the day following his keynote.

The Hodge-Bretz debate was not quite what its instigator had planned. Hodge provided the curious audience with a retelling of his glacial hypothesis. Afterward, Bretz rose and merely suggested that all interested parties drive out across the Palouse Hills to the nearby southern part of the Cheney-Palouse scabland tract. Enroute, Bretz demonstrated with a kitchen sieve how above a certain level the loess was silt only, but below that level the loess contained basalt and granite sand and granules (deposited by the highest flood water). At the former Palouse-Snake divide he ponted out the great bars, the prow-pointed loessial hills, and the spectacular scabland erosion. The demonstration was a clear example of Bretz's inductive approach to the Channeled Scabland problem. The flood hypothesis had emerged from the field evidence, and the field data provided the rebuttal to all critics.

Another hypothesis involving glacial origins for the Channeled Scabland was authored by Charles Keyes (1935). In a cursory description of regional geography, Keyes ascribed both Moses Coulee and Grand Coulee to supraglacial streams that developed along the axes of waning ice lobes. This deductive approach was not substantiated by the field data.

FLINT'S HYPOTHESIS

Probably the most serious challenge to the Spokane Flood hypothesis was that posed by Richard Foster Flint. While a student at the University of Chicago, Flint had been associated with Bretz. Flint's Ph.D. thesis, pursued under R. D. Salisbury's supervision, was completed at Chicago in 1925, when Bretz was an associate professor in the same department. In 1925 Flint became an instructor at Yale. Flint's early research at Yale concentrated on ice stagnation in New England (Flint 1928a, 1928b, 1932, 1933). One of his colleagues in the department at the time was Aaron C. Waters, also an instructor at Yale (1927–1930). Waters (personal communication, 1978) suggested to Flint that he would find much more interesting field examples on the Waterville Plateau of eastern Washington. During the 1930s Flint published a number of papers on the general glacial features of the region (Flint 1935, 1936, 1937), culminating in his extensive attack on the Spokane Flood hypothesis (Flint 1938). Waters recalls that he was so incensed by Flint's published papers that he began his own glacial studies in the region. Unfortunately World War II intervened, and he never completed a contemplated exposé of Flint's work. It was not until 1965 that Weis and Richmond (1965) revealed important errors in Flint's glacial work. Failure to recognize the catastrophic flood effects had led to misinterpretations of former ice margins and to misidentification of sedimentary origins.

Flint's paper on the Cheney-Palouse scabland tract is reproduced in part here as Paper 8. Note that Flint's carefully worded argument cites a considerable amount of field data to suggest "a picture of leisurely streams with normal discharge" (Flint 1938, p. 472). However, it is obvious from Flint's sedimentological descriptions that he was giving most of his attention to the low-energy slackwater facies of the Missoula Flood deposits in the various scabland channels.

One of Flint's most important arguments was that the surface form of the scabland deposits was that of "non-paired, stream-cut terraces in various states of dissection" (Flint 1938, p. 475). It was an idea that Bretz had introduced (Bretz 1923) and subsequently rejected after closer field study. Flint thought that Bretz's revised interpretation of the deposits as constructional bar forms could explain some, but not all, of the field relationships. He suggested that a sequence of channel aggradation by normal proglacial outwash was followed by dissection to leave remnants of fill that occasionally resembled bar forms.

Flint (1938) accepted Bretz's (1928) arguments that the flood gravel often (1) occurred in the lee of island-like areas, (2) had rounded upper surfaces, and (3) exhibited a parallelism of surface slopes with the dip of underlying foresets. He argued that "terraces" had been extensively dissected by a downstream base-level reduction. The "terraces" were preferentially preserved in the lee of island-like areas. In addition, the low precipitation plus the high permeability of the gravel prevented gullying, so the gravel deposits developed rounded slopes by dry creep. Finally, he showed that many of the gravel slopes did indeed truncate the underlying bedding. As specific cases, he argued that Bretz's Willow Creek bar, Staircase Rapids bar, Palouse Canyon bar, Midcanyon bar, and Shoulder bar were all simply terrace remnants. Subsequent studies have shown that three of these bars have prominent giant current ripples on their upper surfaces (Bretz et al. 1956).

Flint traced the coarse scabland deposits downstream into the Pasco Basin. There he found that the deposits changed from sand and gravel to silt and fine sand containing erratic stones. He named the fine-grained facies the "Touchet beds." The deposits had already been described by Bretz (1929, pp. 516–536; 1930, p. 414), who ascribed them to ponded flood water, and by Allison (1933), who ascribed them to water ponded by ice jams. The silts are recognized only to a uniform elevation of about 350 m (1150 ft.). The stratification ranges from rhythmic parallel laminations to cut-and-fill. The included erratic stones are granite, basalt, and other crystalline lithologies. Intense folding, fracturing, and clastic dikes imply slumping and sliding of the water-saturated silt on gentle subaqueous slopes. Flint thought that these relationships were most consistent with a large lake, which he proposed was ponded by a landslide dam or glacier ice in the Columbia gorge. Following Symons (1882) he named this water body Lake Lewis.

At this point Flint had the necessary tools to erect his hypothesis. The proglacial meltwater streams of normal discharge overran the northern margin of the Cheney-Palouse tract. This flow was derived from lobes of ice at the heads of the Cheney-Palouse and Telford–Crab Creek scabland tracts. Flint thought water from Lake Missoula need not be involved. Instead, he observed that the discharge "was less than that of the Snake River today" (Flint 1938, p. 515).

As Lake Lewis rose, the "leisurely" streams that Flint envisioned aggraded, forming a thick fill. This fill blocked preglacial tributaries to the Channeled Scabland, such as the Snake River, and formed marginal lakes that accumulated fine-grained sediments. The steep scarps on the Palouse loess were then cut by lateral planation of the streams flowing on this fill. When Lake Lewis finally drained, the streams gradually incised the fill to form terraces. Moreover, Flint was able to explain the enigmatic notched spurs and slot-like hanging canyons as the result of superposition of streams from the widespread fill rather than a consequence of divide crossing by catastrophic flood water.

Flint argued that the complex of anastomosing channelways cut into basalt was a consequence of erosion by relatively small streams operating on various profiles. He stated that scabland-type erosion should occur wherever rock material with vertical planes of weakness was subjected to stream flow. As an example of such erosion he cited Red Rock Pass, Idaho, a glacial outlet of glacial Lake Bonneville (Gilbert 1890). He also noted the scabland erosion at Twin Falls, Idaho, where the Snake River flows in a canyon nearly as spectacular as the scabland channels. He noted:

> The . . . [basalt] flows yielded to the hydraulic force of the Snake River as similar flows on the Columbia Plateau yielded to the hydraulic force of proglacial streams, yet I am not aware that unusual floods have been held to have affected the upper Snake River (Flint 1938, p. 492).

Flint's discussion was premature. Malde (1968) describes the physiographic relationships at Twin Falls in considerable detail. The alcoves, scabland, and deep erosion cited by Flint were not produced by normal flows of the Snake River. Instead these features resulted from a catastrophic overflow of Lake Bonneville approximately 30,000 years B.P. Flood water was hydraulically ponded in the Rupert Basin by the narrow basalt gorge upstream of Twin Falls. The gorge was unable to pass the entire flood discharge, so an upland channel was carved across the basalt surface, rejoining the Snake canyon near Twin Falls. Erosion on a grand scale took place at this junction and created a canyon segment more than 15 km long, as much as 1500 m wide, and 150 m deep. Malde calculates that the flood removed about 1 to 1.5 km^3 of basalt from this reach. Downstream from Twin Falls Malde (1968) describes gigantic streamlined bars several kilometers long and up to 90 m high. Boulders as large as 23 m in circumference occur on the surfaces on these bars.

In yet another ironic passage, Flint (1938, pp. 504–505) calculated the probable rate of filling for Lake Lewis at the modern discharge of the Columbia River. He stated, "The calculated time, 13 years 1 month, seems grossly inadequate for the deposition of the fill in the scabland tracts." He rationalized his interpretation, however, by referring back to the interpreted filling episode. Bretz's flood theory was so despised that even circular reasoning could be employed to erect an alternative hypothesis.

A careful examination of Flint's (1938) paper reveals that he observed and described the morphological feature that, more than any other, was absolutely incompatible with his elegant theory. On the surfaces of the scabland "terraces" he described an intricate microtopography of anastomosing channels, small depressions, and crescentric channels (Flint 1938, p. 475). In other areas he observed "mamillary undulatory topography." As an example he gave the precise location of the train of giant current ripples on the upper surface of Staircase Rapdis Bar, 3 km north of Washtucna (Flint 1938, p. 486). Although the ripples that he described are somewhat masked by overlying slackwater sediments, Flint (1938, p. 499–500) even stated characteristic ripple magnitude: "The undulations are 20 to 100 feet long, and have amplitudes up to 10 feet. Their axes are generally transverse to the Snake River." How ironic that Flint was the first to describe accurately (without knowing what they were) the very feature that Bretz and others (1956) later presented as incontrovertible evidence for catastrophic flood flows!

THE HOBBS HYPOTHESIS

One final curious example illustrates the strong emotions evoked by the Spokane Flood controversy. W. H. Hobbs, an eminent glacial geologist from the University of Michigan, spent several weeks studying the terrain in southeastern Washington and prepared a paper explaining the landforms as the product of a "Scabland Glacial Lobe." Both Bretz and Flint reviewed the paper for the Geological Society of America, and both recommended rejection. The paper was then submitted to the American Philosophical Society, which had supplied the funds for the study. Bretz again reviewed the paper, and again it was rejected. Although a brief statement of the hypothesis was published (Hobbs 1943), the main manuscript had to be published privately (Hobbs 1947). The author expressed his feelings in the "Foreword" to his paper:

> In the winter of 1942–43 I was listening with much interest to a lecture on the late geological history of the so-called Scabland area which is southwest of Spokane and close to the supposed southern front of the Pleistocene Cordilleran continental glacier. A map projected on the screen showed the scabland to be a lobate area which extended out from the former glacier. Scattered over its surface could be seen dozens of lakes, none of which transgressed its border, an almost sure indication that this lobate area had once been actually covered by a Pleistocene glacier lobe.
>
> Surrounding this lobe on the lecturer's map could be seen a broad apron of gravels, and enveloping the gravels were heavy deposits of silt. These relationships of glacier lobe to outwash and loess duplicated what I had observed in west Greenland. The lecturer explained, however, that the deposits represented upon his map had been laid down by a great flood of water of unknown origin, the "Spokane Flood."
>
> In the belief that my Greenland observations had given me an advantage in interpreting the evidence within the Scabland region, I then and there decided to make a personal study of it on the ground. Although two other very extended studies had already been made of it by Fellows of the Geological Society of America, and their conclusions had been published *in extenso* in its *Bulletin*, the Society provided me with a grant of money which made possible a new study of the area. This field investigation was carried out during two seasons, and the results and conclusions met with unusually enthusiastic general approval when they were presented to the Society in 1945 at its Pittsburgh meeting. Following tumultuous applause in the crowded section the discussion was throughout approving.

The Hobbs paper contains so many hypothetical arguments that one marvels at the limits that were being stretched to find an alternative to catastrophic flooding as the cause of the Channeled Scabland. Hobbs (1947) argued that the scabland was a product of glacial scour and that the Palouse loess was deposited contemporaneous to this glaciation by anticyclonic winds off ice that lay in the various "channels." He interpreted many scabland gravel deposits as moraine remnants modified by glacier-border drainage.

REFERENCES

Allison, I. S. 1933. New version of the Spokane Flood. *Geol. Soc. America Bull.* **44**:675–722.

Bretz, J H. 1923. Glacial drainage on the Columbia Plateau. *Geol. Soc. America Bull.* **34**:573–608.

Bretz, J H. 1928. Bars of the Channeled Scabland. *Geol. Soc. America Bull.* **39**:643–702.

Bretz, J H. 1929. Valley deposits immediately east of the Channeled Scabland of Washington. *Jour. Geology* **37**:393–427, 505–541.

Bretz, J H. 1930. Valley deposits immediately west of the Channeled Scabland. *Jour. Geology* **38**:385–422.

Bretz, J H., Smith, H. T. U., and Neff, G. E. 1956. Channeled Scabland of Washington; new data and interpretations. *Geol. Soc. America Bull.* **67**:957–1049.

Fecht, K. R., and Tallman, A. M. 1978. Bergmounds along the western margin of the Channeled Scablands, south-central Washington. *Geol. Soc. America Abs. with Programs* **10**:400.

Flint, R. F. 1928a. Pleistocene terraces of the lower Connecticut Valley. *Geol. Soc. America Bull* **39**:955–984.

Flint, R. F. 1928b. Eskers and crevasse fillings. *Am. Jour. Sci.* **15**:410–416.

Flint, R. F. 1932. Deglaciation of the Connecticut Valley. *Am. Jour. Sci.* **24**:152–156.

Flint, R. F. 1933. Late-Pleistocene sequence in the Connecticut Valley. *Geol. Soc. America Bull.* **44**:965–988.

Flint, R. F. 1935. Glacial features of the southern Okanogan region. *Geol. Soc. America Bull.* **46**:169–194.

Flint, R. F. 1936. Stratified drift and deglaciation of eastern Washington. *Geol. Soc. America Bull.* **47**:1849–1884.

Flint, R. F. 1937. Pleistocene drift border in eastern Washington. *Geol. Soc. America Bull.* **48**:203–231.

Flint, R. F. 1938. Origin of the Cheney-Palouse Scabland tract. *Geol. Soc. America Bull.* **49**:461–524.

Gilbert, G. K. 1890. Lake Bonneville. *U. S. Geol. Survey Mon. 1*, 438 pp.

Hobbs, W. H. 1943. Discovery in eastern Washington of a new lobe of the Pleistocene continental glacier. *Science* **98**:227–230.

Hobbs, W. H. 1947. *The Glacial History of the Scabland and Okanogan Lobes, Cordilleran Continental Glacier.* Ann Arbor, Mich.: J. W. Edwards, privately printed, 40 pp.

Hodge, E. T. 1934. Origin of the Washington Scabland. *Northwest Sci.* **8**:4–11.

Keyes, C. 1935. Glacial origin of the Grand Coulee. *Pan-American Geologist* **63**:189–202.

Malde, H. E. 1968. The catastrophic late Pleistocene Bonneville flood in the Snake River Plain, Idaho. *U. S. Geol. Survey Prof. Paper 596,* 52 pp.

Symons, T. W. 1882. Report of an examination of the upper Columbia River. U. S. Congress, Senate, 47th Cong. 1st sess. 1882, S. Doc. 186, 135 pp.

Weis, P. L., and Richmond, G. M. 1965. Maximum extent of late Pleistocene Cordilleran glaciation in northeastern Washington and northern Idaho. *U. S. Geol. Survey Prof. Paper* **525-C**:128–132.

7

Reprinted from pages 675–681 and 718–722 of *Geol. Soc. America Bull.* **44**:675–722 (1933)

NEW VERSION OF THE SPOKANE FLOOD[1]

BY IRA S. ALLISON

(Read before the Geological Society, December 28, 1932)

CONTENTS

INTRODUCTION

The basalt plateau of eastern Washington exhibits a peculiar assemblage of physiographic features, including a complex set of interlocking stream channels, abandoned canyons and cataracts, steep-walled buttes, rock-rimmed basins, residual islands of loess between the channels, bar-

[1] Manuscript received by the Secretary of the Society, December 28, 1932.

like deposits of rubbly gravel, and slack-water deposits in certain tributary valleys, which have been described at length by Bretz.[2] He has attributed these features to a catastrophic flood of glacial waters, diverted southwestward across the plateau by a glacial dam in the Columbia River valley on the north edge of the plateau. According to Bretz, these waters were released suddenly and in such volume as to occupy all the channels simultaneously and even to spill across divides into neighboring dischargeways. The torrent of glacial waters thus conceived, Bretz called the Spokane Flood. Critical examination of the area by the writer has revealed new facts regarding the extent of the flood between the plateau of eastern Washington and the gorge of the Columbia River through the Cascade Mountains, and has led him to propose an alternative conception of the flood.

The new data indicate that the flood rose to a consistently high level from the Wallula Gateway to the Columbia River gorge through the Cascade Mountains; that it left slack water deposits both in tributary valleys and on the uplands; hence, that it was virtually ponded. The data also prove that the flood was later than the major physiographic development of the region, as Bretz has continually insisted, so that a method of delaying or even entirely blocking the flow of the flood waters must be sought to explain the ponding. In spite of the obvious difficulties of such an explanation, the writer believes: that the ponding was produced by a blockade of ice in the Columbia River gorge through the Cascade Mountains; that the rise of the Columbia River to abnormally high levels began at the gorge and not on the plateau of eastern Washington; that the blockade gradually grew headward until it extended into eastern Washington; that, as the waters were damned to progressively higher levels,

---

[2] J Harlen Bretz: Channeled scablands of the Columbia Plateau. Jour. Geol., vol. 31, 1923, pp. 617-649;

———: Dalles type of river channel. Jour. Geol., vol. 32, 1924, pp. 139-149;

———: The Spokane Flood beyond the channeled scabland. Jour. Geol., vol. 33, 1925, pp. 97-115, 236-259;

———: The Spokane Flood—a reply. Jour. Geol., vol. 35, 1927, pp. 461-468.

———: Alternative hypotheses for channeled scabland. Jour. Geol., vol. 36, 1928, pp. 193-223, 312-341;

———: Bars of channeled scabland. Bull. Geol. Soc. Amer., vol. 39, 1928, pp. 643-702;

———: Channeled scabland of eastern Washington. Geog. Review, vol. 18, 1928, pp. 446-477;

———: Valley deposits immediately east of channeled scabland. Jour. Geol., vol. 37, 1929, pp. 393-427, 505-541;

———: Valley deposits immediately west of channeled scabland. Jour. Geol., vol. 38, 1930, pp. 385-422;

———: Grand Coulee. Am. Geog. Soc., Special Pub. 15, 1932.

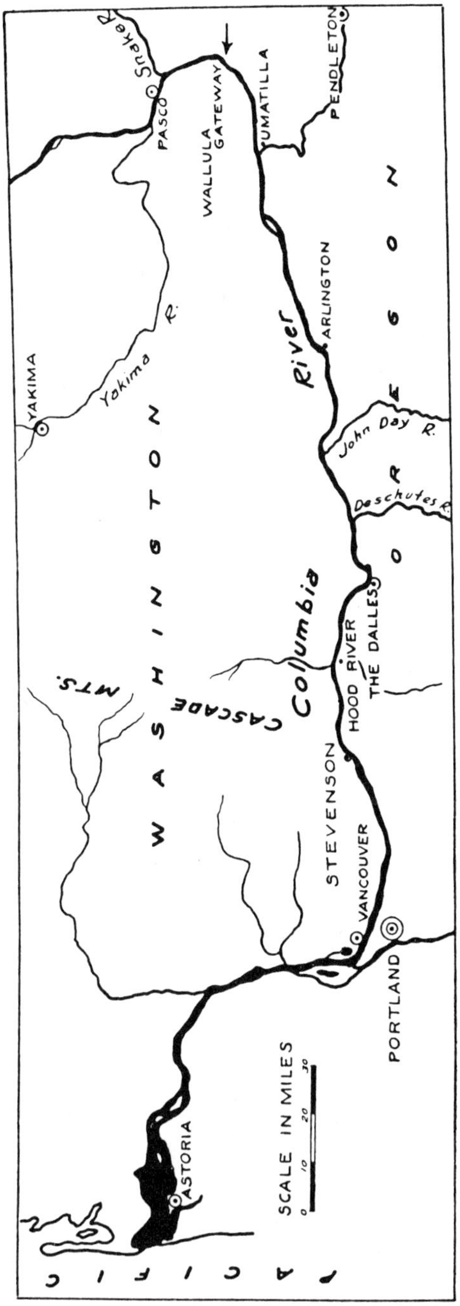

they were diverted by the ice into a succession of routes across secondary drainage divides at increasing altitudes, producing scablands and perched gravel deposits along the diversion routes, distributing iceberg-rafted erratics far and wide, and depositing pebbly silts in slack-water areas. This interpretation of the flood does not require a short-lived catastrophic flood but explains the scablands, the gravel deposits, diversion channels, and divide crossings as the effects of a moderate flow of water, now here and now there, over an extended period of time. It thus removes the flood from the "impossible" category.

With this preliminary statement in mind, let us examine the field data to see how well the mechanism of the flood here suggested, emphasizing the importance of ice, affords a satisfactory interpretation of the facts. For the sake of brevity, this paper will treat only the Yakima Valley in south-central Washington and the area below the Wallula Gateway, shown in Figure 1, for which a considerable amount of new data is available.

FIGURE 1.—*Sketch Map of the lower Columbia River Basin of Oregon and Washington*

The areas treated in this paper are in lower Yakima Valley, and between the Wallula Gateway and Portland, Oregon.

YAKIMA VALLEY

CHANDLER NARROWS SCABLAND

In Yakima Valley (figure 2) the scabland at Chandler Narrows, the local gravel deposits in the lee of projecting spurs or on the east walls of tributary valleys, the mantle of pebbly, gritty silt, and the glacial boulders up to 1100 feet above sea-level, as described by Bretz,[3] are acceptable proof

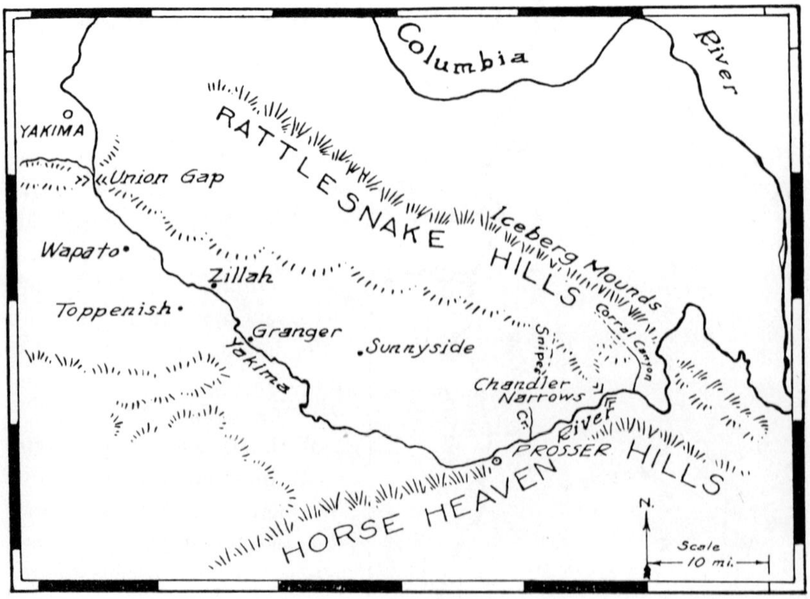

FIGURE 2.—*Sketch Map of lower Yakima Valley, Washington*
Shows the setting of Chandler narrows and the iceberg mounds on the northeast flank of Rattlesnake Hills.

of the invasion by glacial waters from the Columbia River valley. Moreover, the scabland knobs and basins are evidence that the inflowing waters locally had considerable velocity. As noted by Bretz, the current would be "an uphill current, its velocity dependent on a surface gradient only." The writer does not agree, however, that the Chandler scabland requires "enormous velocity over almost the whole wetted perimeter of the Narrows." Instead, it seems more likely that the multiple channels of the Chandler scabland were eroded, not simultaneously, but in succession.

[3] J Harlen Bretz: Valley deposits immediately west of the channeled scabland. Jour. Geol., vol. 38, 1930, pp. 412-422.

On the Prosser (Washington) topographic map, part of which is reproduced in Figure 3, the scabland basin along the east side of Section 7 is oriented in a nearly north-south direction, as is also the smaller one in Section 8. To enter its northern end, the water must have risen to

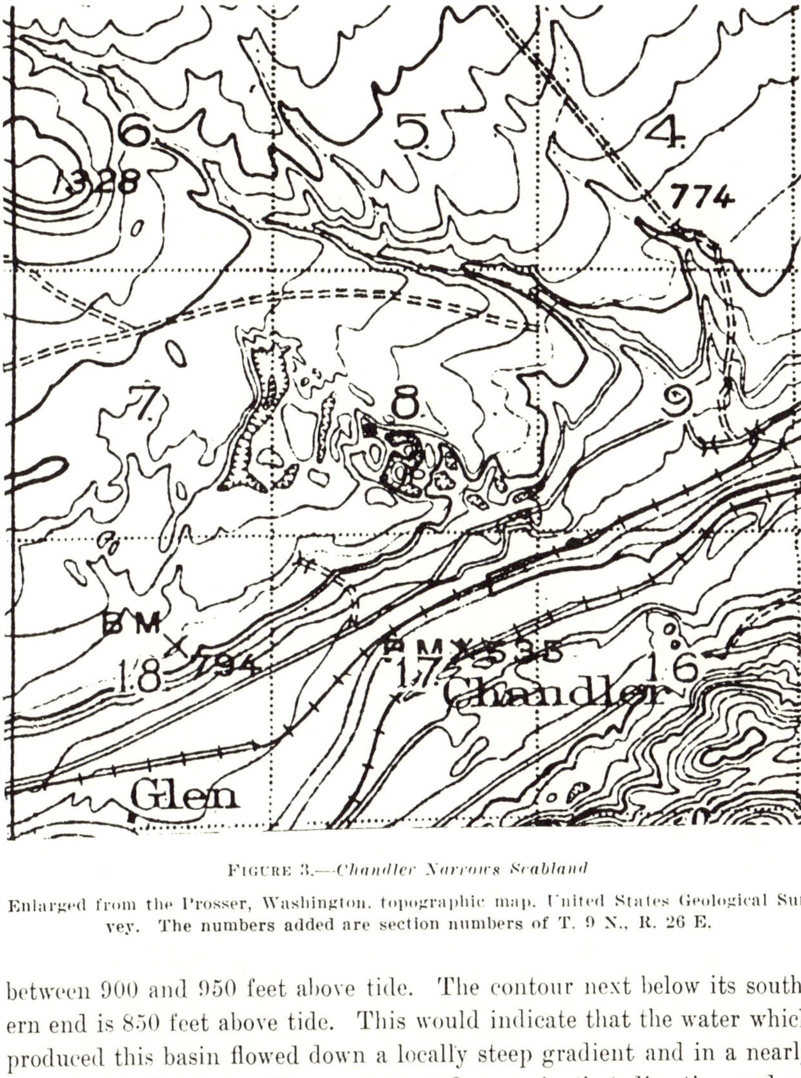

FIGURE 3.—*Chandler Narrows Scabland*

Enlarged from the Prosser, Washington, topographic map, United States Geological Survey. The numbers added are section numbers of T. 9 N., R. 26 E.

between 900 and 950 feet above tide. The contour next below its southern end is 850 feet above tide. This would indicate that the water which produced this basin flowed down a locally steep gradient and in a nearly north-south direction. The concentrated scour in that direction and on that slope is not consistent with the theory of an "advancing wall of water, like a gigantic bore" flowing west-southwestward, 450 feet above the valley bottom, in a valley two miles wide.

The writer therefore proposes another explanation of (1) the localization of scour, (2) the trend of the rock basins or scabland channels, and (3) the locally steep gradient—namely, that these features are due to an abundance of floating ice in the flood waters. That floating ice was abundant is well shown by the abundance of erratic materials, ranging in character from silt to enormous boulders, not only within Yakima Valley but also throughout the flooded area as far down as the Willamette Valley, a tributary to the Columbia River west of the Cascade Mountains. That some of the ice blocks were very large is shown by the size of the iceberg mounds on the northeast flank of the Rattlesnake Hills, one of which is shown in Figure 4, and by the large size of some of the boulders. It is, therefore, not difficult to conceive of a congestion of this floating ice at Chandler Narrows, whether or not we concede high velocity to the approaching waters, for the Narrows lies just beyond a bend in the valley. As soon as an icejam was formed, the water would rise behind the jam and try to find an outlet around the obstruction. On the up-valley (or Yakima) side of the icejam the water-level would be lower, provided that the flood waters from the Columbia River rose faster than the Yakima River, on the other side, could fill the basin behind the icejam, so that any water flowing around the ice might possess high velocity and consequent capacity for erosion. In pouring around the ice obstruction, over rock divides and down the valley walls beyond, the incoming water could well erode such rock basins as are exemplified at Chandler Narrows. These rock basins then become analagous to plunge basins below waterfalls. Of course, icebergs would be carried into these diversion channels and probably would choke them. In that event, the water would rise still higher and pour out of another outlet farther up the slope until it, too, was blocked by accumulating ice. In this way, a large series of channels may be cut—not all at once, but in a definite, though perhaps inextricable, succession. Thus, a greater proportion of the inflowing water would be effective in erosion; the erosive energy would be temporarily and locally concentrated; the total time involved would be greatly lengthened; and no catastrophic flooding would be required. In fact, the writer believes that the scabland buttes and channels of Chandler Narrows could not have resulted from the sort of flood pictured by Bretz, for the reason that such a flood offers no explanation of the localization of erosion ("plucking" of Bretz) of deep, narrow channels alongside steep-walled buttes. The severity of the local relief is better explained by a shifting of the courses of the eroding currents, and such a shifting could easily have been produced if we grant the possibility of an icejam in the Narrows.

[*Editor's Note:* Material has been omitted at this point, including figure 4.]

Cause of the Ponding

In this paper the writer has attributed the slack-water deposits east of the Cascade Mountains to a ponding of the Spokane Flood, for the reason that a general submergence (such as an extended Willamette Sound) sufficient to drown the middle reaches of the Columbia River to the 1100-foot elevation level is inconsistent with the diversion channels, the side-hill dumps of rubble and gravel, the lack of definite deltas in the tributary valleys, the scabland, and other peculiar features of the affected area. The lake produced by this ponding of the Spokane Flood recalls Symons' idea of a glacial lake, which he proposed to call Lake Lewis,[29] and Rus-

[29] T. W. Symons: The Upper Columbia River and the Great Plain of the Columbia. Senate Document No. 186, 47th Congress, 1st Sess., 1882.

sell's idea of a possible glacial dam at The Dalles [30] as the cause of such a lake. Bretz,[31] in 1919, proposed to redefine the term, Lake Lewis, to cover an expanded portion of a supposed submergence of the Columbia River valley east of the Cascade Mountains. In view of Bretz' subsequent papers on the Spokane Flood, his earlier conclusions must be modified. The present paper shows how the data favoring the Spokane Flood and the data suggesting a possible submergence may be reconciled in the idea of a glacial lake made by the ponding of the Spokane Flood. Perhaps the term, Lake Lewis, should be revived for the waters of the Spokane Flood ponded to an elevation between 1100 and 1200 feet above sea-level east of the Columbia River gorge, and the term, Spokane Flood, confined to the phase revealed on the plateau of eastern Washington, and not applied to Lake Lewis and other areas down the river.

The formation within the Columbia River gorge of an icejam, capable of withstanding the pressure of at least 900, and perhaps 1100, feet of water, is a herculean requirement, but the field evidence seems inescapably to demand it. We cannot fall back upon a mere bottleneck restriction to delay the flow, such as Bretz conceived for the Wallula Gateway, because the laminated, somewhat pebbly silts of the valley bottoms are not consistent with that kind of a flood. Moreover, the sidehill deposits and the diversion channels, used by Bretz as evidence of flood, can better be explained by the writer's icejam theory.

Five factors may have contributed toward the formation of an icejam: (1) lava flows in the gorge, (2) freezing of the river, (3) land slides in the gorge, (4) the crookedness of the gorge, and (5) a rapid renewal of uplift of the main arch of the Cascade Mountains. Lava flows, younger than the development of the gorge, occur, for example, in the White Salmon and the Wind River valleys. Underwood Mountain and White Salmon Butte (opposite Hood River) are volcanoes later than the major Cascade uplift and much of the cutting of the gorge. Their lavas, subsequently, were cut off sharply by the Columbia River. If these lavas extended partly or entirely across the river at the time of the flood, they might have offered sufficient obstruction. However, they bear on their surfaces gravel deposits, thought by Bretz to be flood deposits and by the writer to be earlier than the flood, so that, in either event, the lavas are older than the flood. If the lavas had formed a dam, a lake would have formed up-river and would probably have been completely filled with alluvium before the

---

[30] I. C. Russell: A geological reconnaissance in Central Washington. U. S. Geol. Survey Bull. 108, 1893, pp. 25-27.

[31] J Harlen Bretz: The Late Pleistocene submergence in the Columbia Valley of Oregon and Washington. Jour. Geol., vol.. 27, 1919, pp. 489-506.

lavas could have been cut through. The writer tried to fit the valley-bottom silts described in this paper into such an explanation but had to give up the attempt. The absence of even a few remnants of lacustrine deposits in the main valley and the lack of deltas (to be expected under conditions of relatively stable water-level) in the tributary valleys seem to rule out a lava dam. Such a strong and persistent obstruction, however, is given up reluctantly.

The Columbia River freezes over occasionally even in the prevailing mild climate of Vancouver, Washington, and frequently in the colder section east of the Cascade Mountains.[32] If, as commonly believed, the temperature in glacial time was lower than at present, freezing of the river would be easier. If general freezing of the river occurred, a large part, perhaps the greater part, of the ice may have been river ice instead of glacial ice. When the waters backed up sufficiently to reach the ice-sheet, they undoubtedly would lift off and tear away a large number of additional icebergs. The higher they rose, the more ice they would release, thus insuring continuance of the flood by adding to the icejam downstream.

A great landslide area, conditioned by the weak Eagle Creek formation underlying the Columbia River basalt, fronts the Columbia River near Stevenson, Washington, for a distance of about six miles. The basalt cliffs have receded about three miles from the river, but the slides are still active and apparently cause the present Cascades in the Columbia. As these cliffs were undoubtedly nearer the river at the time of the flood, a slide sufficiently large temporarily to dam the river is easily conceivable. A slide two or three hundred feet high would perhaps be enough to initiate an icejam, which in the narrow (generally about one mile wide) and somewhat crooked confines of the gorge (especially above Mosier) might grow headward so as to pile up to about 800 feet near Hood River, to 900 or 1,000 feet near Mosier, and 1,100 feet or more at The Dalles and beyond. A landslide dam would, presumably, be short-lived geologically. It would also allow a relatively rapid decline of the flood, which the scablands, the sidehill deposits, and other phenomena seem to require.

Except for a 45-degree turn near Stevenson, Washington, the course of the Columbia is fairly straight west of Hood River, Oregon. A nearly right-angle turn near Mosier, Oregon, another kink between Mosier and

[32] Press dispatches of December 14, 1932, report an icejam completely filling the Columbia River at The Dalles, Oregon, and another, a mile and a half long, damming the Yakima River near Granger, Washington. The sharp turn of the Columbia River at the east end of the gorge is reported to cause many icejams.

Lyle, Washington, and the angle near The Dalles at which the river enters the gorge through the Ortley Anticline are favorable to the formation and preservation of an icejam. The sheer wall at Mayer Park, opposite Lyle, Washington, much of it 600 to 900 feet high and facing directly up-river, would offer especially noteworthy resistance to the push of a mass of ice; a sharp deflection to the north would be necessary to get by until it was overridden.

Gravels, 525 feet above sea-level at Lyle, 625 feet at Hood River, 650 feet at White Salmon, and 500 feet at Carson, are probably equivalent in age to one or more of the terraces less than 350 feet above sea-level near Portland, Oregon, and they probably owe their greater elevation, in part, to uplift in that portion of the gorge since they were deposited. If such an uplift were very rapid, slack water might have been produced in the Columbia River, but its probability and its adequacy to begin the flood appear doubtful.

For the present at least, the writer favors a landslide as the initiating cause of the flood, with freezing of the river (or the water-ice mixture) and the confinement of the ice in the narrow and somewhat crooked gorge as contributing causes. Admittedly, such an icejam is difficult, but it appears far more feasible than the sudden release of an overwhelming, catastrophic flood on the plateau. Moreover, it has certain advantages in interpretation of the field evidence, in that it gives a reason for the local plucking in the scabland, the upland diversion channels, the perched gravel deposits, the pebbly silts on the upland, the laminated, more or less pebbly silts in tributary valleys, and the coincidence of the upper limits of erratic boulders and silts. Furthermore, it does not require a sudden release of an enormous volume of water, and it extends the time for scabland development. These advantages more than offset the difficulties of the icejam theory.

Summary

In conclusion, it may be noted that the writer agrees with Bretz that there was a Spokane Flood, and that the erratics, the scabland, the diversion channels, and certain perched gravel deposits and sidehill rubble heaps are authentic records of it. The writer's interpretation differs from that of Bretz, however, in many particulars, among them: (1) That the distribution of all glacial erratics is definitely assigned to the flood; (2) that erosion by the flood below Wallula Gateway was local and limited in volume; (3) that most of the gravels near Umatilla, Hood River, White Salmon, Lyle, Carson, and Portland are older than the flood; (4) that the flood attained a uniform level from the Wallula Gateway to the Columbia River gorge through the Cascade Mountains; (5) that this uniformity and the pebbly silts on the uplands and in tributary valleys record a general ponding of the flood; and especially, (6) that ice, rather than mere volume, was the critical factor in the flood. Accordingly, our conceptions of the flood, and particularly its cause, its mechanics, its geologic effects, its volume, and its duration differ widely. Perhaps this revision will make the idea of such a flood more generally acceptable.

8

Reprinted from pages 464–472, 493–494, 496, 502–507, 520–523 of *Geol. Soc. America Bull.* **49**:461–523 (1938)

ORIGIN OF THE CHENEY-PALOUSE SCABLAND TRACT, WASHINGTON

Richard Foster Flint

[*Editor's Note:* In the original, material precedes these excerpts.]

REGIONAL SETTING

The plateau, in central Washington, consists of a thick pile of basalt flows, chiefly of Miocene date, dipping inward centripetally with an average descent of about 25 feet per mile. This dip controls the slope of the plateau surface, whose descent culminates in a faulted fold which forms the Horse Heaven Hills, a massive ridge 1000 to 1500 feet high. The basin at the north base of this ridge extends northwest to Saddle Mountain, north into Esquatzel Coulee and Snake River canyon, and east to the vicinity of Walla Walla. As thus broadly defined, it will here be termed the Pasco basin, from the town of Pasco near its center.

The plateau is surrounded on the west, north, and east, by highlands of crystalline rocks older than the basalt. Some of the highland valleys at the north and east margins of the plateau carry extensive fills of siltstone and shale interbedded with the plateau basalt. These sediments are generally referred to the Latah formation.

In general, the plateau drainage is consequent on the long dip slopes of the basalt. The Palouse River—Rock Creek—Cow Creek system, 75 miles long, is the most conspicuous example. Some streams, however, follow the axes of shallow east-west synclines. Outstanding examples are Snake River between Palouse Canyon and Devils Canyon, and Washtucna Coulee. Finally, the Columbia River, augmented by the Snake, leaves the Pasco basin by a great gap (Wallula Gap), one mile wide, 8 miles long, and in places 1000 feet deep, which it has cut through the Horse Heaven Hills.

The Pasco basin contains the dissected remnants of an early Pleistocene fluvial fill, named the Ringold formation (Merriam and Buwalda, 1917). The surface of the basalt to the north and east is covered with a mantle of fine sand and silt, called "Palouse soil" (Bryan, 1927). This body of sediment is in places more than 100 feet thick, and probably is chiefly of eolian origin. Much of it is now maturely dissected by small streams into a systematic network of small hills and valleys.

The basalt surface, with its dissected cover of "Palouse soil," is cut by a system of valleys and channels, many of which are now dry. They were cut by streams flowing south and southwest on the dip of the basalt, and west through synclinal depressions, and are clearly shown on the latest map published by Bretz (1928b, pl. 5). It is generally agreed that the valleys constitute the normal, preglacial drainage pattern of the plateau, but that many of them were widened, deepened, and modified by the making of subsidiary channels and closed depressions, during their occupation by proglacial meltwater, so that today they are scabland tracts rather than normal valleys. The term scabland (or scabrock) is in common local use, denoting areas of basalt denuded of the customary cover of "Palouse soil."

The scabland tracts vary from almost unmodified preglacial valleys, some of them deep and narrow, in the basalt, to great areas 10 to 20 miles wide, bordering preglacial streamways, stripped of "Palouse soil," seamed with shallow reticulated channels, and interrupted by shallow rock basins, with a maximum relief within the basalt, of perhaps 50 feet. Where the preglacial valleys are modified, the mouths of their lateral tributaries may hang above their floors; and lateral channels, which leave the main channel only to return to it farther down the tract, and hang above the main channel at both intake and outlet, are not uncommon. The hanging channels do not integrate into systematic profiles, but occur at many different elevations. In places, the scabland tracts cross preglacial divides, indicating surfaces of streams that rose high enough to spill over the valley sides that retained them.

The scabland tracts contain quantities of gravel, including glacial erratic stones and till balls, as well as finer-grained sediments, very irregularly distributed both vertically and areally, and are flanked along their margins by steep scarps in the "Palouse soil," facing inward and revealing the work of lateral planation by former streams flowing on profiles corresponding to the scarp bases. These scarps, numerous and conspicuous features of the scabland, will be referred to, for the sake of brevity, as Palouse scarps.

The scabland features, here summarized briefly, are described by Bretz, to whose publications the reader is referred (Bretz, 1923b, p. 618-620, 626-630; 1927a, p. 200-202; 1928b).

A veneer of wind-blown sand and silt, probably postdating the scabland, covers the plateau generally. In some places the sand forms pronounced dunes, as near the northeastern margin of the Pasco basin.

In the eastern part of the plateau, the strongly rolling surface of the maturely dissected "Palouse soil" stands in abrupt contrast with the black cliffs and ragged surfaces of the scabland tracts. The steep Palouse scarps at the margins of the broader tracts resemble wave-cut headlands rising above the surface of a choppy sea. The "Palouse soil" areas are wheat lands, green in spring and pale yellow in summer, brown fallow fields alternating with fields of standing wheat. The extreme northern and eastern parts of the plateau, which have more than 15 inches of rainfall annually, are diversified with yellow-pine timber on shaded slopes and on the scablands. Toward the Pasco basin (the direction of decreasing precipitation) the wheat lands gradually give way to sage-covered sandy plains which, with adjacent scablands, afford sheep pasture in winter. In much of this area, cultivation was formerly attempted, and abandoned ranches and homesteads are numerous. Little rain falls during the summer, and dust storms are frequent, moving northeast into the wheat country where fallow fields contribute more soil material to the wind.

A more complete statement concerning the regional setting is given elsewhere (Flint, 1938).

HISTORICAL SUMMARY

The point at issue is the genesis of the scabland tracts. Although Calkins (1905, p. 42) mentioned fills and terraces of well-rounded basalt gravel "along all the coulees of the wheat lands," and Leighton (1919, p. 34) stated the opinion that the coulees of the plateau were the loci of meltwater stream courses that were filled with detritus which was later dissected, the first extensive discussion was furnished by Bretz (1923a, p. 588; 1923b, p. 620). His thesis was that the meltwater occupied pre-existing valleys on the plateau in such great volume that it overflowed divides and constituted a vast though short-lived flood, the "Spokane flood" (Bretz, 1925, p. 98) which at its maximum had a catastrophic discharge. The flood was held to be 3 miles wide by 600 to 800 feet deep in the lower Snake River canyon, (Bretz, 1925, p. 111; 1928c, p. 662), and one mile wide by 375 feet deep in the western part of Washtucna Coulee (Bretz, 1923b, p. 639; 1928a, p. 317). These streams built great constructional "bars," and then subsided so quickly that these deposits were left almost unmodified by erosion or deposition by the subsiding water (Bretz, 1925, p. 105). Nevertheless, deep central channels were held to have been excavated in basalt while the flood diminished, draining the water away from high-level channels that hang above the deep central ones (Bretz, 1923b, p. 628, 644; 1924a, p. 149; 1927a, p. 210).

The principal lines of evidence on which this concept was based appear to have been these:

1. Great vertical distance, amounting in places to many hundreds of feet, between the floors of preglacial valleys and (a) the bases of Palouse scarps at the valley rims, (b) high-level stream deposits, (c) abandoned

hanging distributary valleys, and similar fluvial features. This relationship is clear and incontrovertible.

2. Definite indications that nonglacial tributaries to scabland tracts were ponded to high levels during proglacial discharge, and that deposits derived from the scabland were stream-built up these tributaries. This evidence is physically similar to the evidence cited in (1).

3. Great breadth of some scabland tracts, amounting to as much as 20 miles in places where the vertical dimension is small.

4. Interpretation of the stream deposits in the scabland tracts as constructional "bar" forms, essentially unmodified since their deposition.

5. Presence, as an integral scabland feature, of ragged basalt surfaces, including anastomosing channels and rock basins, which are held to be the product of subfluvial quarrying, and to have required large vigorous streams for their development (Bretz, 1927a, p. 201), although some basins were conceded to be plunge basins at the bases of extinct waterfalls.

6. Evidence of contemporaneous discharge through all the scabland tracts (although the Grand Coulee tract continued to discharge meltwater after the abandonment of the others). This evidence is well supported.

The concept of the "Spokane flood" was received with general skepticism,[1] but only two authors have entered the problem since the appearance of Bretz's publications. Hodge (1934) published in outline form an interpretation of scabland origin, a complicated sequence of ice advances and drainage changes, involving quarrying of basalt by direct glacial erosion, and cutting of channel complexes in the basalt by stream diversion around blocks of stagnant glacier ice and jams of berg ice, and viewed most of the closed depressions in the basalt as plunge basins.

Although retaining the concept of a "Spokane flood," Allison (1933) showed that the water was ponded to a high level, from the Wallula Gap to the Columbia River gorge through the Cascade Mountains. He attributed the ponding to an icejam of enormous size, originating in the gorge, and by continued accretion growing upstream into eastern Washington, forcing the scabland streams to rise and seek new and ever-higher diversion channels. Allison's recognition of high-level ponded water is an important and valuable contribution to the problem. Equally important is his recognition, along the Columbia River downstream from the Wallula Gap, of an earlier gravel fill, whose eroded remnants were submerged by the ponding and locally veneered with slack-water deposits. It is interesting to note that a part of Allison's evidence of ponding is the same as that cited much earlier by Bretz (1919; 1923a, p. 605) as favoring late-glacial marine submergence of this district.

[1] See, for example, discussions by W. C. Alden, J. Gilluly, E. T. McKnight, G. R. Mansfield, O. E. Meinzer, and J. T. Pardee *in* Bretz (1927a).

THESIS OFFERED FOR DISCUSSION

The Cheney-Palouse scabland tract (Bretz, 1928b, p. 450) was selected for study because it is the largest of the four tracts described by Bretz, and because it was said by him [2] to carry the most convincing evidence of enormous proglacial discharge. The study has yielded facts that support the following synthesis of events (Fig. 1):

1. Meltwater streams of normal discharge, originating in glacier ice, which overran the northern part of the tract, scoured basalt on steep gradients, mainly along preglacial drainage lines.
2. These streams gradually filled their valleys with gravel, sand, and silt, as rising ponded water in the Pasco basin reduced the gradients to about two-thirds of their former values.
3. The streams then dissected the fill, removed the greater part of it, and notched and channeled the basalt extensively, as the ponded water drained away and restored the original gradients, while glacier ice at the head of the tract continued to supply the necessary discharge.
4. These events occurred in the Wisconsin glacial stage, but erosion of basalt by meltwater, accompanied by a lesser fill, may have occurred during at least one earlier glaciation.

The discussion that follows embodies the facts upon which this reconstruction is based. Although the reconstruction is very different from the "Spokane flood" hypothesis, and although some of the observations cited here run counter to some of those cited in support of that hypothesis, I wish to acknowledge the aid I have received from the extensive information and excellent descriptions contained in Bretz's thirteen papers. The later investigator always benefits from the earlier work of the pioneer. Any reader thoroughly familiar with the literature of the scabland will realize how much any subsequent worker owes to Bretz. He has repeatedly stated that the scabland problem could be solved only by detailed field study, and has shown a cordial interest in my investigation. Prior to a revisit to the field on his own part, he prefers for the present not to indorse unqualifiedly the fill hypothesis presented here, but he states that if the additional field evidence offered by me is acceptable, he considers the fill hypothesis a consistent explanation of the facts in the Cheney-Palouse tract.[3]

PRINCIPAL FEATURES OF THE CHENEY-PALOUSE SCABLAND TRACT

The Cheney-Palouse scabland tract (Pl. 1) is about 100 miles long, extending from Spangle, Marshall, and Waukon in the vicinity of Spokane, to the Pasco basin. Its long axis is oriented northeast-southwest, down the

[2] Bretz, J H.: Unpublished communication.
[3] Bretz, J H.: Unpublished communication.

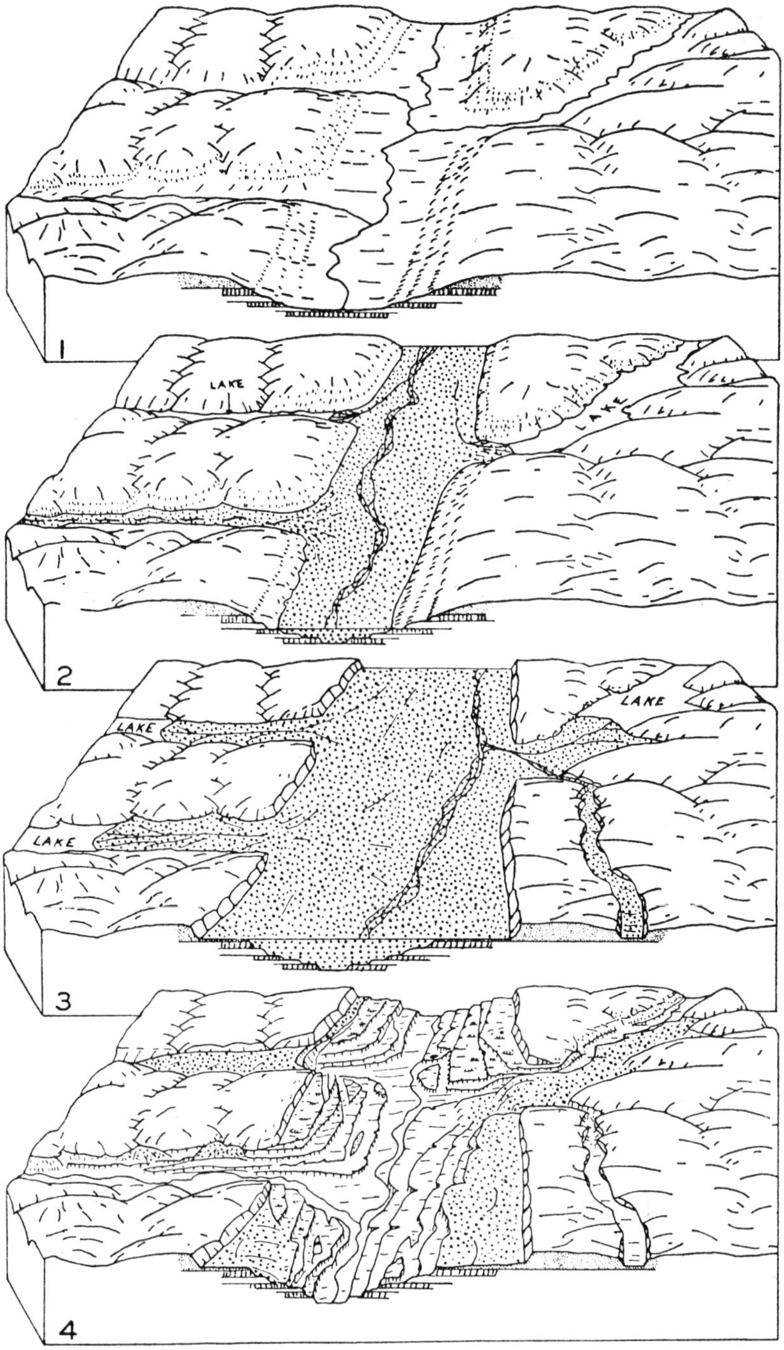

FIGURE 1.—*Evolution of an ideal scabland tract*

(1) Normal preglacial valley with tributaries, cutting "Palouse soil" and underlying basalt; (2) Meltwater stream has eroded valley floor and begun to build a fill, backing up tributaries; (3) Main valley and mouths of tributaries completely filled with stream sediments; lateral planation on fill has cut Palouse scarps; spillover (right) has entered head of small preglacial tributary; (4) Today—dissection has left remnants of fill, chiefly in tributary mouths, and has greatly eroded the basalt, cutting superposed channels across spurs.

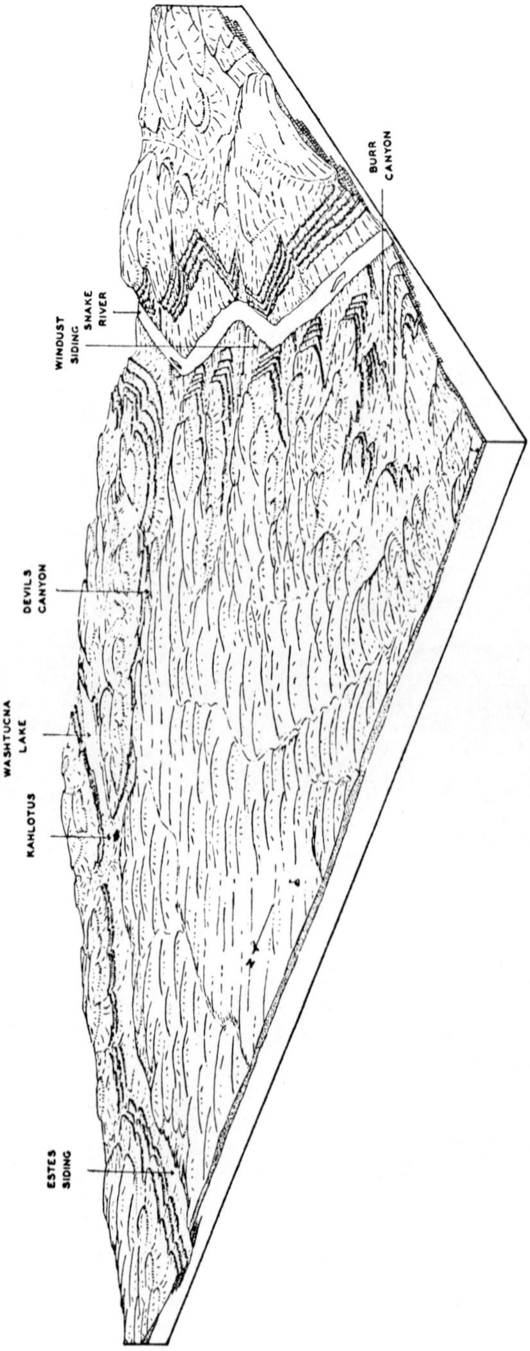

FIGURE 2.—*Snake River canyon and Washtucna Coulee*

Block diagram of an area of about 90 square miles, looking northeast, and showing the Devils Canyon distributary. Base of block represents sea level; highest elevations are somewhat more than 1500 feet.

dip of the basalt. At its southern end it bifurcates into two discharge routes, one consisting of Washtucna Coulee with its continuation known as Esquatzel Coulee, and the other, farther southeast, consisting of the canyon of Snake River. These two routes are connected transversely by both Palouse Canyon and Devils Canyon, which will be discussed later in detail. This great tract consists of a short segment of the preglacial Snake River system, and most of the preglacial Palouse River system, including Rock Creek and Cow Creek, whose headwaters reach to the northern end or head of the tract. The gross pattern of the system is just as it was in preglacial time, but details have been altered by the proglacial streams. The alterations consist principally of the stripping away of "Palouse soil" over wide areas, the cutting of minor lateral channels and rock basins *de novo,* and the trenching of a few preglacial divides.

The long profiles of the streams named are perfectly normal, and although some valley widening has occurred, I agree with Bretz (1923b, p. 646; 1928a, p. 204) that Snake River canyon was nearly as deep in preglacial time as it is now. The same may be said of the valleys of Palouse River and its chief tributaries as far downstream as Hooper, and of their preglacial continuation (Bretz, 1925, p. 98) in Washtucna and Esquatzel coulees, now dry.

The southwestern part of the Cheney-Palouse tract consists of deep and relatively narrow canyons little modified from their preglacial condition. The Snake River canyon scabland, for example, is 2 to 4 miles wide, by a maximum of 800 feet deep (Fig. 2). On the other hand, the northeastern part of the tract consists of broad expanses of roughened basalt surface, with a relief in most districts of considerably less than 100 feet. Throughout its length, the tract is flanked by Palouse scarps, and scarped, island-like residuals of "Palouse soil" are numerous, marking the positions of the more conspicuous preglacial interfluves, with innumerable small valleys in the "Palouse soil" left hanging above the bases of the scarps which now behead them.

SIGNIFICANT FIELD DATA

ADVANCE SUMMARY

The field observations given in the paragraphs that follow, may be briefly summarized thus:

(1) The scabland sediments exhibit relatively small average grain size, good size sorting, fair to good rounding of the larger size fractions, and predominantly shallow-fluvial stratification.

(2) The surface form characteristic of these deposits consists of nonpaired stream-cut terraces with lunate channels; not constructional "bars."

(3) The sediments grade downstream, both stratigraphically and topographically, into parallel-bedded silt and fine sand in the Pasco basin.

(4) Scarps and benches stream-cut in the "Palouse soil", near the vertical upper limit of scabland, do not fall into a single long profile, but are multiple, recording streams cutting laterally on various profiles.

(5) The scabland carries notched spurs, small slotlike canyons, small dry falls with plunge basins, and hanging distributaries and tributaries in the basalt, which must have been cut by relatively small streams operating on various profiles.

(6) All the foregoing data combine to yield a picture of leisurely streams with normal discharge, aggrading and then degrading, and emptying into standing water in the Pasco basin.

[*Editor's Note:* The detailed presentation of the field data has been omitted at this point.]

SUMMARY

Although, as pointed out during this discussion, some of the features described in the foregoing sections can be explained as the products of a "flood," the combination of them cannot be so explained. On the contrary, the evidence submitted points to streams of normal discharge, which lowered their profiles through deposits that they themselves had built. This view demands a relative rise of baselevel followed by a relative decline. We may now turn to the Pasco basin, where evidence of such deleveling, definitely related to the scabland fill, is present.

RELATION OF SCABLAND FILL TO DEPOSITS IN THE PASCO BASIN

ADVANCE SUMMARY

The scabland fill grades downstream into silt and fine sand containing erratic stones. This fill occupies the Pasco basin, and the Columbia basin as far west as the gorge through the Cascade Mountains. Chiefly because of its stratification and uniform upper limit, this fine-grained fill is believed to be lacustrine. It is thought to record a lake ponded by a landslide dam or glacier ice in the Columbia gorge, the rise and fall of the lake surface providing a fluctuating baselevel for the scabland streams, causing them first to deposit a fill and later to dissect their deposits.

TOUCHET BEDS

General features.—The Pasco basin, and continuously with it the Columbia basin downstream as far as the gorge through the Cascade Mountains, carries a deposit of buff- to cream-colored silt and fine sand with erratic stones. Because this deposit is believed to be a stratigraphic unit, and because it is particularly well exposed in the vicinity of Touchet, it will be referred to in this paper as the Touchet beds. This material is younger than the "Palouse soil," upon which it rests, and grades both stratigraphically and topographically into the scabland fill. It ranges in thickness from a barely appreciable veneer up to more than 100 feet in single exposures. Topographic relations suggest that in places it is considerably thicker. Its grain size is related to the mouths of the scabland tracts, being coarse (sand and gravel) near them and fine (medium and fine silt) away from them. It carries erratic stones ranging from sand grains to boulders, including many non-basalts, and these also are larger and more numerous near the mouths of the tracts than away from them. The coarser phases are invariably well stratified, showing cut-and-fill (fluvial) bedding and subflat parallel bedding such as might have been made in shallow ponded water (Pl. 3, fig. 1). In some places the finer phases show delicate subflat parallel lamination with local indications of rhythmic deposition, but more commonly they show little or no stratification (Bretz, 1929, p. 540). The Touchet beds differ from the "Palouse soil," however, in being less compact, with few exceptions [12] "brighter" in appearance (lacking ferruginous stains), free from conspicuous accumulations of caliche in the subsoil zone, and in the fact that it carries erratics. Moreover, its vertical distribution and topographic expression differ from those of the "Palouse soil." It has not been recognized above an elevation of 1150 feet, and is usually thin and patchy above 1100 feet. But at lower elevations, where sufficiently thick, it is sharply dissected, exhibiting sloping, plane-topped interfluves between V-gullies which contrast with the mature expression of the valleys and interfluves in the "Palouse soil."

[*Editor's Note:* The detailed description of the Touchet beds has been omitted at this point.]

[12] For example, an 18-foot section exposed in the SW cor. Sec. 3, T 9 N, R 31 E, shows coarse-sand zones limonitized just above their contacts with underlying silt layers, which are relatively "tight".

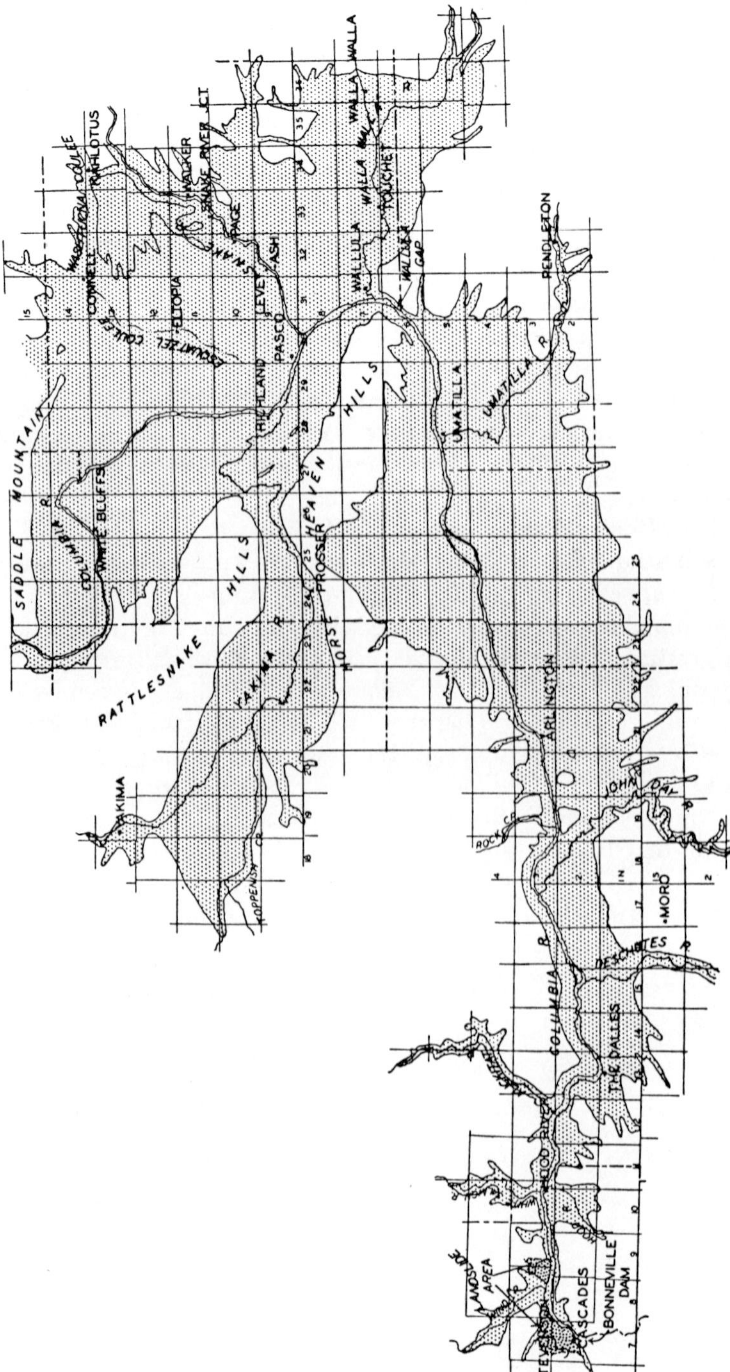

FIGURE 5.—*Approximate area covered by Touchet beds and erratics*

Shaded area shows those parts of the Pasco basin and the Columbia Valley west to Stevenson, which lie below the 1150-foot contour. Strong stippling shows areas of modern landslides in the Columbia gorge. Townships are numbered for reference.

DISCUSSION OF THE PONDING

Historical.—Several published works have recorded ponded water in the region between the mouths of the scabland tracts and the Columbia gorge.

1. Symons (1882, p. 108, pl. 26) recognized a former lake in the Pasco basin, which he named Lake Lewis. The evidence consisted of silt, (rafted?) boulders, and exposures of the Ringold formation.

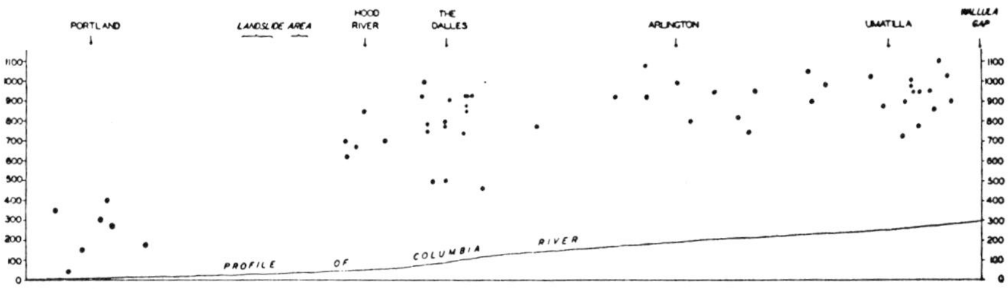

FIGURE 7.—*Erratics between Wallula Gap and Portland*

Profile representing about 150 miles along the Columbia River, with vertical dimension exaggerated 200 times, showing longitude and elevation of erratics accurately located by Bretz (1919), Piper (1932, p. 128), and Allison (1933; 1935). The erratics occur throughout a wide zone along both sides of the valley, and have been brought south or north into the plane of the profile. The absence of low-level erratics from the profile east of Hood River does not represent the true situation. Erratics are present at lower elevations, but probably they were not regarded as critical. This is a compilation of published data from outside the region which is the subject of the present paper. Silt localities are not shown.

2. Russell (1893, p. 26-27) recognized Symons's Lake Lewis, noted that its maximum depth was "nearly a thousand feet", and that it was ephemeral, and suggested that it was either a lake ponded by a dam, perhaps glacial, in the Columbia gorge, or an arm of the sea introduced by temporary subsidence.

3. Bretz (1919) held Lake Lewis to be an arm of the sea, but later (1925) considered the silt and erratic stones to be deposits of the "Spokane flood", their upper limit declining downstream.

4. Hodge (1931) held the erratics to be lacustrine, but thought they antedated the present Columbia canyon.

5. Piper (1932, p. 129) believed the erratics near The Dalles were ice-rafted on a ponded Columbia River.

6. Allison (1933) showed that the upper limit of silt and erratics is essentially uniform throughout its extent, and held that ponding was the result of a dam caused by landsliding and a jam of floated ice near Stevenson, in the Columbia gorge.

Fossils.—Constant search for fossils in the Touchet beds has resulted in only two finds. A road-cut in the NE 1/4 Sec. 13, T 8 N, R 29 E, near Kennewick, at elevation 500 feet, in a fine-sand phase of the deposit, yielded several calcareous shell-like forms which have been identified[15] as the fossil fruits of *Lithospermum ruderale* Douglas, a common plant in the dry soils in that district today. As the plant is terrestrial, the fruits must have been washed into place, and hence do not provide critical information as to whether the Touchet beds are lacustrine or marine.

A more significant find was made in a road-cut in the E ½ Sec. 30, T 7 N, R 34 E. Here, a fine-sandy silt facies of the Touchet beds yielded diatoms. These were isolated by P. D. Krynine, and were kindly examined by K. E. Lohman, of the United States Geological Survey, who identified them as *Melosira granulata* (Ehrenberg) Ralfs, *M. ambigua* (Grunow) Müller, and *M. solida* Eulenstein. These determinations have led Lohman to the opinion that the inclosing silt was deposited in a fresh-water lake.[16]

Possible dam in the Columbia gorge.—The fossil evidence favors the first of Russell's hypotheses. On the evidence so far at hand, I favor the concept of a ponding caused by damming in the Columbia gorge. The best evidence is the areal distribution of the silt and the vertical distribution of the erratics. The high-level silt occurs only upstream from the central part of the Columbia gorge, and the erratics, although reaching consistently high elevations (up to 1150 feet) east of the gorge, are nowhere higher than 400 feet in the region west of the gorge (Fig. 7).

A dam in the Columbia gorge might result from (1) warping or faulting, (2) local volcanic accumulation (rejected by Allison, 1933, p. 719), (3) local glacier ice, or (4) landsliding, with or without the cooperation of icejams, as suggested by Allison. There is no direct evidence of the first two, and the facts thus far known make it very unlikely that ponding of the required depth could have been brought about by either of these agencies. Glacier ice and landsliding, however, appear to be valid possibilities.

A valley glacier, emanating from Mount Hood, descended Hood River valley to within 5 miles of the Columbia in a late glaciation, and to within 3 miles of the Columbia during an earlier glaciation (Bretz, 1919, p. 495). Bretz regards it as unlikely that this ice tongue could have blocked the Columbia to a high elevation without leaving local traces. On the other hand, until definite evidence to the contrary is brought forward, the hypothesis of a glacier-ice dam ought to be retained as a possibility.

The data on a possible landslide dam are as follows: The Oligocene or Miocene Eagle Creek formation, consisting of weak siltstones, shales, volcanic ejecta, conglomerates, and flows, is exposed in a north-south anticline so that it crops out through 25 miles along the north bank, and

[15] By Mr. L. C. Wheeler and Dr. Ivan M. Johnston, through the kindness of Professor M. L. Fernald.

[16] Lohman, K. E.: unpublished communication.

through 14 miles along the south bank of the Columbia, between Hood River and Skamania (Hood River quadrangle).[17] This mass of weak strata, at least 2700 feet thick at Table Mountain, is overlain by about 1500 feet of basalt.

The contact lies at about 2000 feet A.T. in the crest of the anticline on the north side of the river, but descends southward with the pitch of the fold. At Red Bluffs and at Bald Mountain, both north of the river, extensive sliding has occurred, with development of landslide topography beneath two areas aggregating several square miles. At Red Bluffs, the cliffs have receded 3 miles from the river, which has been crowded over against the south bluffs by the slides. Recent slides which apparently dammed the Columbia to a depth of 100 feet have been described by Lawrence (1936, figs. 4, 5). Extrapolation on the contact of resistant basalt on the weak Eagle Creek formation at Table Mountain, shows that if extensive landslides had occurred when this cliff was much nearer the Columbia than it is now, the resulting debris could have dammed the Columbia up to the 1150-foot contour. This is admittedly a large requirement, which can perhaps best be met by successive slides at the same place, or by slow continuous sliding on a gentle slope. By either of these means, it seems possible that the Columbia might have been dammed for hundreds of years.

The answer to the dam question will have to follow detailed study of the Columbia gorge. I do not defend any one hypothesis here, because the character of the water body is not vital to the fill hypothesis of scabland origin. The fill mechanism could have operated equally well if the proglacial streams had been graded to a lake with some other kind of dam, or to an arm of the sea. The outstanding requirement is that the water surface shall have risen gradually to, or close to, the present 1150-foot contour, and then have begun soon to fall again. For this water body, I propose to use Symons's term Lake Lewis, without prejudice as to its origin.

Time relations.—If Lake Lewis was an arm of the sea, we have all the time we need for the building of the scabland fill. But if it was a lake without outlet, then the time factor becomes critical. On the latter assumption, the maximum area and capacity of the Lake Lewis basin, assuming an upper limit of 1150 feet, have been determined, and the time required to fill such a basin by the modern Columbia has been computed. The results are as follows:

[17] Important data on the landslide area have been kindly supplied by J Harlen Bretz (unpublished communication); also *see* Allison (1933, p. 720), and Williams (1916, p. 86-93).

Area of Lake Lewis at maximum[18] .... 5916 square miles
Volume of Lake Lewis at maximum[19] .... 1,502,000 acre-feet
Present rate of evaporation from water surfaces (weighted average of 5 stations) (Follansbee, 1934) .... 42 inches per year
Mean annual discharge of all streams tributary to lake area, 1878-1935 (Grover, Canfield, and Parker, 1936, p. 15) .... 200,500 second-feet
Time required for present streams to fill basin[20] .... 13 years 1 month

The calculated time, 13 years 1 month, seems grossly inadequate for the deposition of the fill in the scabland tracts. This calculation, however, is based on present discharge, and we have no data at all on the discharge during the bracket of Pleistocene time in which we are interested. There is as yet no means of determining whether the discharge was greater or smaller then than now, and if so, by what amount.[21] Further, the calculations are based on no outflow at the dam, but various hypotheses, including those of repeated and progressive landsliding, permit lengthening the time indefinitely, by allowing for outflow during rise of the lake surface.

With the facts thus far at hand, it does not seem profitable, therefore, to carry the time calculations any farther. It should be added, however, that the numerous non-paired stream-cut terraces in the scabland tracts, vertically closely spaced, indicate that the ratio of lateral cutting to downcutting by the proglacial streams was high, during dissection of the fill. As the great volume of debris in these streams would have kept their long profiles in approximate equilibrium throughout the greater part of the dissection episode, this high ratio is understandable even though the surface of Lake Lewis may have descended rapidly.

PROFILES OF THE PROGLACIAL STREAMS

Two principal profiles of the meltwater streams are recorded (Fig. 8). The relatively steep profiles of the present drainage ways, which were followed by the meltwater streams just before they were extinguished by the disappearance of the glacier ice from the plateau at the stream heads near Spokane, have an average gradient of slightly more than 20 feet per mile, measured from a point near Cheney to a point near Snake River Junction, a distance of nearly 100 miles. The less-steep profile recorded by the highest fill remnants and Palouse scarps, which was followed by

[18] By planimeter on 100-foot contours.

[19] Computation from planimeter measurements.

[20] Allowing for decreasing rate of rise, and increase in evaporating area, as water area increased.

[21] The analogy of the late-Glacial lakes Bonneville and Lahontan suggests diminished late-Glacial evaporation, owing to increased precipitation or lower temperature or both. The lowered late-Glacial snowline of the Cascade Mountains (Klute, 1928, p. 75; Louis, 1927, p. 221) carries the same suggestion. Furthermore, the gorges across the Coast Ranges in British Columbia having been blocked by ice, the late-Glacial discharge of the Columbia River in Washington should have been drawn from a more extensive tributary area than that furnishing the present discharge.

the streams at the maximum of aggradation, has an average gradient, through the same segment, of about 13 feet per mile. The figures given necessarily include any changes due to warping that may have occurred since proglacial flow ceased, and whose value cannot be determined. However, local deformation is not indicated by the profiles, and regional

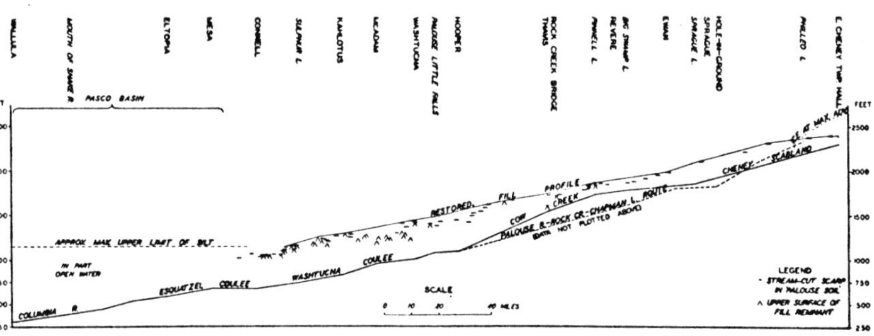

FIGURE 8.—*Vertical relations of Cheney-Palouse tract*

Profiles from East Cheney Township Hall (near Spokane) to Pasco basin, showing main drainage way and a principal subsidiary; limit of glacier ice at maximum advance; positions of Palouse-scarp bases and upper surfaces of fill remnants determined with Paulin altimeters from nearby benchmarks; long profile of fill at its maximum, restored from scarps and fill remnants; and upper limit of silt and erratics (Touchet beds) in Pasco basin. The scarps and fill remnants lie on both sides of the scabland tract, distributed through a zone 5 to 25 miles wide, and have been brought into the plane of the profile.

deformation is not likely to have altered greatly the two profiles relative to each other.

Intermediate positions of the streams during dissection of the fill are recorded by intermediate scarps and terrace remnants, but, as these are not paired, they cannot be correlated. Intermediate positions of the streams during the building of the fill are not recorded; their records are in part obliterated, and in part indistinguishable from those made during dissection.

The stream-cut scarps "hang" above the floor of the Pasco basin, their downstream terminations showing several profiles which are apparently fortuitous, depending on the position of the surface of Lake Lewis at several times when the laterally shifting streams impinged against promontories of "Palouse soil."

Prior to aggradation the initial profiles of the streams were probably somewhat steeper than the latest, post-fill, profiles, because of the excavation in basalt that occurred near the head of the Cheney-Palouse tract. Even if this difference be disregarded, the building of the fill would have had the effect of reducing the initial gradients about 35 percent (from about 20 feet per mile to about 13 feet per mile). Conversely, the disappearance of Lake Lewis would have increased the gradients more than

50 percent above the smaller figure. These changes seem entirely appropriate to have produced the filling and dissection actually recorded.

The reader now has before him all the data essential to an understanding of the fill-and-dissection interpretation of the genesis of the Cheney-Palouse scabland tract. In the following pages are discussed the effects of the fill on eastern tributaries to the tract, the special cases of two distributaries across major divides, and the age relations of the scabland features. A chronologic summary of events, based on the new interpretation, is given in a concluding section.

[*Editor's Note:* The detailed discussions of eastern tributaries, Devil's Canyon, Palouse Canyon, age relations, and correlations have been omitted at this point.]

CHRONOLOGIC SUMMARY AND CONCLUSION

The order of events inferred from the features of the Cheney-Palouse tract and the Pasco basin may be summarized as follows:

1. Advance of glacier ice lobes across the Spokane-Columbia canyon at the northern margin of the plateau, to the divide between drainage north to the canyon and drainage south on the plateau. If the Okanogan Lobe or the Columbia Lobe (Flint, 1936; 1937) reached the plateau before the Little Spokane Lobe, a lake may have been formed in the canyon, and an overflow spill established down the Cheney-Palouse tract, before the encroachment of actual glacier ice south of the divide.

2. Occupation of preglacial valleys and erosion of basalt in them, by the meltwater flow.

3. Rise of Lake Lewis in the Pasco basin, causing aggradation in the tributary valleys to form a thick fill, growing vertically upward and also headward.

4. Cutting of Palouse scarps by lateral planation and development of lateral distributaries from the principal routes as the streams rose on their own fills. Sedimentation in Lake Lewis, progressively overlapping stream deposits in the mouths of tributaries.

5. Draining of Lake Lewis, accompanied by progressive dissection of the fill into cut terraces, together with the cutting of more Palouse scarps and terraces and the erosion of more basalt, chiefly by stream channels superposed from the fill. Offlap deposits built into receding Lake Lewis, most of this secondary sediment later eroded and carried off down the Columbia.

6. Backwasting of the glacier ice margin beyond the divide in the Spokane district, shutting off the supply of meltwater and causing the Cheney-Palouse tract to run dry except for the through drainage of the modern Palouse River-lower Snake River system.

7. During the whole sequence of events, showers of volcanic ash fell at many different times, and the ash was incorporated in the growing fill, and in the deposits of Lake Lewis and the Riparia Lake.

A hypothesis of filling followed by dissection demands no more than one glaciation for the general development of such a scabland tract, but certain features brought out in this discussion seem to indicate that more than one glaciation, with meltwater flow, did, in fact, control the evolution of the Cheney-Palouse tract. The key to the latest fill in the tract is Lake Lewis. If this water body had not come into existence, temporarily providing an abnormally high baselevel, the glacial record in the scabland tract probably would have consisted of a thinner fill in the form of a valley train, slightly dissected by postglacial streams, instead of a thick fill greatly dissected by meltwater streams. Again, if Lake Lewis had maintained its high level until after the withdrawal of the glacier ice from the area south of Spokane, the fill would still be essentially nondissected, except in Snake River canyon, and would form broad valley floors contrasting little with the adjacent surface of the "Palouse soil."

WORKS TO WHICH REFERENCE IS MADE

Allison, I. S. (1933) *New version of the Spokane flood,* Geol. Soc. Am., Bull., vol. 44, p. 675-722.

——— (1935) *Glacial erratics in Willamette Valley,* Geol. Soc. Am., Bull., vol. 46, p. 615-632.

Bretz, J Harlen (1919) *The late Pleistocene submergence in the Columbia Valley of Oregon and Washington,* Jour. Geol., vol. 27, p. 489-506.

——— (1923a) *Glacial drainage on the Columbia Plateau,* Geol. Soc. Am., Bull., vol. 34, p. 573-608.

——— (1923b) *The channeled scablands of the Columbia Plateau,* Jour. Geol., vol. 31, p. 617-649.

——— (1924a) *The Dalles type of river channel,* Jour. Geol., vol. 32, p. 139-149.

——— (1924b) *The age of the Spokane glaciation,* Am. Jour. Sci., 5th ser., vol. 8, p. 336-342.

——— (1925) *The Spokane flood beyond the channeled scablands,* Jour. Geol., vol. 33, p. 97-115, 236-259.

——— (1927a) *Channeled scabland and the Spokane flood,* Wash. Acad. Sci., vol. 17, p. 200-211.

——— (1927b) *The Spokane flood; a reply,* Jour. Geol., vol. 35, p. 461-468.

——— (1928a) *Alternative hypotheses for channeled scabland,* Jour Geol., vol. 36, p. 193-223, 312-341.

——— (1928b) *The channeled scabland of eastern Washington,* Geog. Rev., vol. 18, p. 446-477.

——— (1928c) *Bars of channeled scabland,* Geol. Soc. Am., Bull., vol. 39, p. 643-701.

——— (1929) *Valley deposits immediately east of the channeled scabland of Washington,* Jour. Geol., vol. 37, p. 393-427, 505-541.

——— (1930a) *Lake Missoula and the Spokane flood,* Geol. Soc. Am., Bull., vol. 41, p. 92-93.

Bretz (1930b) *Valley deposits immediately west of the channeled scabland,* Jour. Geol., vol. 38, p. 385-422.

——— (1932a) *The Grand Coulee,* Am. Geog. Soc., Spec. Pub., No. 15, 89 pages.

——— (1932b) *The channeled scabland,* Int. Geol. Congr. (XVI Sess., U. S. 1933), Guidebk. 22: Excur. C-2, 12 pages.

Brown, J. S. (1928) *Ground water in the New Haven area, Connecticut,* U. S. Geol. Surv., W.-S. Paper 540, 206 pages.

Bryan, Kirk (1927) *The "Palouse soil" problem with an account of elephant remains in wind-borne soil on the Columbia Plateau of Washington,* U. S. Geol. Surv., Bull. 790, p. 21-46.

Calkins, F. C. (1905) *Geology and water resources of a portion of east-central Washington,* U. S. Geol. Surv., W.-S. Paper 118, 96 pages.

Campbell, Marius R., and others (1915) *Guidebook of the Western United States, Part A. The Northern Pacific Route,* U. S. Geol. Surv., Bull. 611, 211 pages.

Collins, R. F., and Schalk, M. (1937) *Torrential flood erosion in the Connecticut Valley, March, 1936,* Am. Jour. Sci., 5th ser., vol. 34, p. 293-307.

Columbia Basin Survey Commission (1920). *The Columbia Basin irrigation project,* Columbia Basin Surv. Comm., Olympia, 185 pages.

Flint, R. F. (1935) *Glacial features of the southern Okanogan region,* Geol. Soc. Am., Bull., vol. 46, p. 169-194.

——— (1936) *Stratified drift and deglaciation of eastern Washington,* Geol. Soc. Am., Bull., vol. 47, p. 1849-1884.

——— (1937) *Pleistocene drift border in eastern Washington,* Geol. Soc. Am., Bull., vol. 48, p. 203-231.

——— (1938) *Summary of the late-Cenozoic geology of southeastern Washington,* Am. Jour. Sci., 5th ser., vol. 35 (*In press*).

Follansbee, Robert (1934) *Evaporation from reservoir surfaces,* Am. Soc. Civil Engrs., Tr., vol. 99, p. 710-711.

Gilbert, G. K. (1890) *Lake Bonneville,* U. S. Geol. Surv., Mon. 1, 438 pages.

Grover, Nathan C.; Canfield, G. H.; Parker, G. L. (1936) *Surface water supply of the United States, 1935; pt. 14—Pacific slope basins in Oregon and Lower Columbia River Basin,* U. S. Geol. Survey, W.-S. Paper 794.

Hodge, E. T. (1931) *Exceptional morainelike deposits in Oregon,* Geol. Soc. Am., Bull., vol. 42, p. 985-1010.

——— (1934) *Origin of the Washington scablands,* Northwest Science, vol. 8, p. 4-11.

Jenkins, O. P. (1925a) *The Spokane flood: a discussion,* Jour. Geol., vol. 33, p. 747-748.

——— (1925b) *Clastic dikes of eastern Washington and their geologic significance,* Am. Jour. Sci., 5th ser., vol. 10, p. 234-246.

Kerr, F. A. (1936) *Quaternary glaciation in the Coast Range, northern British Columbia and Alaska,* Jour. Geol., vol. 44, p. 681-700.

Klute, F. (1928) *Die Bedeutung der Depression der Schneegrenze für eiszeitliche Probleme,* Zeitschr. für Gletscherkunde, vol 16, p. 70-93.

Kocher, A. E., and Strahorn, A. T. (1919) *Soil survey of Benton County, Washington,* U. S. Dept. Agric., Bur. Soils, Field Operations for 1916, 72 pages.

Lahee, F. H. (1923) *Field geology.* 2nd ed., New York.

Lawrence, D. B. (1936) *The submerged forest of the Columbia River gorge,* Geog. Rev., vol. 26, p. 581-592.

Leighton, Morris M. (1919) *The road building sands and gravels of Washington,* Wash. Geol. Surv., Bull. 22, 307 pages.

Louis, H. (1927) *Die Verbreitung von Glazialformen im Westen der Vereinigten Staaten,* Zeitschr. für Geomorphologie, vol. 2, p. 221.

Merriam, J. C., and Buwalda, J. P. (1917) *Age of strata referred to the Ellensburg formation in the White Bluffs of the Columbia River,* Univ. Calif. Publs., Bull. Dept. Geol., vol. 10, p. 255-266.

Piper, A. M. (1932) *Geology and ground-water resources of the Dalles region, Oregon,* U. S. Geol. Surv., W.-S. Paper 659, 189 pages.

Russell, I. C. (1893) *A geological reconnaissance in central Washington,* U. S. Geol. Surv., Bull. 108, 108 pages.

——— (1897) *A reconnaissance in southeastern Washington,* U. S. Geol. Surv., W.-S. Paper 4, 93 pages.

——— (1901) *Geology and water resources of Nez Perce County, Idaho, Part I,* U. S. Geol. Surv., W.-S. Paper 53, 85 pages.

Smith, G. O. (1903) *Description of the Ellensburg quadrangle,* U. S. Geol. Surv., Geol. Atlas, Ellensburg folio (No. 86), 7 pages.

Strahorn, A. T., and others (1929) *Soil survey (reconnaissance) of the Columbia Basin area Washington,* U. S. Dept. Agric., Bur. Chem. and Soils, Series 1929, No. 28, 255 pages.

Symons, T. W. (1882) *The upper Columbia River and the great plain of the Columbia,* U. S., 47th Congr., 1st Sess., Senate Ex. Doc. 186, 135 pages.

Trowbridge, A. C. (1911) *The terrestrial deposits of Owens Valley, California,* Jour. Geol., vol. 19, p. 743-744.

Van Duyne, Cornelius, and others (1917) *Soil survey of Franklin County, Washington,* U. S. Dept. Agric., Bur. Soils, Field Operations for 1914, 101 pages.

Williams, I. A. (1916) *The Columbia River gorge: its geologic history interpreted from the Columbia River Highway,* in *Mineral Resources of Oregon,* vol. 2, no. 3, 130 pages.

Part IV

VINDICATION

Editor's Comments on Papers 9, 10, and 11

By 1940 several geologists had developed alternative explanations for the origin of the Channeled Scabland. As in 1927, the major arguments surfaced at one professional meeting. On June 17–23, 1940, the American Association for the Advancement of Science met in Seattle, Washington. The geological section (Section E) included a session, "Quaternary Geology of the Pacific Northwest," held Tuesday afternoon, June 18. The session paper titles suggested that nonflood origins of the Channeled Scabland would be strongly advocated. A post-meeting field trip had been organized during which Flint would be leading a visit to the Channeled Scabland. Bretz had also been invited to participate, but he refused the invitation to confront his adversary. Bretz noted that all of his ideas were in print and that the field evidence would speak for itself. Indeed it did!

The early papers in the session reiterated various hypotheses for the Channeled Scabland. E. T. Hodge (1940) presented yet another retelling of his scenario involving glacial erosion in the scablands associated with complex damming and diversions by river ice. I. S. Allison (1940) presented a synopsis of Flint's fill hypothesis and contrasted it with his own ice-jam theory. In his later published paper Allison (1941) pointed to key shortcomings in the Flint hypothesis. The first shortcoming was that the anastomosing channel patterns and deep rock basins could not have been eroded by "normal" streams. Second, Allison disputed Flint's correlation of the scabland gravels to the Touchet beds, suggesting that the Touchet lacustrine sequence was younger than the gravels. Third, he agreed with Bretz that the peculiar shapes of the scabland deposits required extraordinary processes. The conclusion was that the complex jamming of various channels with ice was the only reasonable explanation for the unusual drainage patterns and depositional features.

PARDEE'S REVELATION

The dramatic event at the 1940 meeting was supplied by a totally unexpected source. Howard A. Meyerhoff (personal communication, 1978) attended the meeting and relates that no one in attendance really suspected that any of the speakers would support Bretz's postulate of a giant flood. The eighth speaker in the session was Joseph Thomas Pardee, who rose to speak on "Ripple Marks(?) in Glacial Lake Missoula."

Pardee quietly described the "ripple marks" at Camas Prairie, an intermontane basin in northwestern Montana. He described their size as "extraordinary," heights of up to 15 m and spacings up to 150 m (Pardee 1940). Pardee's written discussion (Paper 9) also had an understated title, "Unusual Currents in Lake Missoula." His work, dating back to before Bretz's studies, clearly demonstrated that Lake Missoula was the source of catastrophic floods through the Channeled Scabland. He noted that about 2000 km^3 of water were impounded behind a glacial lobe that occupied the basin of modern Lake Pend Oreille in northern Idaho. Actually these relationships had been known for decades. Pardee (1910, p. 376) even credited T. C. Chamberlin with the discovery of the lake strandlines: "Chamberlin conceived the idea of a glacial dam and furthermore tentatively suggested that its location was in the Pend Oreille region with outflow by way of Spokane."

Pardee's new discovery was that this glacial dam had failed suddenly, with a resultant rapid draining of the lake. Evidence for this failure included severely scoured constrictions in the lake basin, huge bars of current-transported debris, and the giant current ripple marks. However, Pardee (1942) did not state the connection to the Channeled Scabland. Perhaps he generously left that point to Bretz.

ACCEPTANCE

In the summer of 1952, Bretz, then nearly seventy years old, returned for his last summer of fieldwork in the Channeled Scabland. The purpose was to investigate new data that had been obtained in surveys for the Bureau of Reclamation's Columbia Basin project. Professor H. T. U. Smith accompanied him, acting in the field as "skeptic for all identifications and interpretations" (Bretz et al. 1956, p. 761). Mr. George E. Neff of the Bureau of Reclamation pointed out many new exposures of flood sediments.

Bretz returned from the 1952 field season with a wealth of new data. The U.S.B.R. had been especially generous in supplying maps, aerial photographs, and sedimentological information. Bretz wrote the extensive report over the next year. He recalls sending a draft to R. F. Flint for criticism. In that paper he, of course, provided a meticulously detailed refutation of Flint's fill hypothesis. Bretz (personal communication, 1978) recalls that the response from Flint was but one sentence, "Scabland is proving to be pretty complicated."

The published report of Bretz's 1952 field work (Paper 10) is noteworthy for its study of the scabland depositional features. Extensive excavations for the irrigation project and new topographic maps proved that the gravel hills called bars by Bretz (1928) were indeed subfluvial depositional bedforms. Most convincing of all was the presence of giant current ripples on the upper bar surfaces. These showed clearly that bars 30 m high were completely inundated by phenomenal flows of water. Numerous examples of giant current ripples were found on the same bars that Flint had interpreted as terraces. Such features could have been produced only by the flow velocities associated with truly catastrophic discharges. Bretz and others (1956) also modified Bretz's earlier interpretations to allow for several episodes of flooding. The theme of their study, however, was that only a hypothesis involving flooding could account for all the features of the Channeled Scabland.

The results of Bretz's 1952 field study were also ably summarized by Bretz (1959). This paper is reproduced here as Paper 11. It covers the same major points as the Bretz, Smith, and Neff (1956) report, but it presents the critical new field data in a shorter format.

One field experience during the summer of 1952 nearly put Bretz's name on the map (literally). While Bretz and Smith were visiting the U. S. Bureau of Reclamation offices in Ephrata, a two-day excursion was organized to show the two visitors various aspects of the Columbia Basin Irrigation Project. At one point the group stopped at the summit of High Hill to view the rugged, unpopulated scabland of the southern Hartline Basin. There, a prominent unnamed coulee was developed in the Trail Lake anticline. Mr. Walcott of the U.S.B.R. announced that the coulee was to be named "Bretz Coulee."

When the final maps of the region appeared (Coulee City 7.5-minute quadrangle of the U. S. Geological Survey, 1965), Bretz found that the coulee had been named "Hudson Coulee." Bretz relates the story of that name (personal communication, 1978):

> The Hudson brothers tried desperately to make a farm on the gravel bar below the cataract cliff facing south from Hartline Basin. The younger brother returned from a store, carrying a rifle bought on credit. The brothers were down to their last cent, and the older one ordered that the gun be returned to the store. Resistance led to a struggle. The younger brother was shot.
>
> The remote life, financial worries, and the unwitnessed murder pushed the older brother to madness. He walked several miles to two hated neighbors, called them to their doors, and shot them on the spot.
>
> The whole township was alerted by daybreak. Who was this demented man planning to kill next? Several days of search found the body of the older Hudson. He had shot himself with the rifle purchased by his brother. Putrifying odors and swarming blue bottle flies betrayed his hiding place in some rugged scabland of the district.
>
> Yes, I'm glad that the U. S. Bureau of Reclamation never succeeded in fixing my name to that coulee.

Although his name was not immortalized in the region's geography, Bretz's hypothesis for the Channeled Scabland, as espoused by Bretz and others (1956), became accepted by nearly all the geological community. Subsequent investigators have mainly added details to Bretz's masterful interpretation (Trimble 1963; Fryxell and Cook 1964; Richmond et al. 1965; Baker 1973).

The second edition of Flint's comprehensive text *Glacial and Pleistocene Geology* does not mention the Channeled Scabland. One reference (Flint 1957, p. 168) is made to "the temporary diversion of the Columbia River through the Grand Coulee," which is cited as a conspicuous example of a large glacier-margin river. The irony remained that this oft-cited textbook, noted for its thorough coverage, actually failed to mention the most spectacular event to occur during the Pleistocene on our planet. Moreover, despite its discrediting and Flint's own reluctance to mention the idea after 1956, the scabland fill hypothesis continued to appear in texts through the 1960s (Thornbury 1969, pp. 390–391). Eventually, however, Flint's tacit acceptance of the flood hypothesis was recorded in the third edition of his text. That volume contains this single sentence (Flint 1971, p. 232): ". . . features, collectively known as channeled scabland, were widely created east of the Grand Coulee by overflow of an ice-margin lake upstream (Bretz 1969)."

REFERENCES

Allison, I. S. 1940. Flint's fill hypothesis of origin of scabland. *Geol. Soc. America Bull.* **51**:2016.

Allison, I. S. 1941. Flint's fill-hypothesis for channeled scabland. *Jour. Geology* **49**:54–73.

Baker, V. R. 1973. Paleohydrology and sedimentology of Lake Missoula flooding in eastern Washington. *Geol. Soc. America Spec. Paper 144,* 79 pp.

Bretz, J H. 1928. Bars of Channeled Scabland. *Geol. Soc. America Bull.* **39**: 643–702.

Bretz, J H. 1959. Washington's Channeled Scabland. *Washington Div. Mines and Geology Bull.* **45**:57 pp

Bretz, J H. 1969. The Lake Missoula floods and the Channeled Scabland. *Jour. Geology* **77**:505–543.

Bretz, J H., Smith, H. T. U., and Neff, G. E. 1956. Channeled Scabland of Washington: new data and interpretations. *Geol. Soc. America Bull.* **67**:957–1049.

Flint, R. F. 1957. *Glacial and Pleistocene Geology.* New York: John Wiley and Sons, 553 pp.

Flint, R. F. 1971. *Glacial and Quaternary Geology.* New York: John Wiley and Sons, 892 pp.

Fryxell, R., and Cook, E. F. 1964. A field guide to the loess deposits and channeled scablands of the Palouse area, eastern Washington. *Wash. State Univ. Lab. Anthropology Rept. Inv. 27,* 32 pp.

Hodge, E. T. 1940. Glacial history of southeastern Washington. *Geol. Soc. America Bull.* **51**:2024.

Pardee, J. T. 1910. The glacial Lake Missoula, Montana. *Jour. Geology* **18**: 376–386.

Pardee, J. T. 1940. Ripple marks (?) in glacial Lake Missoula, Montana. *Geol. Soc. America Bull.* **51**:2028–2029.

Pardee, J. T. 1942. Unusual currents in glacial Lake Missoula, Montana. *Geol. Soc. America Bull.* **53**:1569–1600.

Richmond, G. M., Fryxell, R., Neff, G. E., and Weis, P. L. 1965. The Cordilleran ice sheet of the northern Rocky Mountains and related Quaternary history of the Columbia Plateau. In *The Quaternary of the United States,* Wright, H. E., and Frey D. G., eds. Princeton, N.J.: Princeton University Press, pp. 231–242.

Thornbury, W. D. 1969. *Principles of Geomorphology.* New York: John Wiley and Sons, 594 pp.

Trimble, D. E. 1963. Geology of Portland, Oregon and adjacent areas. *U. S. Geol. Survey Bull. 1119:* pp. 1–119.

9

Reprinted from pages 1570–1571, 1573–1574, 1587–1588, 1594–1599 of *Geol. Soc. America Bull.* **53**:1570–1599 (1942)

UNUSUAL CURRENTS IN GLACIAL LAKE MISSOULA, MONTANA

J. T. Pardee

[*Editor's Note:* The contents listing has been omitted at this point.]

ABSTRACT

The area submerged by Glacial Lake Missoula includes several intermontane basins and constricted interconnecting valleys or "narrows" that drain to a single outlet, the Clark Fork River. A sudden failure of the ice dam that blocked this valley, near the Idaho-Montana State line, caused unusually large and rapid currents through the narrows and wind gaps in the partly submerged rim of Camas Prairie basin. Evidences of such currents include commensurate, but otherwise ordinary, effects of streams confined to rocky channels and the unique giant ripple marks. At its high stage the lake is roughly estimated to have held more than 500 cubic miles of water of which nearly three-fourths was stored above a constricted part of the Clark Fork Valley called the Eddy Narrows. Calculations based on available incomplete data indicate a flow through the Eddy Narrows that reached a maximum of 9.46 cubic miles per hour. Whether the lake was completely drained at that time has not been determined, but a later set of beaches testifies that the basin held a lake soon after the rapid outflow. Apparently the final draining was gradual.

INTRODUCTION

In the Pleistocene much of the Clark Fork drainage basin in northwestern Montana (Fig. 1) was briefly submerged by "Glacial Lake Missoula" (Pardee, 1910). A lobe of the Cordilleran glacier advanced southward along the Purcell Trench into northern Idaho where it occupied the basin of Pend Oreille Lake and crowded up the valley of the Clark Fork River to a position near the Montana State line. The water thus impounded in the Clark Fork Valley rose to a maximum altitude of 4150 feet or about 2000 feet higher than the floor of the valley just above the dam. On the east, south, and west it was held in by mountains, and on the north possible outflow channels were blocked by other lobes of Cordilleran ice.

The ice dam is thought to have failed, permitting a sudden large outflow. The consequent depression of the water surface in the arm of the lake above the dam caused huge and rapid currents in the narrow valleys and passes connecting this arm with the wider depressions in the lake basin farther east. The records of these currents are described. The interpretation of certain gravel deposits and denuded

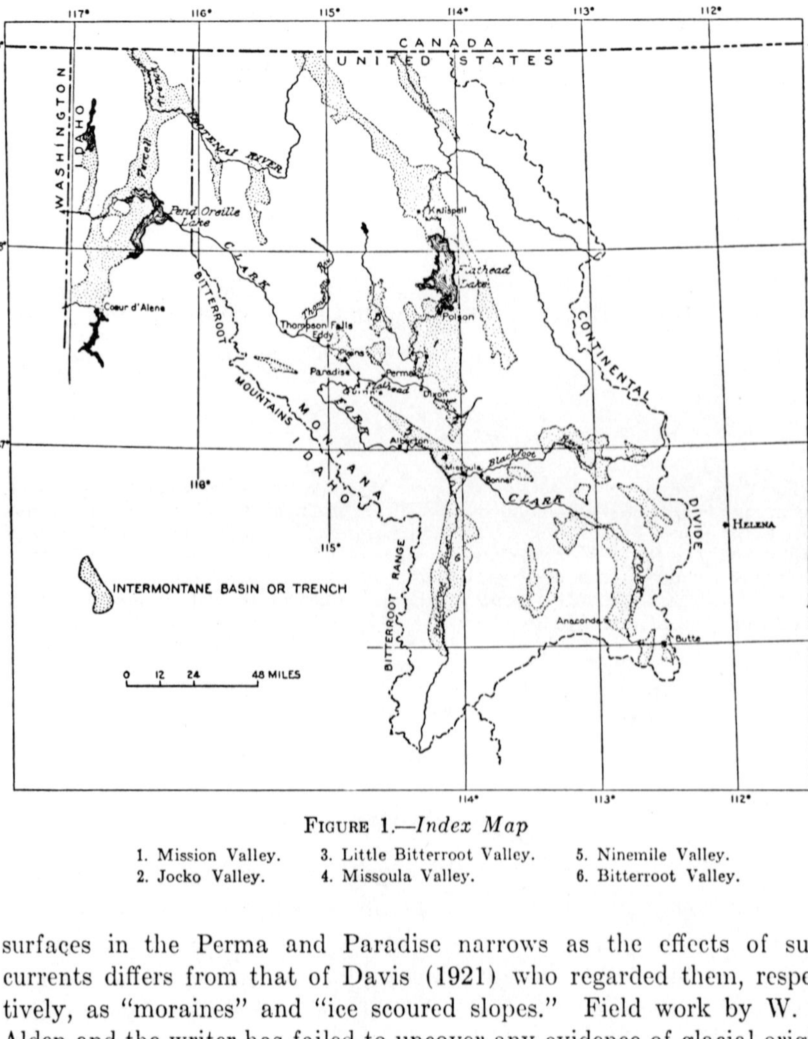

FIGURE 1.—*Index Map*

1. Mission Valley.	3. Little Bitterroot Valley.	5. Ninemile Valley.
2. Jocko Valley.	4. Missoula Valley.	6. Bitterroot Valley.

surfaces in the Perma and Paradise narrows as the effects of such currents differs from that of Davis (1921) who regarded them, respectively, as "moraines" and "ice scoured slopes." Field work by W. C. Alden and the writer has failed to uncover any evidence of glacial origin.

Not long after the rapid outflowing ceased the basin again held a lake. So far it is not clear whether this body was part of the original ponded water whose escape had been halted or a new filling. In contrast to the earlier outrush, final draining seems to have been very slow.

[*Editor's Note:* Figure 2, plate 1, and the acknowledgments have been omitted at this point.]

TOPOGRAPHY

BASINS

Although the region of Glacial Lake Missoula is occupied chiefly by mountains, the topography is diversified by a number of wide valleys (Fig. 1) or intermontane basins.

The basins are enclosed by generally rough and steep mountains standing from 2000 to 4000 feet higher. Exceptionally high rugged ranges west of the Bitterroot Valley (Bitterroot Range) and east of Mission Valley (Mission Range) rise abruptly to serrate crests 5000 to 6000 feet above the adjacent lowlands. Some of the basins are isolated except for narrow outlet valleys that join them in a common drainage system. Others are more closely connected in irregular groups and in certain linear series known as trenches. Most of the individual basins are one-fifth to half as wide as they are long. Their floor areas range from about 50 square miles in Camas Prairie to 400 square miles or more in Mission Valley and consist mainly of gently sloping plains that merge with a medial flat or flood plain or end as terraces above it. Several of these basins were submerged by the central and eastern parts of the glacial lake. Among them the Mission, Jocko, and Little Bitterroot valleys (Figs. 1, 3) form a closely spaced group drained by the Flathead River. Another group that drains into the Clark Fork River above the mouth of the Flathead includes the Missoula, Ninemile, and Bitterroot valleys. Two other relatively small basins submerged in the central part of the lake are of particular interest in its history: Plains Valley (Fig. 9; Pl. 1, fig. 1) along the Clark Fork River a few miles below the mouth of the Flathead, and Camas Prairie, east of Plains Valley and a short distance north of Perma.

NARROWS

Clark Fork River and its tributaries drain the basins through comparatively narrow valleys which in certain places become severely constricted and gorgelike narrows. The 70-mile stretch of the Clark Fork Valley between the Missoula-Ninemile-Bitterroot group of basins and the Plains Basin generally is not more than 1 or 1½ miles wide and is enclosed by steep, rugged mountains 2000 to 4000 feet high. Parts narrowed to half a mile or less occur near Alberton (Alberton Narrows), above Quinn's Springs (Quinn's Narrows), and below the

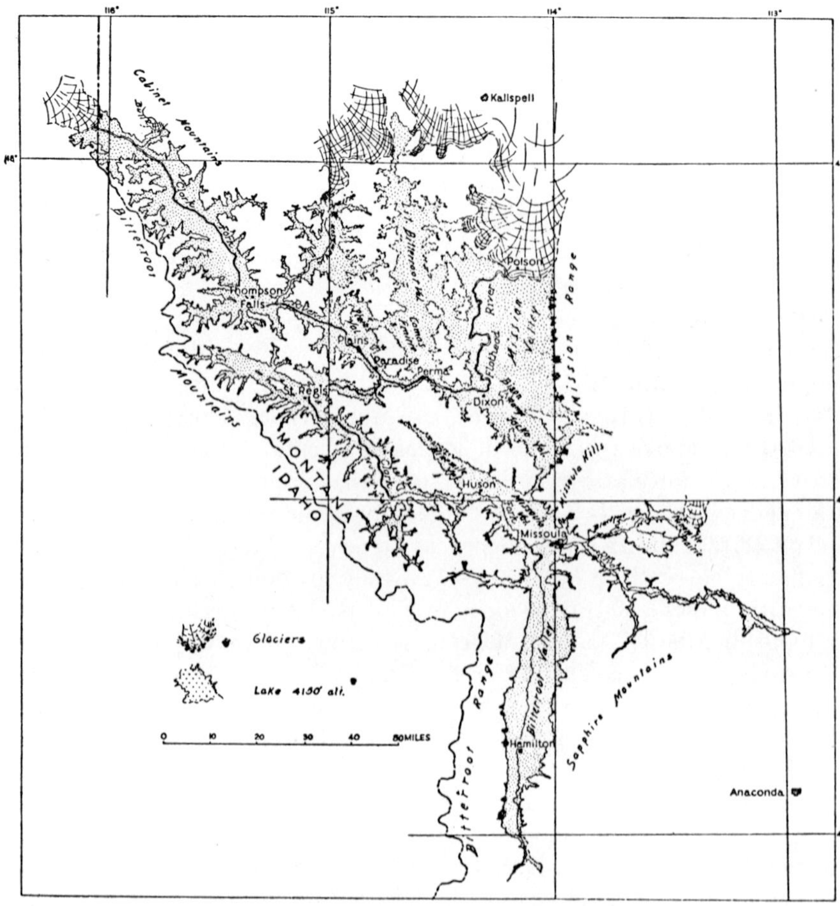

FIGURE 3.—*Glacial Lake Missoula*
High Stage

mouth of the Flathead River from Paradise to Plains (Paradise Narrows). The Clark Fork Valley throughout 10 miles below Plains Basin is again less than a mile wide (Eddy Narrows) between steep, rugged slopes 4000 feet high (Fig. 2).

[*Editor's Note:* Material, including plate 2, has been omitted at this point.]

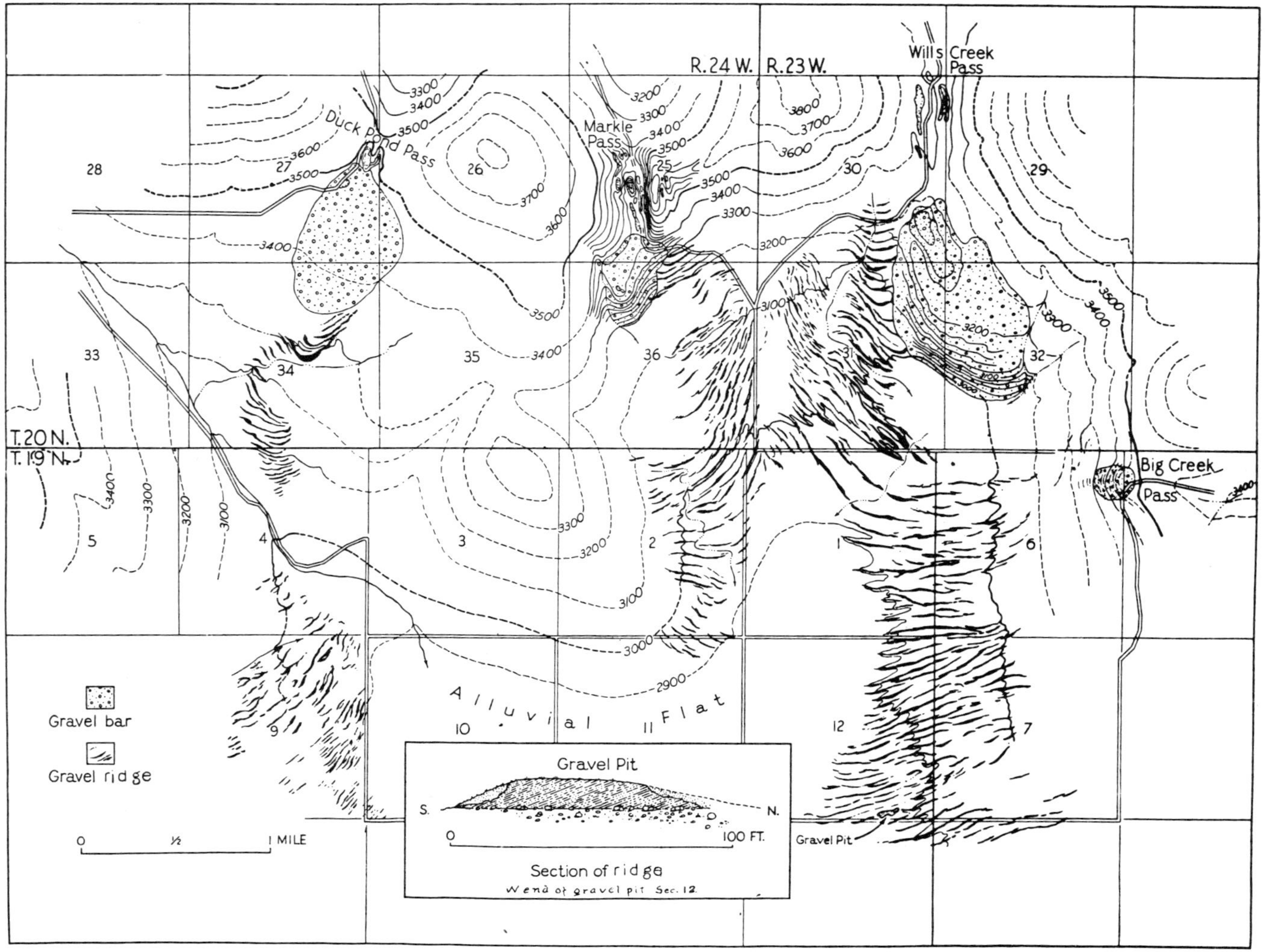

NORTHERN PART OF CAMAS PRAIRIE BASIN

[*Editor's Note:* Material has been omitted at this point. Plate 3 appears on the preceding page.]

Giant ripple marks of Camas Prairie.—Camas Prairie is distinguished by a number of unique ridges that have the form, structure, and arrangement of ordinary ripple marks but are so large that the term "ripple mark" seems inappropriate. It is retained, however, for want of a better one. Except for a few inconspicuous ones below Big Creek Pass, the ridges (Pl. 3) are grouped in belts appended to the other three northern gaps—namely, Duck Pond, Markle, and Wills Creek passes, a relationship suggesting formation by currents entering the basin through those channels. The relative scarcity of similar ridges below Big Creek Pass might be explained by the relatively small size of its stream.

The distribution, linear form, and arrangement of the ripple-mark ridges as mapped were obtained from aerial photographs (by the U. S. Forest Service) of the belt below Duck Pond Pass, and by pace and compass traverses of the belts below Markle and Wills Creek passes. An unmapped part of the Wills Creek Pass belt extends 1 mile or more farther south. The ridges range from 1 foot or less to 50 feet in amplitude and from a few feet to as much as 500 feet in wave length, but the mean height is between 15 and 30 feet and the wave length about 250 feet. Most of the smaller ridges occur toward the lower (basinward) ends of the belts, and the exceptionally high ones near the middle. Toward the lower end of the belt in front of Duck Pond Pass the ridges gradually decrease in size until they fade out completely.

Individual ridges are from a few rods to half a mile or more in length. Their axes are slightly curved, and convex forward (toward the basin). Their crests are rather wide and rounded and with few exceptions their cross profiles are asymmetrical, the southern or downstream sides being the steeper. As exposed in a pit in one of the ridges about 3½ miles south of Wills Creek Pass (near the southeast corner of sec. 12, T. 19 N., R. 24 W.) the gravel is largely subrounded pebbles ranging from 8 to 20 millimeters in diameter. Sandy matrix is rather scant, the mass is loosely compacted, distinctly stratified, and throughout the cross section the dip of the bedding is parallel to the steeper south (downstream) side (Pl. 3). Toward the pass the gravel of the ridges coarsens, but no other exposures of the structure were seen. In comparable situations the ridges in front of Markle and Duck Pond passes are of similar composition and texture.

The giant ripple marks lie on a smooth bedrock surface that rises gently from the interior flat to the bordering mountains, and, according to the direction which they happen to take across this surface, their profiles are sloping or horizontal. Toward the passes they are overlain by the bar-like deposit of coarse gravel at the end of each channel. All appear to have been formed by large currents plunging over the north rim and continuing with decreasing velocity out into the basin. Such currents might have formed when the water surface in Camas Prairie Basin first became lower than the water surface in Little Bitterroot Basin thus causing masses of water to pass over the ridge, chiefly through the gaps. The initial direction and momentum of the descending masses of water plus increased specific gravity due to the fine sediment picked up on the way would tend to hold the currents to the bottom for a considerable distance.

[*Editor's Note:* Material has been omitted at this point.]

AREA AND VOLUME OF THE GLACIAL LAKE

The maximum altitude of the water surface in Glacial Lake Missoula is placed at 4200 feet as indicated by the uppermost of a series of beaches (Pardee, 1910). Repeated checks with the aneroid have indicated that the altitude of the highest definite beach is 4150 feet, although some doubtful beach gravel was found in one place (divide east of Oliver Gulch in T. 21 N., R. 22 W.) near the 4200-foot level. The outline of the lake at the 4150-foot level (Fig. 3) has been prepared with the aid of topographic maps and other surveys that were made after 1910. As thus determined, the area of the lake in round figures was 2900 square miles, and its volume 500 cubic miles. The areas, average depths, and volumes of water held by the different arms of the lake are roughly estimated (Table 2).

DEVELOPMENT OF THE UNUSUAL CURRENTS

Apparently the ice dam failed sufficiently to permit a sudden great outrush of water. The arm of the lake just above the dam was affected first, and there being no obstacle to its free discharge its surface fell rapidly throughout. Beyond Thompson River, however, the outflow was restricted by the Eddy Narrows and successively held back by the several narrows farther up. The lowered water surface in the arm above the dam induced a strong current through Eddy Narrows which lowered the water surface in the Plains Arm. As a result inflows were caused through Rainbow Lake Pass from Camas Prairie Arm and through the

TABLE 2.—*Approximate areas, depths, and volumes of arms of Glacial Lake Missoula*
The water surface was at 4150 feet altitude.

Arm	Area (Square miles)	Average depth (Feet)	Volume (Cubic miles)
CLARK FORK VALLEY FROM MONTANA-IDAHO STATE LINE TO MOUTH OF FLATHEAD RIVER			
Lower Clark Fork (State line to Eddy Narrows)	465	1400	123.3
Thompson River Valley	115	600	13.0
Eddy Narrows	20	1380	5.2
Plains Basin	90	1350	23.0
Paradise Narrows	10	1330	2.5
FLATHEAD RIVER VALLEY			
Perma Narrows	10	1280	2.4
Camas Prairie Basin	95	870	15.6
McDonald Narrows	25	1200	5.7
Little Bitterroot Basin	260	900	44.3
Jocko Basin	105	900	17.9
Mission Basin	575	1100	120.0
CLARK FORK VALLEY ABOVE MOUTH OF FLATHEAD RIVER			
Quinn's Narrows	35	1200	8.0
Saint Regis River Valley	175	600	18.2
Superior Arm (Quinn's Narrows to Alberton Narrows)	120	940	21.4
Alberton Narrows	30	900	5.1
Missoula-Ninemile Basin	245	740	34.3
Bitterroot Basin	405	700	53.7
Hellgate Arm (Clark Fork Valley above Missoula Basin)	50	150	1.4
Blackfoot River Valley	65	440	5.4
Totals	2895 square miles		520.4 cubic miles

Paradise and Perma Narrows from that and other arms farther up (Fig. 9). The consequent lowering of the water surface in the Camas Prairie Arm caused an inflow across the submerged northern rim. In succession the other narrows released outflows from the more distant basins. The subsequent behavior of the currents thus generated was controlled largely by such topographic relations as the distribution and capacities of the different basins, the lengths and cross sections of the narrows, and in particular the "bottle necks" formed by the Paradise and Eddy narrows.

VOLUME AND DURATION OF THE UNUSUAL CURRENTS

Evidently the streams were of unusual size but were short-lived because of limited water supply. No precise data, of course, are available. The following results, based on formulas for small streams, are introduced merely to give some idea of their size.

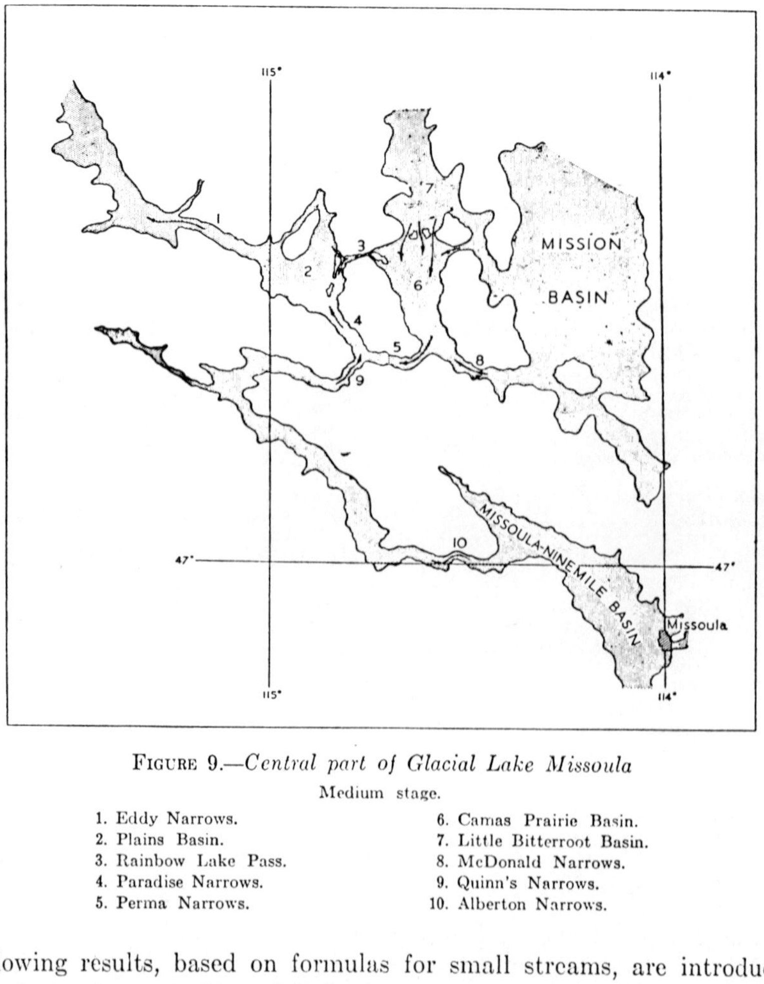

FIGURE 9.—*Central part of Glacial Lake Missoula*

Medium stage.

1. Eddy Narrows.
2. Plains Basin.
3. Rainbow Lake Pass.
4. Paradise Narrows.
5. Perma Narrows.
6. Camas Prairie Basin.
7. Little Bitterroot Basin.
8. McDonald Narrows.
9. Quinn's Narrows.
10. Alberton Narrows.

Eddy Narrows provided a nearly straight sluiceway 10 miles long with no very prominent projections along the sides to hinder the flow. As indicated by its high eddy deposits and the denudation of its sides the stream was at least 1000 feet deep. Available information of the upper limit is not precise enough to permit a satisfactory determination of the surface gradient, which is therefore assumed to have been the same as the average descent of the present river bed—namely, 2.3 feet per mile through the 10 mile stretch of the narrows. The average cross section is about 5,900,000 square feet, and the wetted perimeter 9600 feet. Using these figures with the Chezy formula (Merriman, 1931) for rough stone channels gives a velocity of 52.3 feet per second. According to the formula adopted by Manning (Lindquist, 1934) the same figures would indicate a velocity of 78 feet per second. The velocity of the eddying currents is rather closely related to that of the main stream (Leighly, 1934). As determined by a maximum pebble size in general of 3 inches (76 mm. approximately) the currents that formed the high eddy deposits had (Hjulstrom, 1935) a velocity of about 65 feet per second which may be assumed as about the same as the main current. Incidentally, this figure happens to be the mean of the two other determinations. Accepting it as approximately correct the flow through the Eddy Narrows would appear to have reached a peak of 386,500,000 second feet or about 9.46 cubic miles per hour. As only about 380 cubic miles of water was impounded above the Eddy Narrows the maximum flow could not have continued for long. The maximum discharge of the Mississippi River at Natchez in the flood of February 1937 (Williams and Crawford, 1939) was 2,100,000 second feet or about .05 cubic mile per hour.

Sufficient data are not available for even rough estimates of the flows carried by the other narrows or by the wind gaps around Camas Prairie or by Clark Fork Valley near the ice dam. However, according to data compiled by Hjulstrom (1935) the velocity of the stream that formed the giant ripple marks is indicated at one place (gravel pit in SE¼ sec. 12, T. 19 N., R. 24 W.) by a maximum pebble size of 40 mm. to have been 7 feet a second. This current presumably was at least twice as deep as the ripple marks are high, or about 100 feet, and it may have occupied the full width of the basin floor; obviously no estimate is possible without more data.

AGE OF THE UNUSUAL CURRENTS

Although they have remained exposed since the water subsided, the giant ripple marks and other features produced by the unusual currents show very little effect of weathering. Rock surfaces stripped of their surface mantle remain bare except for a relatively thin new cover. Talus aprons generally do not reach more than one third to half way up a cliff (Pl. 2, fig. 1). Except for later stream trenching in a few places, the giant ripple marks look as though they had just emerged. Where protected from stream erosion the high eddy deposits are likewise unmodified. Few of the many channel depressions in the wind gaps around Camas Prairie Basin have become filled. Large parts of the coarse gravel deposits in Plains Basin and Clark Fork Valley at Thompson Falls have survived attacks by the river. Altogether these conditions suggest an age not greater than Middle Pleistocene.

LAST STAGE OF THE GLACIAL LAKE

Either the rapid outflow was greatly slowed when the lake was partly drained or, if it had become emptied, the basin soon thereafter became, partly at least, refilled, and in this new body of water no unusual currents seem to have developed. The evidence for this delayed drainage or refilling consists mainly of beaches formed on the current-produced features of the preceding outflow.

A series of small beaches are plainly visible on the side of Boyer Bench facing Plains Basin (Pl. 8, fig. 2). From rather faint lines below Locust Hill they become more definite southward where the higher ones, as they turn eastward around the terrace, end in small spitlike embankments. Elsewhere rather faint strand lines are visible on a high eddy deposit along Vanderburg Creek east of Perma (NW¼ sec. 9, T. 18 N., R. 23 W., Fig. 5) and on the basinward side of the coarse gravel deposit in front of Wills Creek Pass (Pl. 3). A small gravel bar continuous with a beach at 3400 feet altitude lies across the lowest part of the stream channel of Markle Pass at its northern entrance. Another blocks the north entrance to a channel about 140 feet higher (Fig. 7). A similar relation between beaches and channel features exists at the northern entrance to Wills Creek Pass.

The highest of the later beaches is at 3540 feet. Whether that figure is the upper limit of the last submergence cannot be positively stated. Where the outflow features are absent present knowledge of the beaches is not sufficient to distinguish the younger from the older.

A relatively slow current during the final outflow may have formed the gravel bar in Plains Basin east of the river near the entrance to the Eddy Narrows. As exposed by a pit along the highway (Pl. 8, fig. 3) the gravel is fine-textured (about 15 to 20 mm., pebble size), clean, and its bedding fore-set. The gravel rests on a platform about 50 feet higher than the river and contains fragments of the varved silt.

SUMMARY

Failure of the ice dam in the Clark Fork Valley near the Idaho-Montana State line permitted the rapid draining of Glacial Lake Missoula estimated to have held 500 cubic miles of water. Rapid falling of the water surface caused huge currents in the narrows between sub-

[*Editor's Note:* **Plate 8, figure 5, and figure 7 have not been reproduced here.**]

merged basins. Current-produced features include the unique giant ripple marks of Camas Prairie, a series of gravel ridges having the form, structure, and arrangement of ordinary ripple marks but of incomparable size. The flow through Eddy Narrows is estimated to have reached a maximum of nearly 9½ cubic miles an hour. After the outflow the basin again held a lake which drained away slowly.

REFERENCES CITED

Alden, W. C. (Manuscript in preparation) *Physiography and glacial geology of western Montana and adjacent areas*, U. S. Geol. Survey.

Davis, W. M. (1921) *Features of glacial origin in Montana and Idaho*, Assoc. Am. Geog. Ann., vol. 10, p. 122-132.

Grover, N. C., et al (1938) *The passage of turbid water through Lake Mead*, Soc. Civil Eng., Pr., Part II, p. 730-796.

Hjulstrom, Filip (1935) *Studies of the morphological activity of rivers as illustrated by the River Fyris*, Geol. Inst. Upsala, Vol. 25, p. 298, Fig. 18.

Leighly, J. B. (1934) *Turbulence and the transportation of rock debris by streams*, Geog. Rev., vol. 24, p. 453-464.

Lindquist, E. G. W. (1934) *On velocity formulas for open channels and pipes*, Royal Swedish Inst. Econ. Res., Pr., p. 18, Stockholm.

Meinzer, O. E. (1916) *Artesian water for irrigation in Little Bitterroot Valley, Montana*, U. S. Geol. Survey, W. S. Paper 400-B, p. 15.

Merriman, Mansfield (1931) *Treatise on hydraulics*, 10th ed. (revised with aid of Thaddeus Merriman) p. 287-288a, John Wiley & Sons, New York.

Pardee, J. T. (1910) *The Glacial Lake Missoula*, Jour. Geol., vol. 18, no. 4, p. 376-386.

Williams, G. R., and Crawford, L. C. (1939) *Maximum discharges at stream measurement stations*, U. S. Geol. Survey, W. S. Paper 847, p. 120.

10

Reprinted from pages 958–960, 965–967, 1043–1049 of *Geol. Soc. America Bull.* **67**:958–1049 (1956)

CHANNELED SCABLAND OF WASHINGTON: NEW DATA AND INTERPRETATIONS

J H. Bretz, H. T. U. Smith, and G. E. Neff

[*Editor's Note:* In the original, material precedes these excerpts.]

Introduction

The plateau character of eastern Washington is determined by the Miocene Columbia River basalt flows. In northern and eastern marginal portions, islandlike hills of much older crystalline rocks project above the general plane surface. Locally, folding and faulting of the basalt have made strongly expressed corrugations, uplifted little-eroded linear tracts, and depressed areas of only incomplete filling. All relief features, whether diastrophic, erosional or depositional, are interruptions of a regional southwestward slope of the plateau basalt flows.

The original low-lying basalt plain received fluviatile and lacustrine sediments—Ellensburg, Latah, early Ringold (?)—before recorded deformation or erosion. After folding and some weathering and erosion, similarly derived sediments were deposited in structural basins and are still flat-lying. The Ringold formation in the Pasco basin is the outstanding record of this time. Considered to be of Pleistocene age solely from its scanty vertebrate remains, it shows no evidence of contemporaneous glaciation.

Overlying all these older formations is an extensive, thick silt mantle, the "Palouse soil." Most of it is clearly of loessial origin, but a detailed history has not yet been deciphered. A rolling to hilly topography, made mostly by stream erosion but some by wind, appears to have existed on this mantle during the time of accumulation of the dust and has subsequently become much more marked. Calcified and reddened zones, even caliche layers, are known deep in the loess. A widespread subsurface caliche was formed after loessial deposition had passed its maximum. It exists also on older rocks where not loess-covered.

The scablands are glacial-river courses where the Palouse and other sediments were swept off the basalt chiefly along preglacial drainage lines, and the basalt itself was notably eroded in places. In these glacial-river courses are extensive stream gravels of the same age and origin. Only the northern edge of the plateau was invaded by the Cordilleran ice sheet.

In the lower reaches of Snake River canyon, in Pasco basin of the Columbia, and in lower Yakima and Walla Walla valleys are widespread lacustrine silts—the Touchet beds—also contemporaneous with glaciation in highlands to the north, east, and west. The Touchet silts record a "Lake Lewis" backwater ponding either from a great blockade in the Columbia gorge across the Cascade Range or from a low stand of this part of the continent and an estuary from the Pacific. "Lake Lewis" is only a general term for possibly several Pleistocene pondings of the Columbia and its tributaries in the southwestern part of the plateau.

The term channeled scabland was first used by Bretz (1923a; 1923b) for the remarkably anastomosing system of glacial-river courses distributed over an area approximately 125 by 125 miles on the plateau.[1] Bretz (1927a) noted that:

"The channeled scablands . . . cover almost 2000 square miles . . . of the plateau. They are elongate tracts, oriented with the gentle dip slope of the underlying basalt flows, mostly bare rock or with a thin cover of coarse basaltic rubble, commonly with canyons in them, and are bounded by steep slopes of the deep loessial soil of the plateau. They constitute a curious anastomosing pattern, the down-dip convergences inherited from an earlier normal drainage pattern and the divergences, equally numerous, produced by crossing of divides of this older pattern. There are hundreds of tracts of the higher loess-covered areas in the scablands, from a fraction of a square mile to many townships in extent, all discontinuous and bounded by the scabland areas. The steep marginal slopes in loess are in striking contrast to the gentle slopes of the older drainage pattern surviving within each isolated tract. Canyons in the scablands are multiple and anastomosing, amazingly so in some tracts, deep canyons and shallow ones uniting and dividing in a labyrinthine fashion about bare rock knobs and buttes unlike any other land surface on the earth. . . . These canyons are interpreted as channels, not valleys, hence the term 'channeled scabland' ".

This definition dealt only with erosional features. But the glacial rivers that "channeled" the scablands also left deposits whose character, the leading theme in this study, is as extraordinary as that of the erosional forms.

The glacial water that made the scabland crossed the preglacial divide along the high northern margin of the tilted plateau, found gradients of 20–25 feet per mile in the preglacial drainage ways leading southwestward down the dip slope, greatly enlarged them, and crossed minor divides among them to make the anastomosis. At some of these crossings canyons 400–900 feet deep were eroded by cataract recession. Empty rock basins more than 100 feet deep and not of plunge-pool origin occur in many places. The glacial rivers which swept the loessial cover[2] off the underlying basalt and left the sharply expressed marginal scarps, descending from the mature topography of the loess "islands" to the essentially bare basalt of the channel floors, are also recorded by great mounds of basaltic debris, chiefly in the Quincy structural basin and in portions of the canyons of Snake and Columbia rivers which discharged westward. Any satisfactory theory for the genesis of this unique assemblage of land forms must take account of all these features and relations.

Four distinctly different explanations, based on field studies, have been proposed during the past 30 years. At least three theories utilize enough apparently supporting field evidence to have been accepted among geologists acquainted with portions of the region, but each has lacked sufficient evidence to eliminate its rivals.

Field study on which this paper is based is the execution of a long-cherished plan by Bretz to restudy the region with the later

[1] After the appearance of 11 more titles by this author and 5 papers by writers who did not accept his interpretation of the origin of channeled scabland, the details of this definition still satisfactorily encompass all erosional and most depositional features of channeled scabland, except for one writer's explanation which has been vigorously rejected by everyone else acquainted with the region.

[2] Where used without qualification in this paper "loess" includes both the thin, fairly recent loess mantling the mature slopes of the "Palouse topography" and the Pleistocene Palouse soil, itself largely loessial in origin.

rival interpretations in mind. The great Columbia Basin Irrigation Project in the western part of the scabland complex has made new data available from the extensive new excava-

land demands more than cursory inspection, more than detailed study of limited portions. One major phase of the record never before adequately treated is the topography, com-

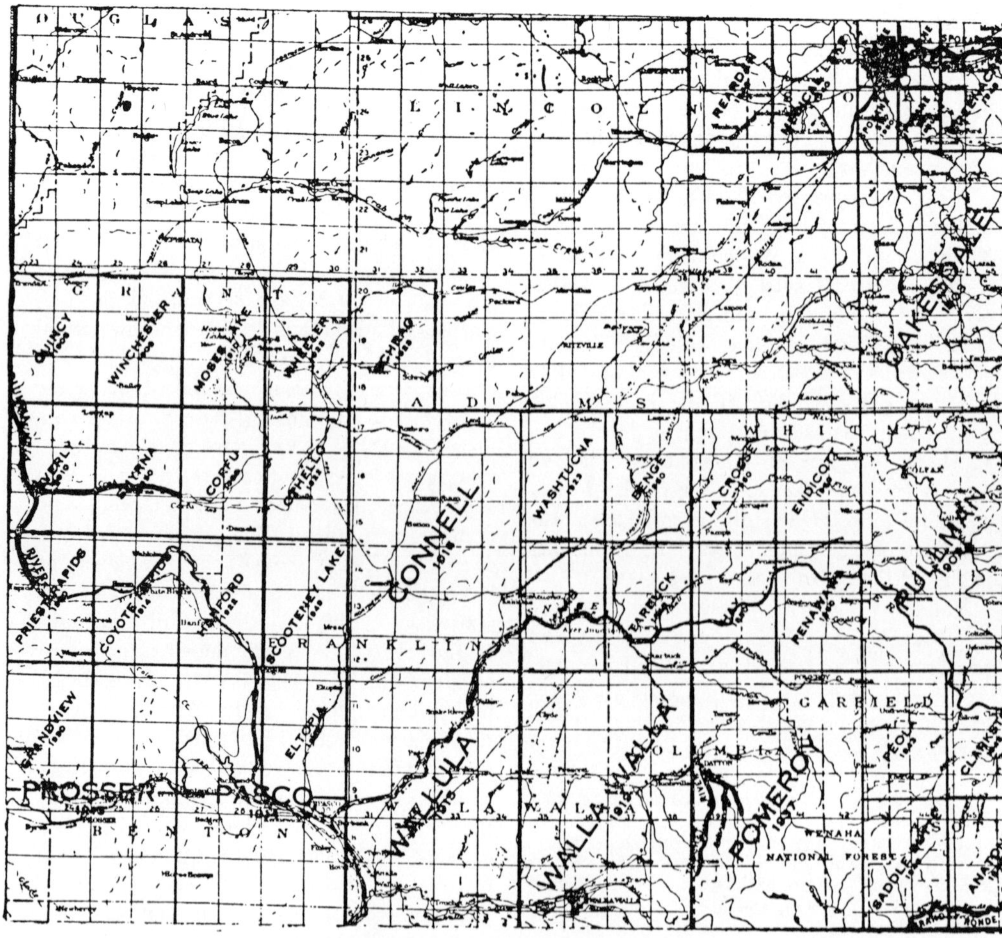

FIGURE 1.—INDEX MAP FOR MOST OF THE QUADRANGLES REFERRED TO IN THIS PAPER
United States Geological Survey.

tions and detailed topographic maps by the U. S. Bureau of Reclamation. Several new quadrangle topographic maps by the U. S. Geological Survey have appeared, and the entire region has been photographed by the U. S. Dept. of Agriculture. The field work was carried out with three men working together as a check on a "ruling-theory" tendency.

The finally accepted theory for the origin of this assemblage must consider the entire area. The elephantine character of channeled scab-

position, and structure of the scabland river deposits, the leading theme of the present paper. A. C. Waters, well acquainted with the scabland, wrote the authors that "the best thing you can do to convince your reader is to beg, cajole, even browbeat him into *really looking* at the region's topographic maps". J. H. Mackin, also well conversant with the field, said that "to understand the scabland, one must throw away textbook treatments of river work".

[*Editor's Note:* Material has been omitted at this point.]

A reader familiar with Bretz's earlier publications will find several revisions of interpretation in this paper, the most striking one being abandonment of the term "the Spokane Flood" and a demand by all three authors for several scabland floods. Obviously the present paper cannot be simply a series of assertions and denials. Field evidence must be presented in sufficient detail to allow the reader to judge of its appropriateness and adequacy for accompanying interpretations.

Summary of Original Interpretations

(1) The Palouse loess of the basalt plateau is older than the scabland. (Accepted by all critics and writers except Hobbs, who thought it essentially contemporaneous.)

(2) The scabland channel complex has been made by glacial water entering and modifying preglacial drainage lines on the loess-covered plateau. (Only Hobbs has a variant interpretation.)

(3) The scabland rock basins have been made by glaciofluvial erosion, chiefly by plucking in the columnar basalt. (Hobbs has differed, asserting glacial plucking; Alden has suggested that they might be collapsed lava caverns.)

(4) The downstream-facing cliffs interrupting channel floors are abandoned cataracts. (Denied by no critic.)

(5) The loessial scarps bordering scabland channels are marginal steepenings made by glacial rivers. (Only G. O. Smith has expressed doubt.)

(6) The scabland gravel deposits are contemporaneous with the channeling. (Flint demurred, believing the channels were made during removal of former complete fluvioglacial valley fills. The critical items here concern the ability of Flint's low-gradient streams of moderate size to detach great boulders from bedrock, to transport them for miles, then to remove the fill and to make the rugged, channel-bottom scabland topography associated with "residual" hills of gravel. Without a contemporaneous "Lake Lewis", whose rise caused the aggradation and whose immediate subsidence caused the erosion, Flint's theory has no logical foundation. Demonstration that the gravel "remnants" are stream-cut terraces is fundamental for his theory, and scabland forms buried under gravel are unallowable.)

Allison postulated a prescabland gravel-terrace system in Lower Snake River canyon, erosional remnants of which were subfluvially shaped by the glacial water from the scablands. (If correct, there must be comparable but unmodified terraces in the canyon upstream from entrance of the scabland rivers. In the terraces modified by the glacial Snake, there should be abundant well-worn Snake River-type pebbles. In unmodified cores of these deposits cementation and/or weathering would be accessory evidence.)

(7) The structure, composition, and topographic expression of these deposits record subfluvial origin as river bars of gigantic size. The undrained fosse depressions among and behind these gravel hills, the midchannel location of many, and the numerous huge current-ripple marks conforming to their shapes are undeniable evidence for bar origin. (Denied by Flint, Gilluly, Hodge, and the sponsors of McKnight. Flint claimed that they are terraced remnants of once-continuous fills, and Gilluly called their long foresets evidence for deltaic origin; Allison accepted them as subfluvial forms, although his concept cannot explain the mid-valley location of many.)

(8) The similar altitude of the upper limit of glacial water in all scabland channel heads and the anastomosing pattern indicate approximately contemporaneous functioning at some time over the entire area. (Strongly questioned by Alden, Gilluly, Hodge, Mansfield and Meinzer, accepted by Flint and Allison. Evidence for contemporaneity at some stage in scabland history must be expected at convergences of channels with separate heads.)

(9) The many minor divide crossings were due to a sufficiently great volume of glacial water to fill pre-existing valleys up to overflow sites. Earlier valley-train deposits may have been a factor in overflow. (Denied by Alden, Flint, Gilluly, Hodge, Mansfield, and Meinzer. Flint specifically shallowed the preglacial valleys with fluvioglacial deposits caused by deep backwater (Lake Lewis) beyond the plateau until spillover into adjacent valleys occurred; this before basalt scabland was made. Allison used local ice jams to make local impoundings and spillovers. There appears to be no third reasonable alternative. Evidence for the flood concept must include demonstration that a general fill never existed and that ice jams left no definite local records of their actual existence.)

(10) Ice-rafted stones and water-transported finer sediments back in valleys, tributary to scabland channels, that did not carry glacial-water discharge were the consequence of back-flooding because of the great volume and depth in the main channels. (Denied or ignored by all but Allison and Flint.)

(11) Bars, high-level channels, tributary valley-mouth deposits, and a "Portland delta" record the flood waters descending the Columbia beyond the plateau, crossing the Cascade Range and entering a Columbia valley estuary. (Allison, 1933, and Hodge, 1934, denied the Portland deposit as a delta; Piper, 1932, denied the flood origin of tributary mouth deposits near The Dalles, Allison and Flint believed a dam (landslide? Mt. Hood glacier? ice jam?) blocked the Columbia Gorge across the Cascades. High-level, berg-rafted, glaciated boulders have been floated to position in lower Columbia and Willamette valleys but perhaps are not all contemporaneous with scabland making. The dam must remain hypothetical until definitive evidence for its location is found on the walls of the gorge.)

(12) Only a great flood could have caused simultaneous discharge of the three great cataracts and one enormous high-gradient spillway out of Quincy basin. (Simultaneity denied or ignored by all opponents of the flood hypothesis, and the four accordant initial discharge levels explained by Meinzer as a fortuitous leveling up by subsequent earth movements. The only other alternatives are that ice jams blocked three of the four dischargeways at a time, in a rotating sequence, or that the basin once had an alluvial fill on which the glacial stream shifted from one dischargeway to another.)

(13) Upper Grand Coulee's dry canyon was made by cataract recession completely through a preglacial divide. (Denied by no critic.)

(14) Water for scabland making came from the bursting of the glacial dam impounding Lake Missoula in western Montana. (Rejected by all except Pardee who discovered evidence for torrential emptying of a large part of the lake. There is little field evidence at or near the dam site to prove a catastrophic bursting, but the lake water certainly discharged westward. Acceptance of evidence for scabland floods entails acceptance of the strong probability of a direct cause-and-effect relationship. (Since this was written, Alden has indicated a "perhaps" acceptance. See footnote 43.)

(15) The scablands east of Grand Coulee were ascribed to a pre-Wisconsin (Spokane) glaciation and Grand Coulee's completion to flood discharge during the Wisconsin glaciation. (Hodge announced three different glaciations during scabland development, while Flint claimed that the Spokane drift is Wisconsin in age.)

Advance Summary of New Evidence and Interpretations

Changes and elaborations in Bretz's earlier interpretations include:

(1) tentative identification of some gravels, both high-level and buried, as remnants of prescabland normal outwash;

(2) recognition of a heavy caliche on the loess and older deposits in part of the region as the marker for immediately prescabland topography;

(3) discovery of giant current-ripple marks on bar deposits in nine different channel-ways, distributed in more than a dozen localities ranging from the northern to the southern, and from the eastern to the western sides of the complex;

(4) establishment of the river-bar origin of the discontinuous mounded gravel deposits in scabland channels from aerial photographs

taken by the U. S. Agricultural Adjustment Administration, the U. S. Production and Marketing Administration and H. T. U. Smith, from the detailed topographic maps of the U. S. Bureau of Reclamation, and from many extensive new excavations;

(5) recognition of the occurrence of several successive floods, the earliest recorded only in the Columbia valley, the second involving perhaps the entire area, later ones restricted to channels already deepened, and the last one again limited to the Columbia valley;

(6) recognition that Touchet silts were not contemporaneous with scabland making.

[*Editor's Note:* Material has been omitted at this point.]

Summary

(Bretz)

Bretz's Flood Hypothesis (1923–1932)

The greatly scarred linear tracts of denuded basalt interlaced in a complicated pattern and elongated with the dip slope of the otherwise loess-covered plateau of eastern Washington have been made by glacial floods entering a maturely developed normal drainage pattern. The adjective "channeled," first proposed by Bretz (1932b), and used in a dozen later papers implies that an enormous volume of water transformed the invaded valleys into gigantic river channels. Such great quantities could continue to arrive only for a brief time, and a large volume of ponded water, capable of being abruptly released, is needed for this interpretation. Lake Missoula's volume of 500 cubic miles (Pardee) back of a glacial dam with one buttress most vulnerably situated for a catastrophic failure is argued as the source. Even then, repetition of the cataclysmic flooding seems indicated by the amount of work done by the short-lived glacial rivers. This paper presents new field evidence believed to be consonant only with the interpretation of repeated floods. Its leading theme is that the associated mounded gravel deposits have constructional forms, are gigantic river bars.

But the demand for unprecedented volume has seemed outrageous to conservative geologists, and other hypotheses have been advanced to explain the extraordinary erosional and depositional features of the Washington scabland. Two of them possess sufficient apparent applicability to have secured adherents

Allison's Ice-Jam Hypothesis (1933)

Allison (1933) argued that the scabland phenomena resulted from a damming in the Columbia River gorge across the Cascade Range, more than 100 miles downstream from the plateau scabland. His dam was initiated by a landslide and composed of berg ice and river ice. Once started, it grew upstream by successive increments, eventually ponding the deep major valleys that bound the plateau and allowing glacial stream water of normal volume to cross low places in divides of the plateau's preglacial drainage system. The complexity of the scabland pattern required many other ice jams to provide for divide crossings farther back up the plateau slope. Water bypassing these ice jams in "run-around" lateral channels gashed into the blocked valleys' slopes to make high-lying scabland and remodeled older terraces to form "perched" bars.

Allison admitted that his Cascade Gorge dam must have been 900–1100 feet high but did not point out any field evidence for its precise location nor for its great "run-around" bypass channels in the gorge walls. Nor has he dealt adequately with the resultant flood down the Columbia (1932) below the dam when that great obstruction finally failed. He never located any smaller dams back on the plateau although he wrote about the excellent records left by stranded bergs in the widest ponding of glacial water in the region, the Pasco basin.

Allison's greatest dam lengthened upstream along the Columbia for 100 miles or more above its inception in the Cascade gorge. It thickened as it lengthened, and many diverted lateral strands of the glacial Columbia cut notches and channels across interfluves between entering tributary valleys and made gravel and rubble "dumps", "deltas", "fans", and "bars" in water ponded in these valleys. Several may occur at different altitudes in one locality, but, taken as a whole, they fall into a general descending profile for this valley stretch (Bretz, 1925). Eventually this ice jam was so thick that the Columbia became a series of ponds in these tributary valleys, standing at about 1100 feet A.T. for the entire distance (Allison, 1933, p. 722). No diversion channels were made at this altitude; only pebbly silts and berg-rafted erratic boulders attest the maximum flooding. How the glacially swollen river traversed this 100 miles of individual lateral pondings on no gradient is unexplained. How the diversion channels and gravel deposits near the main valley escaped modification while the dam was growing to full thickness is a puzzle for the deeper part of this long stretch of Lake Lewis was then almost solidly full of ice. The high-lying silts and rafted boulders may not date from the episode of diversion channels and gravel deposits. Leaving out the 1100-foot records, Allison's picture for this 100 miles of Columbia Valley is appealing but lacks any positive records of the ice blockade itself or of a downvalley rush of water when it failed.

Search was made at Washtucna Coulee, Esquatzel Coulee, Lower Crab Creek, Upper Crab Creek, and Palouse Canyon, where divide crossings or lateral canyons seemed to suggest detours caused by ice jams in the glacial streams. No favorable constrictions and no records of stranded bergs were found. Instead, the upper limits of the glacial rivers, recorded in steepened slopes of the bordering loess, possess uniformly descending profiles along such glacial river courses.

Local ice jams may have occurred in the making of the scabland but no evidence for any has been found. Allison implied that downvalley floods from the inevitable bursting of such dams made the great mounded scabland gravel deposits (remodeled older terraces) which he recognized had largely constructional forms.

Allison also theorized on a deep fluvial fill in the capacious Pasco basin. This basin was undoubtedly occupied by glacial water up to the limits specified, but the only deposits which might conceivably record his detrital fill are huge, elongated mounds, 100 feet or more above the bottoms of closely associated depressions, both forms streamlined like the bars he identified in Snake Canyon, but lying out in the middle of a river valley more than 15 miles wide. The largest depression contains a closed basin $4/10$ mile long and more than 50 feet deep. He found only one terracelike gravel deposit in the basin to fit his theory.

Flint's Fill Hypothesis (1938)

Flint (1938) proposed that the many divide crossings and the abundant scarps in loess-covered tracts bordering the scablands had resulted from deposition in the valleys invaded, a common procedure of laden glacial streams but continued here until low divides were crossed and the plexus was outlined. Raising of base level to produce this deposition was presumably because of a dam in the Columbia gorge, formed probably by a great landside or a local valley glacier pushing down into the gorge. This dam was as inferential as Allison's. Flint's glacial streams across the plateau were shallow and "only small pieces of glacier ice" ever floated down them. During wastage of the dam, the glacial streams gradually removed their fills to leave remnants (15 per cent of original deposits—Flint) which he called terraces modified by creep and slopewash; "in various states of [subsequent] dissection." Every high-lying minor unit of the plexus was as well cleaned out as the larger, deeper ones. Wide-swinging meanders were a feature of the dissection on the broader (10–20-mile) tracts. The rugged basalt scabland, including the numerous and spectacular rock basins, was made largely during this removal.

This Columbia gorge dam must have wasted away very slowly, for the nearly complete removal of the fill in scabland routes was all done with moderate quantities of glacial water while also the valley bottom basalt was being scarified (by meandering streams!) into the butte-and-basin topography characterizing scabland.

Flint's interpretation of the gravel deposits associated with the scabland routes as "non-paired, stream-cut terraces", cannot be accepted. Many of the "remnants" are ungullied hills of gravel standing out in mid-valley, possessing stream-lined forms, having associated closed fosse depressions between them and valley walls, containing bouldery accumulations for the transportation of which his "leisurely" streams never could be held competent, displaying internal structures of long fore-sets parallel to the slope for the entire height of the deposit, and carrying gigantic current ripples on their summits and channel-side slopes. Allison, in a later paper (1941), recognized these gravel piles as subfluvially shaped individual forms made in favored places.

There are divide crossings inexplicable by Flint's rising fill (Drumheller, Othello, Devils Canyon, Palouse Canyon); there were contemporaneously functioning spillways out of basins (three cataracts and Drumheller from Quincy basin, Dry and Long Lake coulees out of Bacon syncline) which, even with adequate fills, required far greater volume of water than Flint allowed.

Flint reported that many loessial scarps margining the scabland river routes were "multiple" and argued that the benched slopes recorded stages in removing the fill, overlooking any alternative possibilities.

The extraordinary pattern of islandlike, scarp-margined, prow-pointed residual hills of loess isolated in wide areas of scabland and cut through by narrow, subparallel accessory glacial river channels was to Flint the consequence of his widespread aggradation. But in the cleaning out of all the filled valleys by lateral planation, no meander scars were ever left in the loessial scarps.[43] Furthermore, Flint's mechanism "breaks down utterly" (Allison, 1941, p. 56) when catharsis of all the high-lying minor divergent strands is required of it.

Flint's arrow went widest of its mark at the place where Allison specified an ice jam in Snake Canyon at the junction of the Cheney-Palouse scabland river with the larger valley. For this district, Flint listed several dry falls and rock basins, stressing their small dimensions (too small to be shown by the 40 ft. contour intervals of the subsequently published topographic maps) as typical. Unfortunately, he missed seeing dry falls 100 to nearly 300 feet high, empty recessional gorges with closed depressions in them more than 100 feet deep, and rock basins equally deep on the rocky summit of the preglacial divide where a width of 9 miles was swept clean of its loess cover and 9/10 of its 80-square mile area has no remnants of his fill.

Flint hypothesized a debris fill completely blocking Snake Canyon just beyond this ravaged divide summit. To make a lake (Riparia) to account for high-lying, berg-rafted erratic boulders farther up the Snake, the fill had to be 800 feet thick. The outlet of the lake had to be around the southern low periphery of the fill. In the subsequent dissection of this dam (during the slow wastage of his Columbia gorge dam), Snake River unaccountably abandoned the low marginal bypass route, where it should have become entrenched, and returned to its buried preglacial gorge. The fill is almost all gone, and the exhumed, rugged basalt topography forbids any concept of a lateral shifting to bring the Snake back.

Character of New Evidence (1952)

The new evidence for torrential discharge comes from detailed topographic maps (10-foot and 2-foot contour intervals) by the U. S. Bureau of Reclamation; from new topographic maps by the U. S. Geological Survey; from 175 miles of main irrigation canals excavated, in large part, in scabland gravel deposits; from dozens of large, recently opened gravel pits; and from aerial photographs not available during any earlier study.

The aerial photographs reveal giant current ripples on mounded gravel deposits interpreted as gigantic river bars, essentially unmodified subfluvial constructional forms.

Long, deltalike fore-set bedding parallels the slopes and extends throughout the body of many of these bars. The large quantity of transported boulders in some bars and their huge size are impossible products of Flint's streams.

Records of repeated debaclelike floods have been found well distributed throughout the whole scabland complex. The western part of the scabland (Grand Coulee, Quincy basin, adjacent Columbia valley) has had at least seven floods, five down Grand Coulee and two affecting the Columbia valley alone. These are accounted for by repeated burstings (and reconstructions) of the glacial ice dam of Lake Missoula in the mountains of western Montana. Altitudes and topographic relations between the lake and the plateau are in consonance with the idea. Striking records that the lake was abruptly emptied (giant current ripples, severely scoured salients, huge barlike deposits, Pardee 1942) strongly support that interpretation.

Flood Correlations

(Bretz)

Correlations of all the postulated floods are not yet feasible. The latest flood across Quincy basin came from Grand Coulee alone, only after that deep canyon had been cut com-

[43] Before one commits himself as skeptical of the interpretation of the present study, he should examine the Benge, Haas, LaCrosse and Starbuck quadrangle maps. A topography probably without close approach to parallelism in his experience with maps or terrain is depicted on them. Some of the "islands" (minor prescabland interfluves) have had one side so largely removed that little valleys draining to the opposite side have been beheaded.

pletely across the plateau's high northern edge and only when the Okanogan lobe, holding an advanced position across the Columbia valley, prevented escape of the glacial water around the northwestern part of the plateau. This event is generally considered to be of late Wisconsin age. At that time only deeper Drumheller and Othello channels and Lower Crab Creek operated downstream from Grand Coulee. Backfilling in the mouth of Upper Crab Creek and perhaps in lower Washtucna Coulee was contemporaneous with it.

Bretz has argued that all channels in the scabland complex functioned contemporaneously until Grand Coulee's great gash was completed. With this interpretation, Flint has agreed (1938, p. 46). But the new evidence of deepening of some channels during a succession of floods renders that concept invalid for later occupations. The idea is not yet to be dismissed for the earliest recorded flooding. It is based on altitudes of upper flood limits at the channel heads on the high northern edge of the tilted plateau. About a dozen definite channel heads are distributed along about 85 miles of this divide, and the upper-limit altitudes (highest scabland, loessial scarp bases, and ice-rafted erratics floated back in minor valleys in the loess) lie between 2450 and 2550 feet. A.T. (Bretz, 1928b, Pl. 5). Although these are not precise markers, there is no cumulative decrease in their altitudes either east or west. Part of this range may well be ascribed to aneroid errors.

The conclusion that the first recorded flooding used all channels can be avoided if the sites of the western channel heads were originally sufficiently lower than the eastern ones to take all the first flood water and (2) in some subsequent flooding, either these heads were closed by an advanced edge of the Cordilleran ice sheet so that higher eastern channels were initiated or eastward down-tilting had occurred since the preceding flood. This sequence can be reversed by blocking the lower western channel heads at the time of an early flood with an expanded Columbia lobe so that only the higher Cheney-Palouse route operated and then in later floods which found a shrunken Columbia lobe, bypassing that eastern route. Postscabland east-west tilting would again be necessary to bring all channel-head upper limits into the same horizontal plane. An eastward down-tilting, however, would remove the present difficulty with altitudes at Mica.

The channeled summit of Steamboat Rock in upper Grand Coulee carries glacial striae and a moraine ridge, both clearly later than the channeling (Bretz, 1932, p. 35). A silt-covered bouldery ridge on the coulee floor close to the cliffs of Steamboat is perhaps morainic. The eastern margin of the Okanogan lobe apparently has invaded the coulee head since cataract recession passed the Rock. If this completely closed the coulee, it must have caused renewed discharge of flood or normal meltwater through the deeper of the eastern scabland channels. Such an episode must have been brief, for very little retreat of the Okanogan lobe would again open Grand Coulee. A flood down Columbia valley west of the scablands followed the last Grand Coulee discharge, its record indisputable in West Bar's giant current ripples and the Beverly bar. Bursting of the Okanogan lobe's dam (Waters, 1933, p. 617), perhaps as a consequence of another Lake Missoula dam burst, may well have supplied the water for this latest glacial river flooding of the Columbia Plateau.

The writers do not believe that successive Lake Missoula emptyings are to be correlated strictly with successive Pleistocene glacial episodes. Dam failures probably came when increasing hydrostatic pressure in front of the Pend Oreille lobe coincided with weakening of the ice front. Decreased alimentation of the lobe would accompany sub-periodic climatic warmings when wastage would also be excessive. Repeated failures[44] may have occurred during any one glacial cycle. This would be comparable to recorded modern glacial lake bursting in Alaska, British Columbia, the Himalayas, the Alps and elsewhere (Kerr, 1934; Klebelsberg, 1948; Stone, unpublished).

The steep scarps in weak loess can hardly be survivals from early Pleistocene glacial floods. If, therefore, they were made in relatively late Pleistocene time, either there were no earlier Lake Missoula pondings and outbursts or the

[44] Pardee's giant ripples (1942) record a lake surface at least 3450 feet A.T. when the dam burst.

scarps record a late flood greater than any earlier ones.

It is significant that the closed basins of the scabland are largely unfilled. Their making was not followed by a dwindling flow of debris-laden meltwater incapable of keeping them flushed out. Flow seems to have ceased abruptly. For Grand Coulee's basins, this may have been caused by an abrupt failure of the Okanogan lobe dam but not for the hundreds of other empty basins in channels of all ages in the complex. If the scabland streams were supplied by lake water, instead of meltwater directly from the ice, they would have had only a minimum load. If they were supplied by a catastrophic failure of the Lake Missoula ice dam, an abrupt cessation would be expectable.

Ordinary valley trains along main scabland channels may have been deposited before flooding or in intervals between some floods. But, if so, these were largely destroyed by succeeding floods. None was made after the last flood to traverse any one channel.

All features of the Columbia plateau scabland are too intricately interrelated to dream of unlike agents or markedly different sequences for different parts. If any theory is correct for the Cheney-Palouse tract, it is correct also for Crab Creek, Grand Coulee, Quincy basin, Othello Channels, Pasco Basin, and the Columbia Valley farther downstream.

However plausible the fill-and-cut or ice-jam theories may be in any particular place, they must be equally applicable everywhere in the complex. The writers are confident that only the flood theory can successfully meet this test.

Recapitulation of Inferred History

(1) Miocene basalt flows (Columbia River formation) spread widely over, and almost completely buried, a mildly rugged topography of ancient crystalline rocks. Along the eastern and northern margins, bounded by highlands of these rocks, isolated peaks and ridges projected through the basalt and tongues of basalt penetrated back into deep valleys in the highlands.

(2) Lacustrine and fluvial sediments (Ellensburg, Latah, Early Ringold) were spread over at least parts of the region.

(3) Snake and Columbia rivers established courses across the basalt plain.

(4) Uplift with deformation down-tilted the region southwestward, warped and locally folded and faulted the basalt. These structures are strongly expressed in present topography. Snake and Columbia rivers held antecedent courses across some of the local uplifts.

(5) Weathering and erosion occurred.

(6) Later (?) Ringold sediment (Pleistocene lacustrine) was deposited on the eroded surface in the largest of the warped basins (Pasco), in southwestern part of the region.

(7) Loess was deposited on the plateau concomitant with continued erosional development of Columbia, Spokane and Snake valleys. Weathered zones and caliche layers in the loess record pauses during accumulation. Caliche developed also on basalt and sedimentaries where not loess-covered.

(8) Glaciation occurred. Canadian Cordilleran ice reached the northern edge of the plateau perhaps two or three times during the Pleistocene.

(9) Glacial water crossed the plateau when the deep Columbia valley along the high northern edge was blocked by lobes from northern highlands. Perhaps valley-train gravels were deposited in valleys across the plateau, their remnants now lying at or above upper limits of scabland gravels, or buried beneath them.

(10) Damming of Clark Fork and tributary valleys in Montana by a Cordilleran lobe in Purcell trench formed Glacial Lake Missoula. The Columbia valley was dammed by a similar lobe in Okanogan trench, presumably during Wisconsin time.

(11) The Lake Missoula dam failed before water had risen to the lowest col in Bitterroot Range. The lake's highest surface was about 4200 feet A.T., its depth close to the dam was 2000 feet. A recorded catastrophic emptying of Clark Fork portion demands similar emptying of the entire water body down nearly to 2500 feet A.T. This water escaped southwestward down Spokane Valley toward heads of scabland channels. Upper limits of channel heads range about 50 feet on either side of 2500 feet A.T.

(12) The great volume of glacial-lake water discharging down the dip slope of the plateau eroded its pre-existing minor valley system,

chiefly in loess, into a network of scabland buttes and rock basins in subjacent basalt. In deeper or broader places, deposits of basaltic and caliche gravel grew into huge bars. Some *de novo* divide crossings made by the flood became cataracts hundreds of feet high; Grand Coulee is the outstanding example. Depth and velocity (because of volume) enabled these rivers to quarry large columnar masses of essentially fresh basalt and transport them for miles. Giant current ripples left on bar summits indicate depth of rivers to have been considerably greater than heights of bars.

(13) Repeated reconstruction of dam with repetition of dam failures and consequent floods took place.

(14) Deepening of some channels in basalt during later floods allowed scarping of bars already formed and growth of lower bars close to scarp bases. These also have great current ripples. In Quincy basin, adjacent Columbia valley, and environs, seven successive floods are recorded, five of them while the Okanogan lobe dam forced the discharge to cross the plateau by way of scabland channels.

References Cited

Alden, W. C., 1953, Physiography and glacial geology of Western Montana and adjacent areas: U. S. Geol. Survey Prof. Paper 231, p. 1-193

Allison, Ira S., 1932, Spokane flood south of Portland, Oregon (Abstract): Geol. Soc. America Bull., v. 43, p. 133-134

—— 1933, New version of the Spokane flood: Geol. Soc. America Bull., v. 44, p. 675-722

—— 1941, Flint's fill-hypothesis for channeled scabland: Jour. Geology, v. 49, p. 54-73

Anderson, A. L., 1927, Some Miocene and Pliocene changes in northern Idaho: Idaho Bur. Mines and Geology, Pamph. 18

Bretz, J H., 1923a, Glacial drainage on the Columbia Plateau: Geol. Soc. America Bull., v. 34, p. 573-608

—— 1923b, Channeled scablands of the Columbia Plateau: Jour. Geology, v. 31, p. 617-649

—— 1925, The Spokane flood beyond the channeled scablands: Jour. Geology, v. 33, p. 97-115, 236-259

—— 1927a, Channeled scabland and the Spokane flood: Wash. Acad. Sci. Jour., v. 17, p. 200-211

—— 1927b, The Spokane flood; a reply: Jour. Geology, v. 35, p. 461-468

—— 1928a, Bars of channeled scabland: Geol. Soc. America Bull., v. 39, p. 643-702

—— 1928b, Channeled scabland of Eastern Washington: Geog. Rev., v. 18, p. 446-477

—— 1928c, Alternative hypotheses for channeled scabland: Jour. Geology, v. 36, p. 193-223, 312-341

—— 1929, Valley deposits east of the channeled scabland: Jour. Geology, v. 37, p. 393-427, 505-541

—— 1930a, Lake Missoula and the Spokane flood (Abstract): Geol. Soc. America Bull., v. 41, p. 92

—— 1930b, Valley deposits west of the channeled scabland: Jour. Geology, v. 38, p. 385-422

—— 1932, The Grand Coulee: American Geog. Soc. Spec. Publ., no. 15

Bryan, Kirk, 1927, The Palouse soil problem: U. S. Geol. Survey Bull. 790-B

Bucher, W. H., 1919, On ripples and related sedimentary surface forms: Am. Jour. Sci., v. 47, p. 149-210, 241-269.

Campbell, M. R., 1915, Guidebook of the Western U. S., Part A, The Northern Pacific Route: U. S. Geol. Survey Bull. 611, p. 142

Culver, H. E., 1937, Extensions of the Ringold formation: Northwest Sci., v. 11, no. 1., p. 57-60

Davis, W. M., 1921, Features of glacial origin in Montana and Idaho: Assoc. Am. Geog. Annals, v. 10, p. 75-147

Eakin, G. R., and Honkala, F. S., 1952, Cenozoic history of Missoula Valley, Missoula, Co., Montana (Abstract): Geol. Soc. America Bull., v. 63, pt. 2, p. 1361

Eardley, A. J., 1938, Yukon channel shifting: Geol. Soc. America Bull., v. 49, p. 343-358

Flint, R. F., 1936, Stratified drift and deglaciation of Eastern Washington: Geol. Soc. America Bull., v. 47, p. 1849-1884

—— 1937, Pleistocene drift border in Eastern Washington: Geol. Soc. America Bull., v. 48, p. 461-524

—— 1938, Origin of the Cheney-Palouse scabland tract: Geol. Soc. America Bull., v. 49, p. 461-524

Harding, H. T., 1929, Possible supply of water for the channeled scablands: Sci., v. 69, p. 188-190

Hobbs, W. H., 1943, Discovery in Eastern Washington of a new lobe of the Pleistocene continental glacier: Sci., v. 98, p. 227-230

—— 1947, The glacial history of the Scabland and Okanogan lobes, Cordilleran continental glacier: Privately printed, J. W. Edwards, Ann Arbor, 36 p.

Hodge, E. T., 1934, Origin of the Washington scabland: Northwest Sci., v. 8, p. 4-11

Jenkins, O. P., 1925, The Spokane flood; A discussion: Jour. Geology, v. 33, p. 747-748

Kerr, F. A., 1934, The ice dam and floods of the Talsekwe, British Columbia: Geog. Rev., v. 24, p. 643-645

Klebelsberg, R. V., 1948, Handbuch der Gletscherkunde und Glazialgeologie: Springer Verlag, Wien, v. 1, p. 212-218

Lane and Eiden, 1940, Sand waves in the lower Mississippi River: Western Soc. Engineers, v. 45, p. 281-291

Leighton, M. M., 1919, Road-building sand and gravel of Washington: Wash. Geol. Survey, Bull. 22, p. 200

Lupher, R. L., 1944, Clastic dikes of the Columbia Basin region, Washington and Idaho: Geol. Soc. America Bull., v. 55, p. 1431-1462

Matthes, G. H., 1948, Macroturbulence in natural stream flow: Am. Geophys. Union Trans., v. 28, p. 255-262

McKnight, E. T., 1927, The Spokane flood: a discussion: Jour. Geology, v. 35, p. 453–460

Nobles, L. H., 1952, Glacial sequence in the Mission Valley, Western Mont. (Abstract): Geol. Soc. America Bull., v. 63, pt. 2, p. 1286

Page, B. M., 1939, Multiple Alpine glaciation in the Leavenworth area, Wash.: Jour. Geology, v. 47, p. 785–815

Pardee, J. T., 1910, The glacial Lake Missoula, Mont.: Jour. Geology, v. 18, p. 376–386

—— 1922, Glaciation in the Cordilleran region: Sci., v. 56, p. 686–687

—— 1942, Unusual currents in glacial Lake Missoula, Mont.: Geol. Soc. America Bull., v. 53, p. 1569–1600

Piper, A. M., 1932, Geology and ground-water resources of the Dalles region, Ore.: U. S. Geol. Survey W. S. Paper 659, p. 130

Raisz, Erwin, 1941, Land forms of the Northwestern States (Map): Cambridge, Mass., The Author

Schwennesen, A. T., and Meinzer, O. E., 1918, Ground water in Quincy Valley, Wash.: U. S. Geol. Survey Water-Supply Paper 425, p. 131–161

Thiel, G. A., 1932, Giant current ripples in fluvial sediments: Jour. Geology, v. 40, p. 452–458

Trimble, D. E., 1950, Joint controlled channeling in the Columbia River basalt: Northwest Sci., v. 24, p. 84–88

Waters, A. C., 1933, Terraces and coulees along Columbia River near Lake Chelan, Wash.: Geol. Soc. America Bull., v. 44, p. 763–820

University of Chicago; Chicago 37, Illinois; University of Massachusetts, Amherst, Mass.; U. S. Bureau of Reclamation, Ephrata, Washington

Manuscript Received by the Secretary of the Society, May 24, 1954

11

Reprinted from *Washington Div. Mines and Geology Bull.* **45**:7–57 (1959)

WASHINGTON'S CHANNELED SCABLAND

By J HARLEN BRETZ

Unparalleled in the whole wide world, scarcely even approached by any other landscape of similar origin, are the Channeled Scablands on the Columbia Plateau in eastern Washington. Four contrasting theories have been proposed for this unique topography, at least three of which cannot be correct. None of the four can point to specific processes now in operation, nor to any combinations of observed processes, that yield more than feeble comparisons for the extraordinary features of the Channeled Scablands. Understanding the Scablands involves imagination and courageous departure from accepted views. Two of the four theories possess this qualification. Two are conventional and inadequate. Although the same agencies are called for in each theory; i. e., glacial ice and glacial rivers, the differences among these theories are in time and space relations and in quantities.

Visualize a fairly plane tract, a plateau surface 125 miles across. Cover its basaltic rock foundation with a thick, powdery soil, the famous Palouse wheat soil. Bound this part of the Columbia Plateau on north, south, and west by the deep valleys of the Spokane, Snake, and Columbia Rivers. Tilt the plateau down toward its southwest corner. Allow normal rainfall for many thousands of years to collect into streams that flow along the direction of tilt, small streams converging to make larger ones. Admit that these plateau-born streams must make valleys in the deep soil—converging valleys. Permit some part of them to erode down into the bedrock.

Now change the climate so that the British Columbia and northern Washington highlands become buried in glacial ice. Thicken that ice until it buries most mountain peaks and spreads slowly toward the south. Visualize the advancing margin reaching and then crossing the canyons bounding the north side of the plateau. Realize that the field evidence records the advance as reaching only a few tens of miles out on the plateau, the ice melting at that margin and its meltwater escaping along the larger streamways in the thick soil, deepening and widening them. Conceive that the arctic climate continued for a long time, so that ice continually flowed southward, only to waste away along the northern margin of the plateau, its meltwater using the minor valleys across that plateau. Nobody adequately conversant with the area questions any of the above requirements for a successful theory for the origin of the Channeled Scabland.

Now examine a map showing the extent to which the old drainage system became invaded by meltwater streams (pl. 1). Compare that pattern with any other drainage system on the earth. The map

[*Editor's Note:* Plates 1, 2, 3, and 4 have not been reproduced here.]

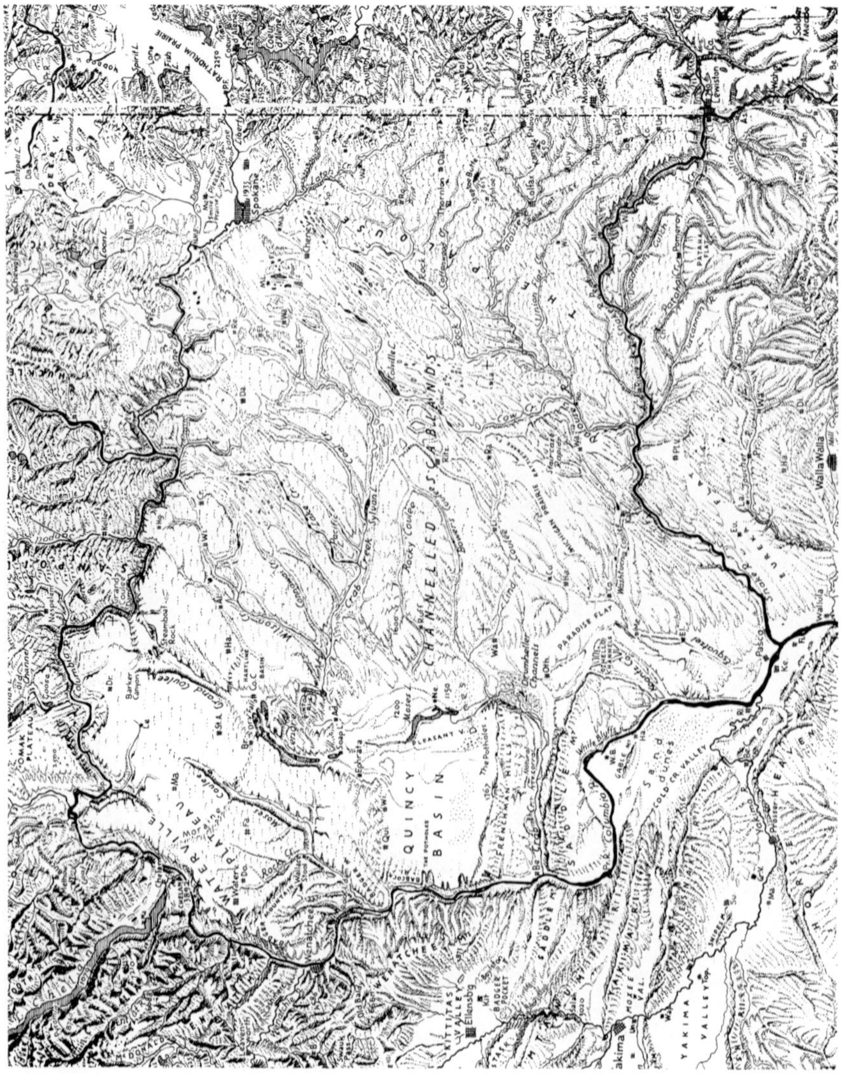
Spokane
CHANNELLED SCABLANDS
QUINCY BASIN
WATERVILLE PLATEAU
PARADISE FLAT
Walla Walla
Wallula
Ellensburg
KITTITAS VALLEY
YAKIMA VALLEY
Rocky Coulee
Barker Canyon
Cow Cr.
Lewiston

shows a feature that never existed before the meltwater invasion occurred, a feature that no other comparable drainage system in the world possesses. Many streamways of this pattern *diverge* downstream, only to *converge* again. A network of drainage channels was made out of the original treelike converging pattern. Travelers by air report that from altitudes of, say, 20,000 feet, the scorings left by the glacier-born streams look about as definite as the depiction in plate 1. The network is vivid to the man in the air because the deep soil was cut through, and many channels were eroded into the black basalt beneath. These long attenuated strips and belts of bare or nearly bare rock outline the unique pattern of Washington's Channeled Scabland.

At this point, theory-makers go four different ways almost as divergent as the cardinal points of the compass. Before one can critically examine and judge any one of the four theoretical routes indicated on our imaginary guidepost, he must know the general facts that demand explanation.

GENERAL FEATURES OF THE CHANNELED SCABLAND

Anastomosis.—This outstanding feature has been outlined above. In the 50 or more single glacial streamways that split downgrade to make 2, about 25 definitely involve crossing a divide of the preglacial drainage pattern. Some overflowed the transgressed divide "broadside" from one channel to the other. Most of them were spillovers from the side of a large channel to the head of a smaller one. Among those divergences, the majority ended by rejoining the larger streamway they had left. But the discharge of more than a dozen offshoot meltwater strands did not mingle with water from the parent stream until after reaching the Columbia southwest of the complex.

This anastomosis (or plexus or network or complex) involved divide crossings as much as 300, even 400, feet above present bottoms of the parent plateau river course. Many of the pre-existing valleys headed miles south of the ice margin, and only by these spillovers could they ever have obtained the stranded erratic boulders derived from bedrock far north of the plateau. Only berg ice could have carried such boulders across the violated divides.

None of these divide crossings south of the known limit of the ice sheet can be explained by direct glacial damming. Here is a challenging problem for the hydrologist and the geomorphologist.

This interlaced glacial river pattern with its divide crossings has been explained as the consequence of:

(1) enormous flooding in the preglacial valleys (Bretz, 1923a-1956),
(2) damming of these valleys by jams of berg and river ice (Allison, 1933),
(3) filling of the early valleyways with gravelly outwash from the glacial margin (Flint, 1938), and

(4) a glacial advance (the "Scabland Glacial Lobe") far beyond the limits accepted by everybody but the author of that theory (Hobbs, 1943).

The proposed floods, ice dams, and ice sheet have vanished, of course. The great gravel deposits of the third theory are less readily disposed of, yet they also are gone.

The flood theory shocked and displeased many geologists, especially when it demanded that every channelway be flooded simultaneously. This demand was based on a near-coincidence of altitude of upper ends of all channels heading at the margin reached by the ice sheet.

Rock basins.—Another outstanding feature of the Channeled Scabland is the existence of hundreds of undrained basins along the channelways. Under the semi-arid climate of the lower southern and southwestern part of the area, these rock-walled and rock-bottomed basins are empty or contain alkaline lakes. Swamps and fresh-water lakes are numerous in the northeast. Depths of the basins may exceed 100 feet. One (Scooteney Lake, Othello Channels, fig. 21), on the very summit of a channeled divide, is 135 feet in depth. Admitting the existence of the rock basins (without an extinct waterfall at the upstream edge to explain them as plunge basins), one doubter of the flood theory suspected that they might be collapsed lava caverns lying in channelways only by coincidence. Other skeptics minimized their magnitude or ignored their existence. The fourth theory which, because of serious internal and external disharmonies, has never had followers, saw these rock basins as glacially plucked. The flood theory accepts plucking but considers that the tremendous volumes and velocities of its rivers, operating on the closely and vertically jointed basalt, made plucking possible in favorable places.

Scarps in loess.—As readily recognized as the rocky ledges and cliffs of the glacial river routes are steep margining bluffs above the scabland basalt. They are developed in the overlying Palouse soil which is actually a thick dust deposit, known as loess. Much of the Channeled Scabland is dotted with "islands" of this loess, also sharply outlined by steep bluffs and commonly with sharp prows pointing upstream. Indeed, all the loess-covered tracts enclosed by the channel divergences and convergences are "islands." Above scabland level the islands have only the gentle slopes of the prescabland topograpy, definitely untouched by glacial rivers, definitely much older in erosional development.

That the steep bluffs were made along the margins of glacial rivers is accepted by all.

Butte-and-basin topography.—Where the loess was swept off wide scabland tracts, the bare or nearly bare basalt is in some places (Palouse-Snake divide, and Drumheller and Othello channnels, figs.

18 and 21) amazingly scarified with anastomosing channels and rock basins surrounding buttes and small mesas of rock. Such tracts may have a relief of 100 to 300 feet, the erosion logically imputable to no agency other than glacial rivers. Two such tracts measure 9 miles across. The fill theory interprets such wildernesses of labyrinthine canyons as eroded into subjacent bedrock during removal of the theoretical gravel cover by glacial streams "no larger than Snake River of today." A better term for this theory is fill-and-cut. The flood and ice-jam theories presuppose an abundance of water for doing the work.

Cataracts.—Dozens of extinct cataracts and cascades exist along the glacial river courses. They constitute the Scabland's most impressive scenery. They range from a quarter of a mile wide and a few tens of feet high to 3½ miles wide and 400 feet high (Dry Falls in Grand Coulee, figs. 2, 3, and 4, and pl. 2). They commonly possess plunge pool basins, with or without lakes. Like the Niagara prototype of a retreating cataract, they migrated upstream by undercutting from the points of inception and thus left recessional gorges or canyons up to 5 miles wide and 900 feet deep (Upper Grand Coulee, fig. 8). The maximum retreat has been about 20 miles.

A repeated criticism of the flood theory has been the obvious impossibility of maintaining any tremendous discharge long enough for the cataract retreat recorded in these longer gorges. No theorist or critic has denied their waterfall origin, but significantly located examples have unfortunately been overlooked and their views invalidated thereby.

Three most challengingly located extinct waterfalls (Potholes, Frenchman, and Crater, figs. 13-17) were developed where glacial meltwater escaped across the rim of a broad basin (Quincy, fig. 20), whose bottom was too low to be eroded down to bedrock, and spilled over high cliffs bordering the Columbia River valley. The challenge lies not only in the great width of the discharging streams (therefore, huge volume) but, most important, in the fact that their channel upper limits are so nearly the same as to make almost impossible any rational concept of a sequence in operation. (See p. 26.) The flood theory has made these three contemporaneously operating cataracts one of its most strongly emphasized arguments. The rival theories have failed to give them a hearing.

Broad gravel deposits.—Plate 1 shows two large areas and many smaller ones as gravel-covered. All such places were sites of relatively slack current because of decreased gradient or actual ponding or overloading of the stream in shallow marginal situations. Exposures in some of these scabland gravel deposits reveal foreset bedding that dips in the direction of current through a maximum of more than 200 vertical feet (mouth of the Palouse River, fig. 28) and thus apparently records a growing delta front or a growing

FIGURE 2. Fall Lake, great cataract group, Lower Grand Coulee, looking northeast from Vista House in Dry Falls State Park. The lake is 380 feet below the camera. The cataract cliff has a maximum height of 400 feet above the lake, whose basin is 80 feet deep. The white area near the right side of the view is a desiccated lake flat in the eastern of the two recesses of Dry Falls. (Photograph by Frank Guilbert)

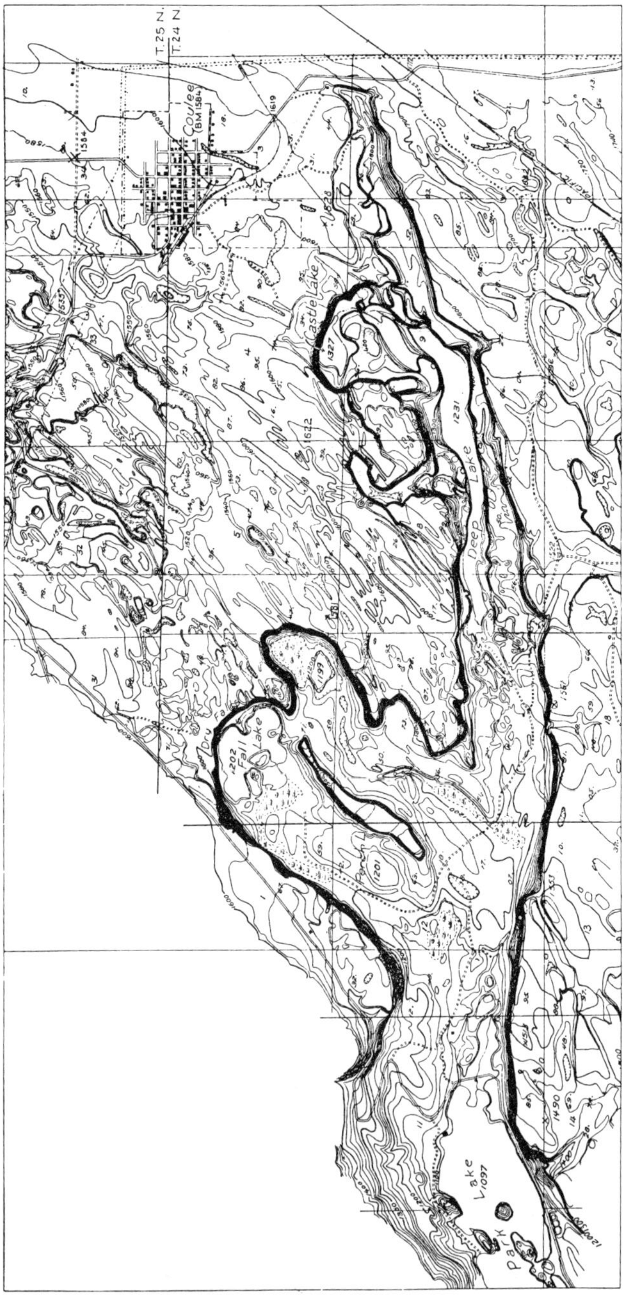

FIGURE 3. The great cataract group in Hartline Basin. Contour interval in T. 24 N. is 20 feet; in T. 25 N. it is 5 and 20 feet, changing on the 1,560-foot contour. (Reduction of a part of sheet 4, Grand Coulee topographic map, U. S. Geological Survey; scale approximately 1 inch to the mile)

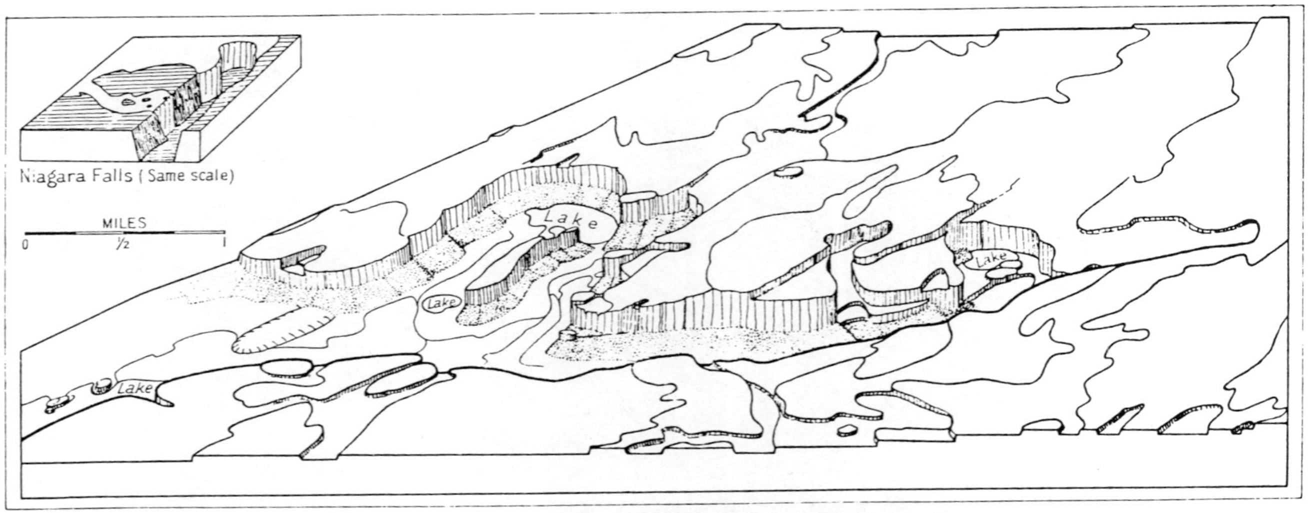

FIGURE 4. Block diagram of the great cataract group, looking north. Compare with topographic map (fig. 3) and the photograph (pl. 2). The walls of Niagara Falls and gorge, having no talus, appear higher proportionately than those of the cataract with talus halfway up the original cliffs. (From American Geographical Society Special Publication No. 15, p. 11)

giant river bar. The flood and ice-jam theories can take such inferences in their stride. The other two find different interpretations, such as morainic gravel deposits or a denial of the continuity of such inclined stratification through large vertical intervals. Imperfect exposures can mislead a geologist into incorrect inferences, especially if he already has a pet theory. This comment applies to all four ways shown on our imaginary guidepost!

Bars.—Proponents of the fill-and-cut theory point to hills of outwash gravel here and there along the streamways as remnants surviving later glacial stream erosion. Defenders of the theory of enormous flooding call these hills giant river bars, averring that they were built up where they stand. But the flood theorists are challenged to show an adequate source for all that water. It admittedly never could have come from daily summer melting along the ice margin. Adherents of the ice blockade theory incline also to call these gravel hills river bars. They get the necessary local flooding by abrupt failure of dam after dam.

The fourth theory ignores the stream deposit character of the gravel hills, erroneously calls them "mutilated moraines" and thus has to bring the ice sheet far south across the plateau. As we linger at the fourway guidepost, this theory seems the least in accord with the field evidence.

Giant current ripples.—A number of publications on the four theories appeared before the U. S. Department of Agriculture had completed its photographic coverage from the air, although vertical views had already been made for the U. S. Geological Survey in mapping Upper Grand Coulee. These U. S. D. A. overlapping verticals allowed stereoscopic viewing of any part of the region. A truly startling discovery resulted from study of photographs of some of the scabland gravel surfaces. The photographs showed a recurring pattern of parallel ridges and swales oriented at right angles to the known direction of flow in a dozen or so of the glacial rivers. Because of a strong suspicion that these corrugations were current-made ripple marks, several examples, widely separated on the plateau (Palouse-Snake junction, Upper Crab Creek, and Columbia Valley opposite Crescent Bar, figs. 24 and 27-31) were examined in the field in 1952. The ridges were so low and widely spaced that no earlier study had more than noted their existence. Their significance now became apparent, for every one of the ridges was 2 to 4 times as steep on the downstream side as on the upstream. This, plus the approximate parallelism, the dying away of the ridges toward the margin of the glacial streams, and the uniform spacing, was, to all geologists who to date (1956) have commented on the published photographs, convincing evidence that these ridges and separating swales were gigantic current-made ripple marks in the gravel bed of the glacial streams. Sizes justify the term "gigantic."

Ridges stand as much as 15 feet higher than the swales and are spaced from 175 to 300 feet apart. The larger the ridges, the wider the spacing. Some groups are nearly a mile across. They are clearly constructional forms made in favored places at the close of deposition of the gravel.[1]

When certain gravel hills, called bars in the flood and ice-jam theories and erosional remnants in the fill-and-cut theory, were found to have these gravel ridges not only on the summits but also descending the lateral slopes of the current-elongated deposits, the flood and ice-jam theories gained perhaps their strongest supporting evidence. No comment on this interpretation has yet come from confirmed skeptics of those theories.

Backwater Deposits.—Least conspicuous of all significant features of the scabland are deposits made in preglacial valleys that entered the complex from areas untouched by the northern ice sheet or its meltwater. Turbulent water flowing *up* the lower stretches of these valleys, especially along the east side of the scabland, is recorded by deltaic and bar gravel accumulations at entrances to the denuded scabland. Foreset bedding inclines upvalley, erratic berg-drifted boulders occur far up the valleys, and a largely unstratified and unsorted pebbly silt mantles the gentler valley slopes. These give a record of glacial water as high as, but no higher than, the bases of loess scarps margining the scabland. Because of the southwestern slope of the plateau, the altitudes reached by these records decrease southward.

Proponents of the flood theory argue that the upvalley foresetting in silt, sand, and gravel can mean only that an adequate current flowed *against* the valley bottom slope, that it could not have been a through current and that the gradient which caused the current was the slope of the intruding water's surface. Thus an adequately rapid rise, a flood with turbulence enough to suspend and transport the sediment, is recorded. It must have been a very brief experience. Perhaps a succession of floods occurred.[2]

The ice-jam and fill-and-cut theories quietly neglect adequate consideration of the evidence for strong upvalley currents back out of scabland.

---

[1]The steep, downstream slope of any current ripple, large or small, is initially determined by debris dropped in the lee of the crest. Inclined bedding at the angle of stability results, and as the ripple grows the downstream-dipping strata come to constitute the structure of the entire ripple. This is beautifully shown in exploratory excavations made in 1958 by the Corps of Engineers, U. S. Army, at the site of the proposed Monumental Dam in Snake River just below the mouth of Devils Canyon. Here the giant ripples on a large scabland gravel bar stand 25 feet or more above their associated troughs and are spaced roughly 400 feet apart. The exploratory pits show that stratification of the gravel in the ripple ridges consistently dips downvalley at an angle of 25° ± for the full thickness of a ridge.

[2]The most extensive records of this back flooding of tributary valleys from the scabland are in the Snake River canyon above Riparia, in the Walla Walla River valley and the Yakima River valley.

Scabland of different ages.—In the dozen papers (1923 to 1932) supporting the flood theory, there is only one suggestion that such flooding occurred repeatedly. Not until the 1952 studies, based largely on data from new excavations, new aerial photographs, and new topographic maps of the great Columbia Basin irrigation project, did the idea of recurring floods find adequate field evidence to support it.

The evidence lies in a variety of relations among the river channels and their bars and in scabland buttes and basins definitely more weathered in the older, higher channels that were abandoned after main channels had become sufficiently enlarged to contain later floods. By this picture, the anastomosis was at its maximum in the earliest recorded flooding, and the latest flood was confined to Grand Coulee alone. An adequate presentation of the evidence for successive deepening of main channels and abandonment of some earlier, higher lying river routes will exceed the space that can be devoted to it here. A few cases will be considered in following pages.

This section on "General Features of Channeled Scabland" constitutes a prologue only. The appended list of references contains only the important titles that deal primarily with our topic—the origin of Washington scabland. From them are taken what detailed descriptions and interpretations seem desirable in this bulletin. A complete account of the scabland's features and a complete history are yet to be written. Indeed, a complete history awaits further field studies also, and from them will come additions to, emendations of, and perhaps subtractions from the treatment bound between these covers. All too briefly for a thoroughly critical appreciation and appraisal, we shall now look at some outstanding places in the Channeled Scabland.

GRAND COULEE

The longest and deepest of all scabland canyons, with the lowest floor at the head and the widest and highest dry falls cliff in midlength—this is Grand Coulee (Bretz, 1932; Bretz and others, 1956). It has the longest gorge made by cataract recession, a record of a former cataract twice as high as its existing Dry Falls, the largest lake (Lenore) and the most saline one (Soap), the most elaborate distributary system in its lower terminal portion, the greatest accumulation there of pebble and cobble gravel and the largest stream-rolled boulders, and it functioned from the beginning of scabland history to become at last the only surviving flood dischargeway. It merits its name, Grand Coulee.

Grand Coulee is a tandem of two canyons, separated by an uncanyoned portion in midlength (pl. 3). The Upper Coulee, whose floor is largely covered now by the Equalizing Reservoir, is 25 miles long and has essentially continuous walls approximately 800 to 900 feet high. It leads southward out of the Columbia River valley at

Grand Coulee Dam, its head a great notch cut in that valley's wall. The floor of the notch is 650 feet above the river. Often described as a Glacial Columbia River course, it came into being because the northern ice sheet pushed a lobe (Okanogan) across onto the plateau and thus blocked the preglacial Columbia farther downstream. Glacial dams at this place were the cause of all scabland river invasions of the plateau.

The detour at Grand Coulee found no pre-existing valleyway to follow for the full length of the Upper Coulee. It is the grandest example of a divide crossing in the whole complex.

The tilted plateau of our introductory picture has some marked departures from a simple plane; huge wrinkles or upfolds of the basalt caused by deep-seated earth movements long before the scabland history began. Grand Coulee is intimately related to one of these, the Coulee monoclinal flexure, a steep warping up of approximately 1,000 feet toward the northwest.

The initial spillover along Upper Grand Coulee was across the higher side of the flexure as far as that upper canyon extends. Encountering the steep slope of the monocline, the glacial river must have originally cascaded about 800 feet down, to spread out considerably in the broad uncanyoned portion where Coulee City and Dry Falls State Park are located. The spectacular Dry Falls are a part of the head of the Lower Grand Coulee canyon. The entire head is the Great Cataract.

This Lower Coulee follows the monoclinal flexure almost to Soap Lake at the mouth of Grand Coulee into the broad, gravel-filled Quincy Basin. It is eroded *in* the bent and broken basalt flows, following an impossible course for any preglacial stream; hence is, like the Upper Coulee, wholly of flood origin. Hogback islands in Lake Lenore and tilted flows (figs. 6 and 7) along the highway from Dry Falls to Park Lake and in other places repeatedly demonstrate this fact along about 10 miles of Lower Grand Coulee's 17-mile length. The great gravel deposits of Quincy Basin represent only a third or a fourth of the estimated 11 cubic miles of rock excavated in making Grand Coulee and its smaller distributary coulees (Dry, Long Lake, Jasper, Lenore, and Unnamed). All that enormous task was accomplished by glacial floods during the scabland-making. Most of the debris was carried on through and beyond Quincy Basin.

Upper Grand Coulee's excavation began at the 800-foot cascade a little north of Coulee City. This cascade shortly steepened to become a waterfall and, as such, it retreated upstream until it finally cut through the divide into the preglacial Columbia Valley, thus destroying itself and leaving the great notch by which water, which is pumped up about 280 feet from Franklin Roosevelt Reservoir back of the Grand Coulee Dam, now enters the Equalizing Reservoir. Thence by a canal using eastern distributary coulees, this water for

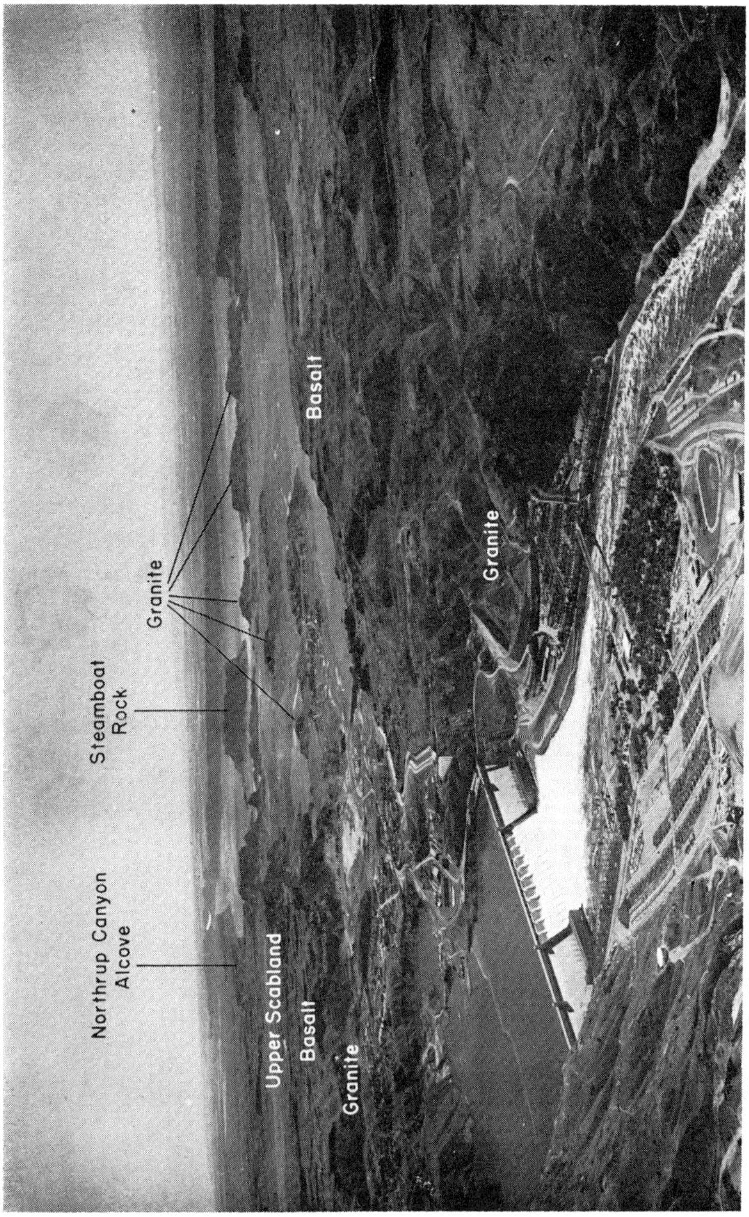

FIGURE 5. Grand Coulee Dam and head of Grand Coulee. (U. S. Bur. Reclamation photograph)

irrigating more than a million acres bypasses Dry Falls and the Lower Coulee to reach Quincy Basin.

The evidence for this interpretation of the origin of Upper Grand Coulee is manifold. A plunge basin at the foot of the initial fall, found by the U. S. Bureau of Reclamation, contains at least 300 feet of gravel below the level of the scabland floor immediately downstream in the Coulee City uncanyoned tract. The glacial river above the falls was relatively shallow and much wider than the gorge it made. Subsidiary lateral falls operated successively and were successively abandoned as the great cataract retreated. Northrup Canyon (fig. 9), on the east side, contains one subsidiary dry cataract as wide as Niagara and three times as high. Barker Canyon, on the west side, is but a little smaller. Steamboat Rock (figs. 5 and 10), 880 feet high and a square mile in area, stands isolated out in the Upper Coulee as a "Goat Island," where for a time the cataract was divided into two. Steamboat's broad summit is as high as the tops of the Coulee's adjacent margining cliffs and is typical scabland.

FIGURE 6. Lake Lenore, Lower Grand Coulee, looking northward. The lake basin has been excavated in the elongated belt of tilted basalt flows which constitute the Coulee monocline. The line of cliffs is on the high side of the flexure, and the ravines that preglacially descended the monoclinal slope have all been left hanging by the glacial river's erosion along the tilted flows. A very similar profile of gabled headlands exists along the east side of lower Moses Coulee. (U. S. Bur. Reclamation photograph)

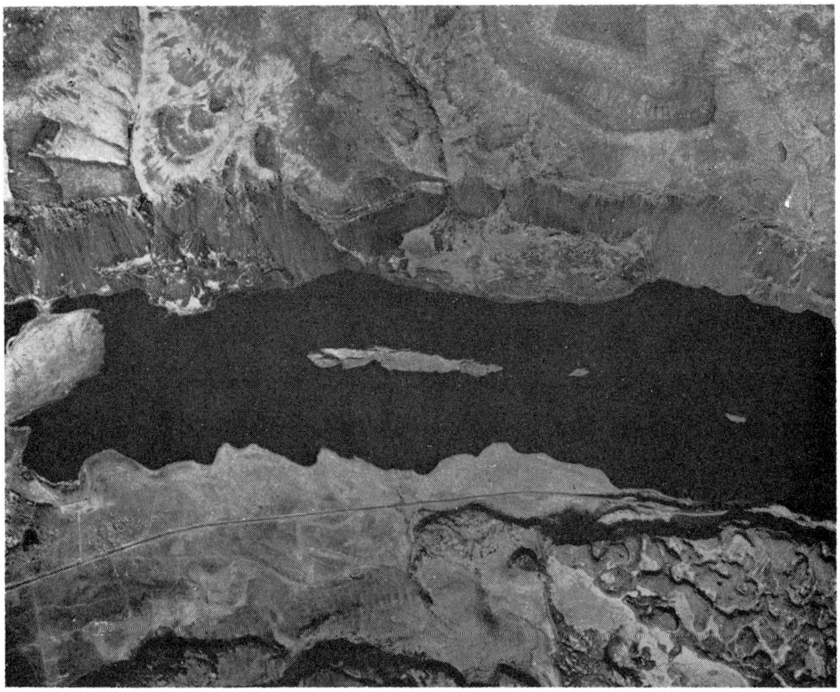

FIGURE 7. Lake Lenore, Lower Grand Coulee, and its hogback islands along the axis of the eroded Coulee monocline. Cliff summits on left stood above all flood water and carry small preglacial drainageways now hanging between their gable-like truncated divides. Cliffs on east side of lake are considerably lower, and their summits were floodswept and made into scabland. Dip of basalt in islands is 45°, whereas the flows shown in all bordering cliffs are nearly horizontal. (U. S. Dept. Agriculture photograph)

FIGURE 8. Scabland floor of Grand Coulee in Hartline Basin. The southern end of the western wall shows along the left edge of the picture. Above it are normal erosional gullies, hanging far above the scabland channel floor. The relief among these channel buttes and basins is more than 50 feet. This vertical aerial photograph was taken at an altitude of about 11,000 feet. (Grand Coulee Series, 33, U. S. Geological Survey)

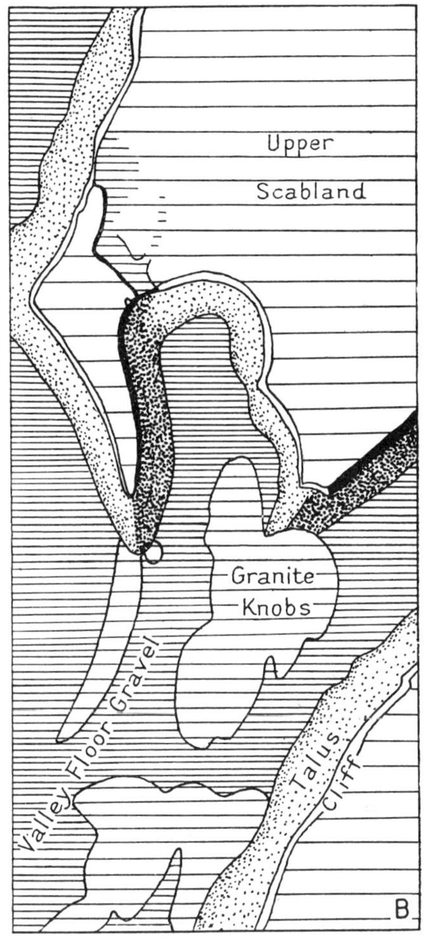

FIGURE 9. Middle alcove of Northrup Canyon group, Upper Grand Coulee.

A. The cliffs facing north and east are in shadow. Grand Coulee floor and cliff show in the northwest corner. Summit scabland occupies most of the northern half of the picture and a little of the southeastern corner. The main Northrup Canyon crosses from the southwest corner to the middle of the east side. The alcove is in the center. Note channels leading to it from the brink of the Coulee wall. Granite hills form the angle between the alcove and the main Northrup Canyon. (Photograph is mosaic of vertical aerial photographs, Grand Coulee Series, Nos. 112-116, U. S. Geological Survey)

B. Line drawing of area shown in *A*.

FIGURE 10. Northwest side of Steamboat Rock. The cliff possesses a hanging valley niche, identical with those margining the upper coulee where summit scabland occurs on each side. It was not made by postglacial drainage from the summit, nor can the valley be interpreted as a preglacial valley and the Rock as a preglacial divide. The channel enters *from the air* on the eastern side. (Photograph by Simmer)

This vanished Upper Coulee cataract, named for Steamboat Rock, found granite hills buried beneath the basal flows north of Steamboat (figs. 5 and 10). Lacking the close vertical joints of basalt, the granite was less readily attacked by the cataract's plunge, and its hills still stand, somewhat reshaped, on the broader floor of this part of the Coulee. A part of the unusual widening here is probably caused by these numerous unyielding pre-basalt hills.

Some excellent exposures of gravel-bar deposits have been made in relocating the highway along the Upper Coulee. Foreset bedding is very obvious and in places records great eddies in the lee of rock shoulders, the foresets dipping *upstream*.

Grand Coulee's system of distributary canyons begins at the broad uncanyoned place in midlength and spreads out over a width of 15 miles before entering Quincy Basin. Very respectable scabland characterizes this system of distributary canyons. One cataract in the group (Unnamed Coulee, fig. 12) is 150 feet high and had three retreating alcoves spanning a head width of more than a mile when the last glacial flood subsided. There is no channel leading to it. The discharging water arrived in a broad sheet across a minor preglacial divide.

FIGURE 11. Bar in Grand Coulee.
A. Looking east. Crest of bar is 150 feet above Coulee floor.
B. Looking south. Crest of bar is 50 feet above the fosse between it and the cliffs.

FIGURE 12. Unnamed Coulee. Oblique aerial photograph, showing unchanneled scabland leading to the northernmost cataract alcoves. Shadows fall to the right. Looking northeast. (Photograph by H. T. U. Smith)

QUINCY BASIN AND ITS OUTLETS

Grand Coulee water, entering Quincy Basin via Soap Lake, Dry Coulee, and Long Lake Coulee, was joined by the discharge of 10 to 15 glacial rivers from the east and northeast. (It is difficult to count rivers that repeatedly divide and reunite.) About half of these were supplied directly from Columbia Valley across the high northern margin of the plateau, the dominant preglacial divide. Half were born of overflow out of the largest scabland tract which the scarred plateau possesses, the enormous Cheney-Palouse tract.

All these rivers eroded their bedrock valley floors, and some garnered great quantities of gravelly and bouldery debris from undercutting their channel walls. The structural Quincy Basin was filled nearly to the brim with this gravel; approximately 500 square miles of it.

The "brim" had irregular heights, and the flooded basin overflowed in four different low places, gravel spilling over with the escaping water and cutting into the rim to a maximum depth of 300 feet. Three of the spillways created great waterfalls (Bretz, 1923a, p. 547; Bretz, 1928b, p. 464; Bretz, 1928c, p. 329; Bretz and others, 1956, p. 984) into the preglacial Columbia Valley on the west —scenic features almost unequalled outside of the scabland. One of these cataracts (Potholes, figs. 13, 14, and 15) was 1½ miles wide, 400 feet from brink to bottom of plunge basin, and its recessional gorge became 2 miles long before the flow ceased. U. S. Highway 10 climbs along a winding route up another of these extinct cataracts (Frenchman Springs, fig. 16) from Vantage bridge to the floor of Quincy Basin. The Great Northern railroad utilizes the notch made by the third (Crater, fig. 17) to climb from the Columbia River to the Basin floor. Plunge-pool basins at the foot of each of the twin Potholes cataracts were dry until leakage and waste irrigation water from the West Canal filled them. The larger one now contains a lake 125 feet deep and 1½ miles long (fig. 14).

Notches for channels to these three cataracts were cut 20 to 90 feet deep in Quincy Basin's western rim, and correspondingly decreased the Basin's capacity for retaining a gravel fill. The most challenging fact, however, is that each spillway began at very close to the same altitude. According to the flood theory, they were contemporaneous in origin and two of them ran neck and neck in the downcutting whereas one cataract was early abandoned.

Alternative interpretations have been offered in an endeavor to make these spillways sequential and thus avoid such unbelievable quantities of water at one time. One suggestion is that earth movements, after one cataract had been formed, depressed another place in the rim (and then another one) so that discharging water was drawn off from one cataract to make another. Then, of course, more

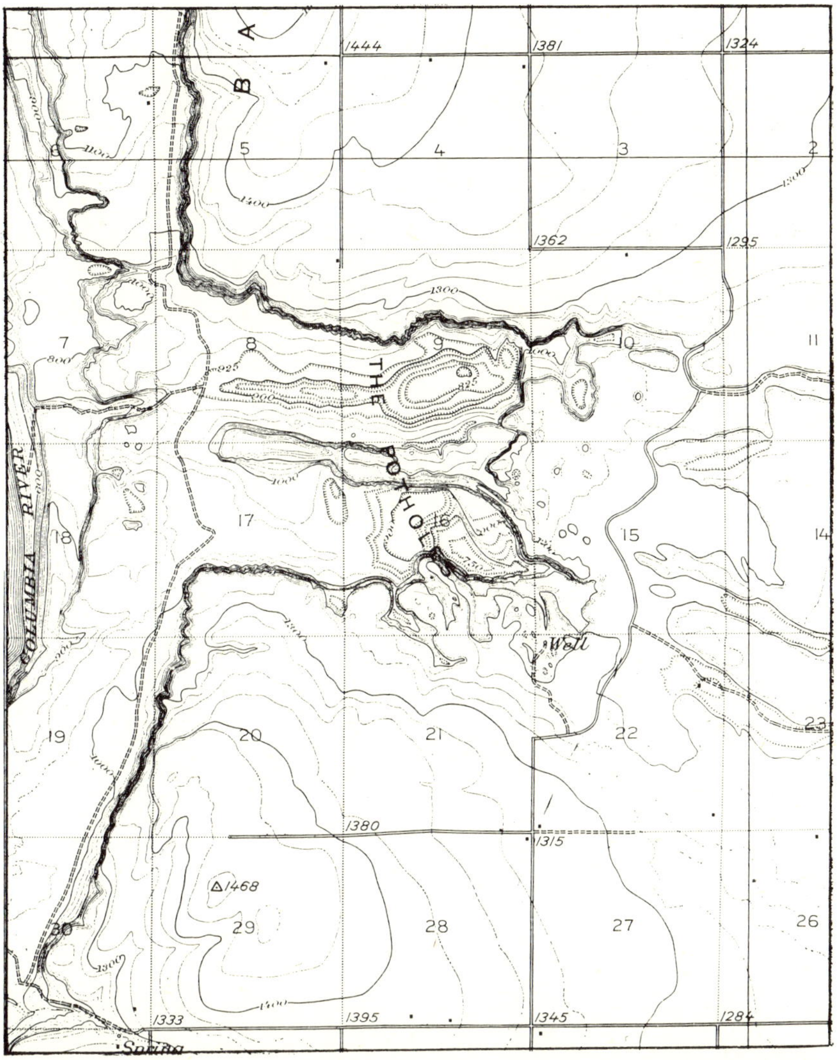

FIGURE 13. Topographic map of The Potholes cataract, one of three cataract spillways from the west side of Quincy Basin. (From U. S. Geological Survey Quincy quadrangle map)

FIGURE 14. Northern alcove of The Potholes cataract in 1954. Water is waste and leakage from West Canal about 2 miles distant and 75 feet higher than cataract brink.

Magnitudes involved: Lake in plunge basin is 125 feet deep. Top of bar on north side of alcove is 50 to 75 feet above lake surface and 175 to 200 feet above bottom of plunge pool. Bar fosse is largely filled with talus. Cliffs back of bar are, including talus, 300 feet high. Scarp of gravel terrace along the Columbia is 150 feet high, and its summit is approximately 250 feet above river. Nearer cliff, across which lake discharges, is 150 feet high above gravel terrace.

Sequences involved: Cataract originated approximately at the near end of the plunge basin. Its total retreat was nearly 2 miles. Tip of the ridge separating the two alcoves is about an eighth of a mile back from the prescabland line of cliffs. Basalt bench below this cliff (part of bench south from Crater cataract) is gashed by glacial-river chutes, recording cascades rather than cataracts and presumed to be of subfluvial origin. Gravel of the terrace overlaps a little on scabland of this bench.

Correlations proposed: Gravel terrace has right altitude above the Columbia to be correlative with West Bar and Bevery Bar. The inconspicuous but recognizable, very bouldery, flattish mound buried under this gravel is a dump from early cataract retreat, comparable to that at Vantage bridge. Scabland beneath terrace gravel is older than the terrace. Cascade chutes are later than the initial Potholes cataract. They and the bar north of the plunge basin may date from a late flood out of Quincy Basin. (U. S. Bur. Reclamation photograph P-222-177-35237)

FIGURE 15. Southern alcove of The Potholes cataract, looking east into Quincy Basin. Outline of lake in plunge basin is determined in part by gravel bars made in last stage of cataract's existence. Two lakes in channel floor leading to the falls are in rock basins of that channel, capacity of which has been increased by dams at near ends. (U. S. Bur. Reclamation photograph 40207)

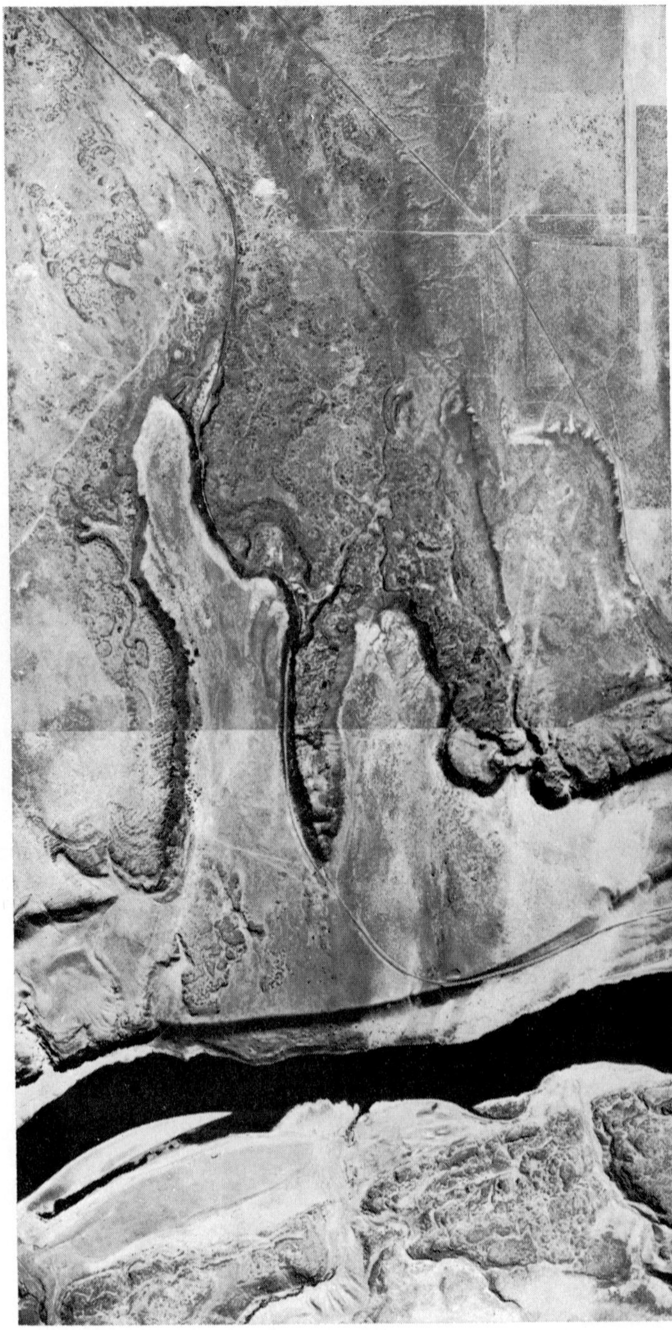

FIGURE 16. Frenchman Springs cataract and Columbia River. Vantage bridge on U. S. Highway 10 is just out of the photograph (at bottom). Highway climbs diagonally northward up a postglacial scarp cut in scabland gravel, swings eastward, and then makes another diagonal climb up the south wall of the northern alcove of the double cataract to the floor of Quincy Basin. Plunge basins exist below the cataracts but have not yet (1956) received leakage and waste water from the West Canal as have the comparable basins in The Potholes cataract. (U. S. Dept. Agriculture photographs AAR-8F-187 and 9F-42)

FIGURE 17. Crater cataract and channel, looking east into Quincy Basin. The beginning of a double fall, like The Potholes and Frenchman Springs, has left a spur or peninsula which the Great Northern railroad utilizes for part of its climb from Columbia Valley to Quincy Basin. The plunge basin, Willow Creek draw, is filled with boulders below the cultivated surface shown in the photograph. (U. S. Bur. Reclamation photograph 40208)

earth movements had to occur to bring all channel upper limits to their present close correspondence. After that, no further movements would be permissible.

A second idea is that a westward-sloping gravel plain once existed in the Basin up to these upper limits. On it, escaping water behaved as it does on a delta; i. e., it developed shifting distributaries, three in number. This procedure could produce a sequence.

A third idea is here proposed for the ice-jam theorists to consider; namely, that the three channels became successively blocked by berg and river ice to make a sequence.

For none of these ideas is there any supporting field evidence. For one, there is positive evidence against it. Solution of the problem of the three western cataract channels becomes still more difficult, except for the flood theory, when the fourth spillway is considered. Although 35 to 50 miles distant, the elevation at its upper limits are the *same* as those of the cataract channels! And the width at its upper limits is 9 miles!

FIGURE 18. Drumheller Channels topography. Looking southward from O'Sullivan Dam. (U. S. Bur. Reclamation photograph DX222-117-142)

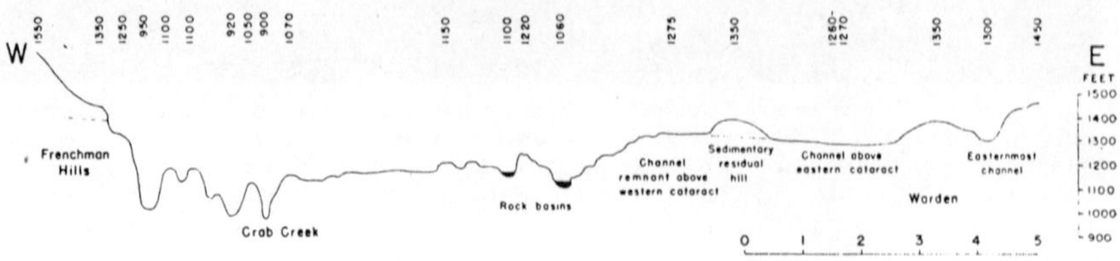

FIGURE 19. Profile across Drumheller Channels. Looking northward. Profile follows a slightly zigzag course across the head of Drumheller Channels, between Warden and the nose of Frenchman Hills anticline. The easternmost channel has been but slightly deepened or scarped, and the Warden cataract channel floors have been cut but little into basalt. The three deep notches, by one of which Crab Creek escapes from Quincy Basin, were made simultaneously. No preglacial drainage course across the anticline could have been deeper in basalt than the 1,100-foot flat between Crab Creek and the two rock basins. (From Geological Society of America Bull., v. 67, p. 976)

This enormous spillway, Drumheller Channels (figs. 18 and 19) in the extreme southeastern corner of the Basin (Bretz, 1923a, p. 595; Bretz, 1923b, p. 636; Bretz, 1928b, p. 466; Bretz and others, 1956, p. 975) was deepened more rapidly than the western channels and eventually captured all their share of the flood water. Its deepening continued through successive floods until it had been cut 300 feet into a previously intact divide. Drumheller is the most spectacular tract of butte-and-basin scabland on the plateau. It is an almost unbelievable labyrinth of anastomosing channels, rock basins, and small abandoned cataracts. Only one channel in the plexus, the route now followed by Crab Creek, has a continuous gradient across

Drumheller's 50-square-mile area; and this route almost surely has rock basins leveled up by the creek's sand and gravel in postglacial time. The average descent across the tract is between 30 and 50 feet per mile. Drumheller was a gigantic cascade rather than a unit waterfall, because the basalt beneath had been flexed, broken, and faulted in one of the plateau's strongly accented upfolds, the Frenchman Hills anticline. Although higher and earlier rock-bound channels were abandoned as the deepening progressed, no marked central channel and no dominant cataract developed.

Drumheller has an erosional record of at least three successive floods. Two are recorded in a cataract ledge, with a separating "Goat Island," 3 to 4 miles southwest of Warden. Alongside it is a gorge-like channel 150 feet deeper. Obviously the making of that gorge was later than the functioning of the cataract and brought its

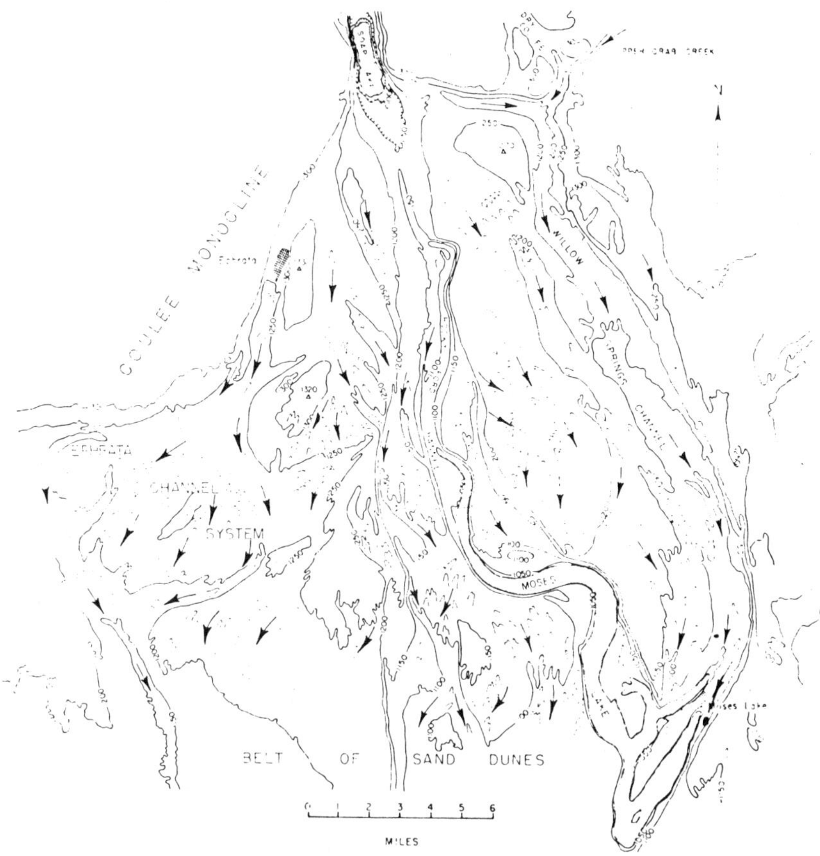

FIGURE 20. Quincy Basin distributary channels. Showing the three channels leading from Soap Lake across the gravel fill and the $\wedge$-shaped heads of minor channels. (From U. S. Bur. Reclamation topographic map R4-5720, sheet 1. Contour interval, 50 feet)

history to an end. The recessional gorge below the Warden cataract is cut off by, and hangs 100 feet up on the side of, the later gorge.

The topography of Quincy Basin's gravel fill is intimately related to the downcutting in Drumheller. Three distributary channels, Ephrata, Rocky Ford, and Willow Springs (fig. 20), reaching bedrock in but few places, cross from the mouth of Grand Coulee to the head of the great Drumheller cascade. Their excavation followed the initial filling of the Basin with waste from its dozen contributing rivers. Concentration of flow and erosion in the gravel succeeded the earlier deposition because of deepening in the Drumheller part of the Basin's rim. Many of the gravel hills and ridges are also identified as bars made during this channel development in gravel. Only one of the channels (Rocky Ford, containing Moses Lake) has any of the characters of a normal river course superimposed on its torrent-made form. These are imputed to a postglacial river from Grand Coulee and Crab Creek, now vanished because of increased aridity.

LOWER CRAB CREEK AND OTHELLO CHANNELS

The further course of the Drumheller discharge presents more difficulties for any conventional interpretation. The water divided, for a time at least, in an elongated downwarp of the basalt (the Othello basin or syncline), a few miles to the south. Part flowed westward along the north limb of Saddle Mountain, another anticlinal fold, and followed the Othello structural valley between this mountain and Frenchman Hills to reach the preglacial Columbia Valley at Beverly. Scabland marks the entire course and is worthy of more attention than can be given here (Bretz and others, 1956, p. 990 +).

Another part of the Drumheller discharge continued southward to cross the preglacial water parting made by the eastern tip of the Saddle Mountain anticline. Othello Channels (Bretz, 1923b, p. 636; Bretz, 1928c, p. 214+; Bretz and others, 1956, p. 995+) (fig. 21), resulting from the discharge across this divide, is a small copy of Drumheller. Its largest rock basin, now containing Scooteney Reservoir, is 135 feet deep, and the entire group of basins and channels is about 3 miles wide. As with Drumheller, this Othello discharge failed to clean off all the sedimentary rock that topped the preglacial divide. Each complex has an isolated residual hill of the weaker rock, entirely surrounded by the wilderness of bare basalt hills and interlocking channelways.

Basalt boulders so large that blasting was necessary before earth-moving equipment could handle them are strewn over the 8 miles of channel between the place of divergence and the entrance into Othello proper. They could have come only from Drumheller, and they must have traveled 10 miles or more on a gradient not exceeding 15 feet per mile. Volume alone could produce the required velocity in the water that carried these boulders.

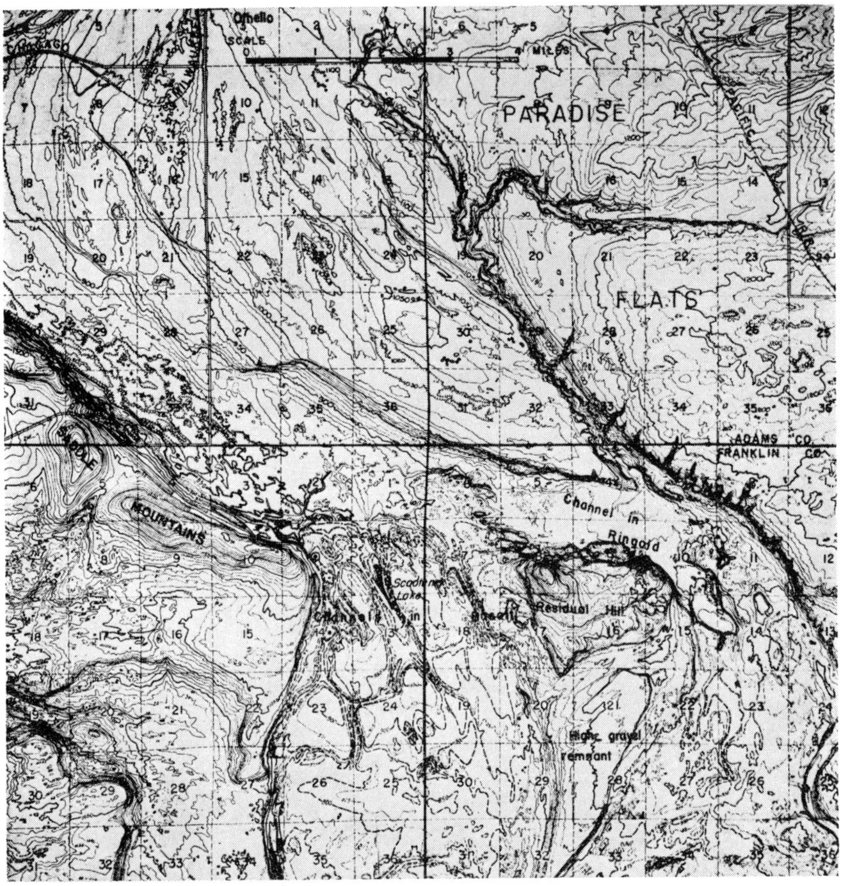

FIGURE 21. Map of Othello Channels. Interval for continuous contours, 10 feet; for dashed contours, 25 feet. (From U. S. Bur. Reclamation topographic map R4-5720, sheet 2)

Part of Othello's discharge reached the Columbia Valley via the Koontz Coulee group of channels across the Ringold plain, a flattish upland determined by a superbasalt sedimentary formation. The Koontz channels did not, in their deepening, reach basalt. Three successive glacial water discharges are recorded on the Ringold plain. Part of the Othello discharge joined, at Connell, a distributary strand of flood water from the great Cheney-Palouse glacial river. This is the only place where any Grand Coulee-Quincy Basin water united with the eastern discharge before reaching the Columbia River valley.

MOSES COULEE

In addition to the two main channel systems of the scabland complex, a third and truly magnificent scabland channel was made west of the Grand Coulee system. Moses Coulee (Bretz, 1923a, p. 600+; Bretz, 1923b, p. 634; Bretz, 1928a, p. 673+; Bretz, 1928b, p. 463; Bretz and others, 1956, p. 988) is a smaller copy of Grand Coulee, with a great ragged compound cataract (The Three Devils, fig. 22) in midlength, beetling cliffs 800 to more than 900 feet high, and a welter of butte-and-basin topography in a central tract. The lower stretch of Moses Coulee is deeply aggraded with gravel, so it has no rock-basin lakes. It has two features lacking in Grand Coulee: a record of a preglacial valley for the lower third of its 50-mile length, and a massive transverse glacial moraine in its northern third. Waterville Creek, the precursor of lower Moses Coulee, now spills some hundreds of feet down into the canyon from a hanging valley that is but little altered from its preglacial condition. Minor tributary valleys on the upland margining lower Moses Coulee also hang far above the great coulee's aggraded floor.

The Withrow moraine crossing of Moses Coulee (fig. 23) is unique in all scabland. It is a very strongly expressed ridge or series of

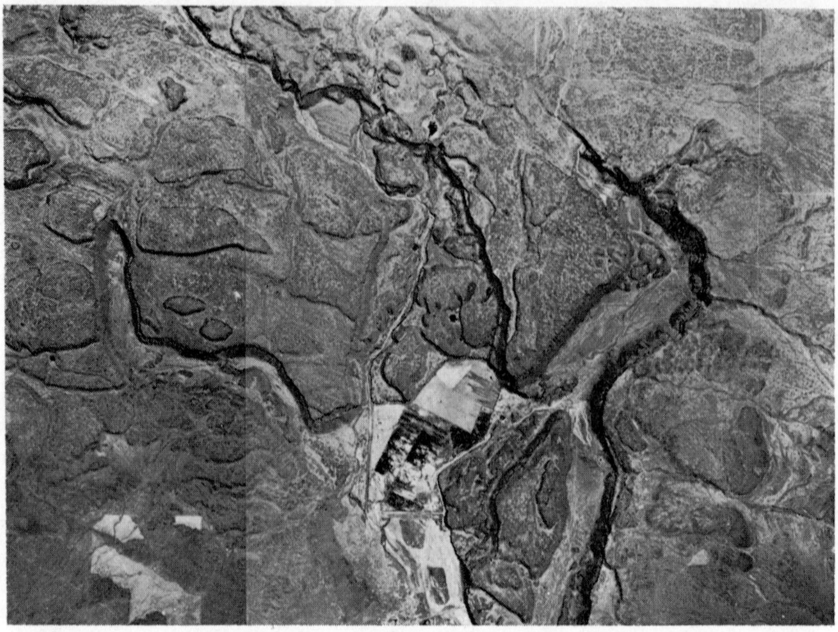

FIGURE 22. Compound cataract in mid-length of Moses Coulee. Three recessional gorges reach back upstream into the broad scabland. The one to the east (right) was started by a double cataract whose members united to leave a "Goat Island" comparable in origin to Steamboat Rock in Grand Coulee. Apparently all three gorges and their cataracts were contemporaneous. (U. S. Dept. Agriculture photograph AAQ-3F-50 and 51)

ridge hills of glacial drift marking the extreme southern limit ever reached by the northern ice sheet on this part of the plateau. Its course is approximately at right angles to that of the Coulee, and its descent into that great gash proves that Moses Coulee is an older feature. In the Coulee, the moraine is an irregularly mounded deposit that in places is more than half as high as the Coulee is deep and that completely conceals the eastern cliff. This latest glaciation of this part of the plateau has destroyed former scabland connections with the head region of Moses Coulee.

Inescapable conclusions from the study of Moses Coulee are (1) that early floods made and used it, and (2) that no flood water issued from the margin of the Okanogan lobe when the Withrow moraine was built. The second conclusion foreshadows an important question which we are approaching: the source of these tremendous quantities of water. The floods were not the direct product of climatic melting. A reservoir of adequate size so situated that it could be catastrophically emptied is indicated.

FIGURE 23. Edge of moraine on Waterville Plateau west of Grand Coulee. Cultivation is limited almost entirely to the smooth unglaciated surface on the left, and the hilly moraine still has its original cover of sage and other brush. The curvilinear dark pattern beyond the moraine is a drainageway still largely in brush. (U. S. Dept. Agriculture photograph AAQ-7F-129)

UPPER CRAB CREEK

Superlative showings of two flood-induced scabland features occur in this torrent-modified preglacial stream valley. One is a group of splendidly proportioned and significantly located gravel bars (pl. 4). The other is a spread of giant current ripple marks on these bars (fig. 24) (Bretz, 1928a, p. 668; Bretz and others, 1956, p. 980+). Neither find possible explanation by any alternative to the flood theory.

Five of the bars are shown by 2-foot contours on maps of the U. S. Bureau of Reclamation. All are hills of gravel with down-valley foreset bedding shown in every pit in them. Three are situated on the valley floor, almost or completely surrounded by bottom land and are streamlined in conformity with the bedrock valley

FIGURE 24. Bar No. 1 in Upper Crab Creek valley, looking southward. Vegetation pattern outlines the current-ripple marks. Fosse, with cultivated field, lies between bar top and foreground scabland. Another cultivated tract, in lower left, is in the blocked mouth of a prescabland gulch. (Photograph by H. T. U. Smith)

outlines. All reach 100 feet or more above adjacent bottom flats. One (fig. 24 and pl. 4, Bar No. 1), which is built against a scabland cliff, still is highest out away from the cliff. This one damned two small preglacial tributary valleys without completely backfilling their mouths. One of the tributaries has succeeded in cutting through the bar dam, the other and smaller one is still a closed basin back of the bar. Crab Creek, in adjusting its course around and among these bars, has not modified their outlines and has had to use the smaller of the two channelways margining the largest of the bars shown in plate 4. The Great Northern railroad uses the larger channelway.

The concept of bar growth to such magnitudes, and its corollary of a river deeper than the heights of the bars, affronted many uniformitarian-minded geologists, and an earnest effort was made to explain these gravel hills as terraced remnants of a former complete fill. This alternative may look plausible with regard to some scabland gravel deposits and, indeed, appears applicable to Quincy Basin's early, high-level gravel. It cannot apply to the Upper Crab Creek features for the reasons very briefly stated above.

Should these reasons fail to appeal, there remains the second group of features, the giant current ripple marks. Each of these bars possesses giant ripple marks on the summit and on the slope toward the larger of its bounding channels. All ripple ridges that were examined in the field are asymetrical, the steeper side facing downstream. Groups of larger ripple marks have wider spacings between the ridges. The U. S. Bureau of Reclamation map (pl. 4) does no more than suggest their existence, in part because the topographer did not know just what he was sketching. Aerial photographs, both vertical and oblique, (fig. 24) present the convincing evidence that these hills are great bars on whose last gravel accretions the generating great river left its large signature.

A short distance northeast of the bars shown in plate 4, Wilson Creek glacial river entered the larger Crab Creek torrent. In a vertical range of about 150 feet there are 2 terrace-like flats on a gravel deposit margining the tributary river's course. They are strikingly marked by giant current ripples up to 15 feet or more in height. Furthermore, the bottom of the Wilson Creek valley here is similarly marked. Erosion scarps separate the 3 ripple-marked surfaces.

This ensemble is taken to record 3 successive floods, the 2 erosional trenchings of earlier, higher gravel made possible by deepening of Drumheller Channels in the Quincy Basin's rim.

It should be said, in fairness, that these Upper Crab Creek and Wilson Creek bars have never been studied in the field by the founders of the rival theories for scabland genesis.

CHENEY-PALOUSE SCABLAND TRACT

A scabland flood route 75 miles long and in places more than 20 miles wide extends southwest from the vicinity of Cheney to Snake River canyon at the junction of the Palouse River, a little downstream from Riparia. Plate 1 shows that glacial water using this route spawned 10 prominent westward spillovers to the Grand Coulee-Quincy Basin system. All but 2 of these were preglacial divide crossings. More than 75 loessial islands are shown, and mapping on a larger scale would show more.

This is the type region where rivers with discharge "less than that of Snake River today" (Flint, 1938, p. 515) supposedly (1) built up a complete gravel fill from 90 feet thick to as much as 800 feet (in the Snake River canyon) in order to spill across preglacial divides westward out of the tract; and then, (2) reversing their behavior, eroded away almost all of the hypothetical great fill and bit deeply into subjacent basalt to make the challenging scabland topography. They are described as "leisurely" rivers (Flint, 1938, p. 514).

The Cheney-Palouse tract is also an excellent place to try out the ice-jam theory for originating the west-flowing distributary channels and the detailed channeling that isolated the loessial islands. Both the ice-jam and gravel-fill theories limit their water supply to ordinary melting of the lobe of glacial ice spanning the head of this tract. However, the bursting of local ice-jam dams could release short-lived local floods whose records might be misinterpreted as catastrophic events involving the entire scabland complex.

The Cheney-Palouse scabland tract has been superposed on a well-recorded preglacial drainage system from the head near Cheney to Hooper and Washtucna, 60-odd miles distant. Several semiparallel creek valleys converged to join Palouse River in the southern part of the tract. The Palouse itself entered the tract from unviolated eastern country about 40 miles down along the length of the tract.

FIGURE 25. Palouse Falls, 185 feet high.

From Hooper to the Snake, another 10 miles, the glacial water did an extraordinary thing. Although it sent a large distributary strand down Washtucna Coulee to join some of the Othello discharge at Connell, most of its water took a short cut *across* the Palouse-Snake divide south of Hooper and Washtucna.

There is no more significant area than this divide for testing rival theories. Streamless Washtucna Coulee was a major plateau river course of the preglacial Palouse, that did not enter the Snake until nearly 70 miles farther southwest. South of Hooper was a bedrock divide 10 miles across and, with its capping of loess, at least 350 feet higher than the preglacial Palouse valley which the glacial water left to overwhelm the divide. Here are 80 square miles of splendidly shown scabland on the divide summit. Here is the spectacular new Palouse River canyon cut 400 feet deep through the divide, leaving the old valley along Washtucna Coulee with nothing more than two alkaline lakes. (At midlength in Palouse Canyon at Palouse Falls State Park the river spills 185 feet in the clear.) Here is H U Ranch dry cataract (fig. 26), 280 feet from brink to bottom of plunge basin, and that basin 1½ miles long in the bottom of the cataract's recessional gorge. Here are 2 rock basins on the very summit of the preglacial divide, a mile apart and 80 and 120 feet deep. Here are streamlined, prow-pointed and steeply scarped loessial islands 150 feet high. These are all de nova products of the glacial overflow whose volume had been repeatedly reduced by distributaries all the way down along the Cheney-Palouse scabland tract.

Removal of the hypothesized gravel fill of this immense tract was theoretically done by "meandering" glacial streams. They left no meander scars in any bordering loessial bluffs from Cheney to the Snake. The long narrow subidiary channels separating many of the long streamlined loessial islands are no more a record of meandering streams than are the streamlined gravel bars with giant ripple marks (Staircase Rapids bar, fig. 27) (Bretz, 1928a, p. 649+; Flint, 1938, p. 878; Bretz and others, 1956, p. 1000). The scabland features just enumerated are utterly impossible erosional forms for such streams to have made.

By the ice-jam theory, the major divergences to distributaries and around islands required blockades many miles across and as much as 400 feet thick where the deep preglacial valleys still constituted trenches in the complex. The quantity of floating ice required for these dams gives one pause even if he thinks such dams would hold the head of water back of them. The failure, both by defenders and attackers of that theory, to find any topographic record of such dams or of adequate bypass channels while the dams were growing, makes the idea very speculative at best.

The "Shoulder Bar" (figs. 28-31) (Bretz, 1928a, p. 657+; Flint, 1938, p. 981+; Allison, 1941, p. 70; Bretz and others, 1956, p. 1020+) is a square mile of scabland gravel on the *south* side of the Snake nearly opposite the mouth of the Palouse and 500 feet above the valley bottom. It completely blocks the mouth of a small tributary from the south, so that in post-scabland time this stream has had to find a new route to the Snake. The long foresets of this gravel dip *northward, into* Snake Canyon. Its broad flattish summit and southern slopes are completely covered with giant current ripple marks (fig. 31). It therefore belongs to the group of constructional land forms left by the scabland rivers; i. e., gravel bars.

This huge mass of scabland gravel on the south side of Snake Canyon must have been carried across from the north side. The

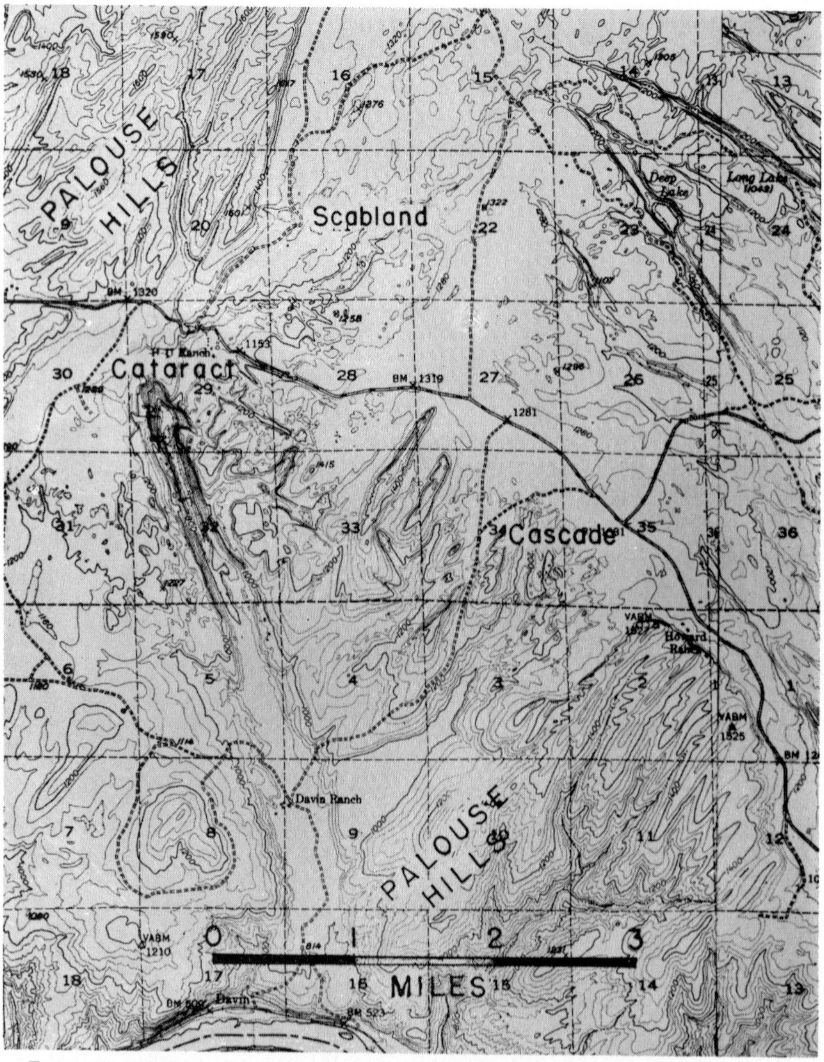

FIGURE 26. Vicinity of H U Ranch cataract. Western part of Palouse-Snake divide scabland. Loessial topography in northwestern and southeastern parts of area, loessial islands in central part. Largest island is 1 mile long and 500 to 1,000 feet wide, and its scarps are 150 feet high. Deep Lake gorge is tributary to Palouse Canyon a mile beyond eastern edge of map. (Part of Haas and Starbuck topographic maps, U. S. Geological Survey. Contour interval, 40 feet)

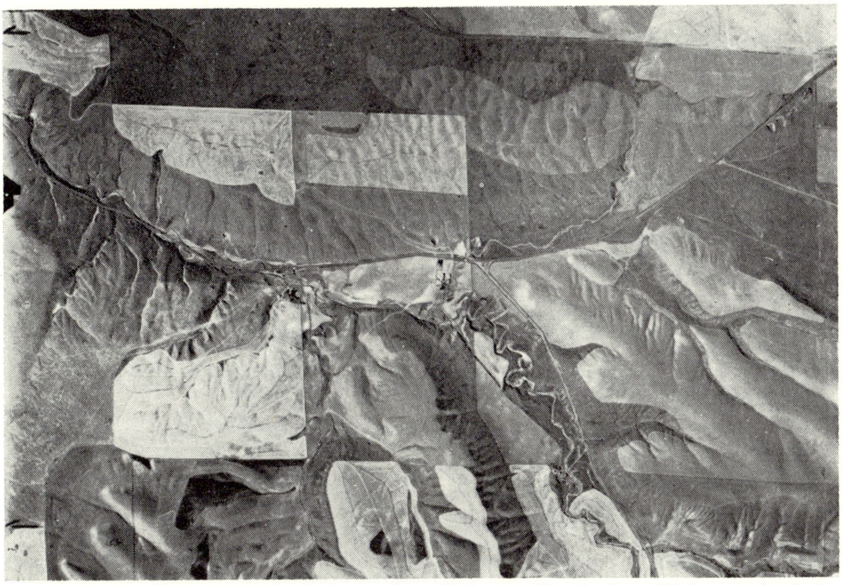

FIGURE 27. Giant current ripple marks. Staircase Rapids bar. Valley east of the ripple-marked surface was partially blocked by bar growth, and the tortuous stream meanders are on a silt flat. Drainage pattern east of creek is that of Palouse Hills topography. (Production and Marketing Administration photograph AAP-26-188)

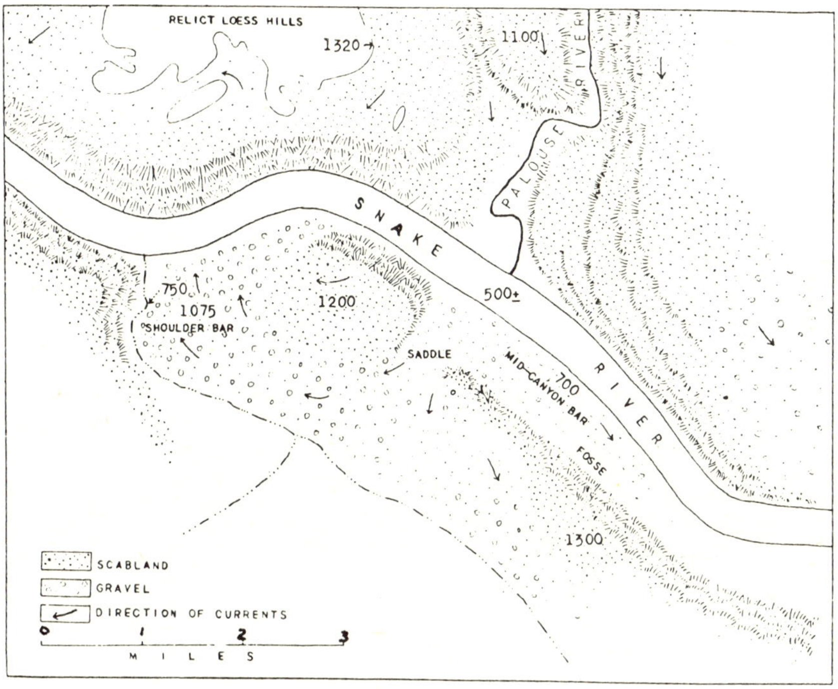

FIGURE 28. Sketch map of region at mouth of Palouse River. (From Geological Society of America Bull., v. 39, p. 656)

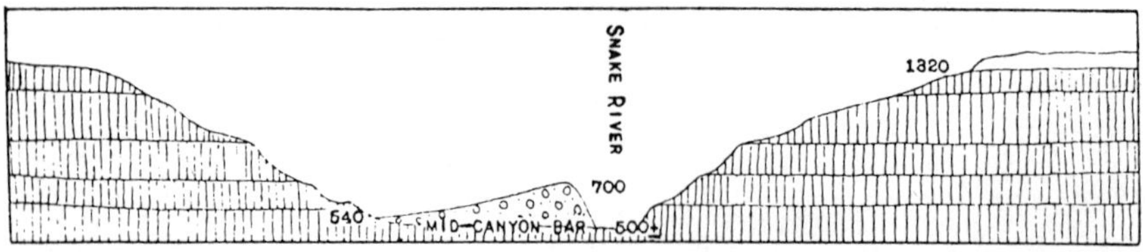

FIGURE 7. - *Cross-section of Snake River Canyon through the Mid-canyon Bar*

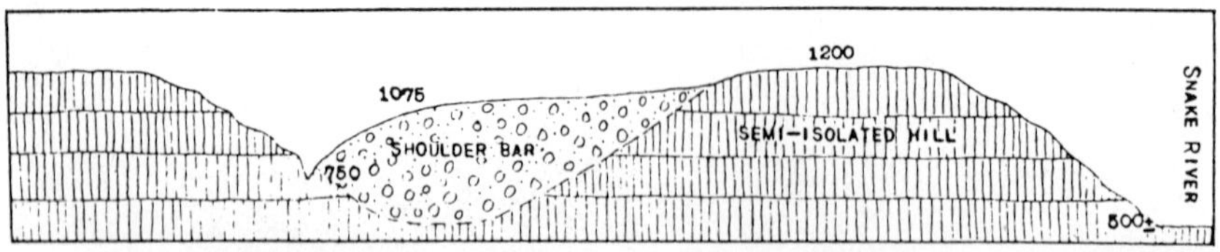

FIGURE 29. Cross-section semiparallel with Snake River Canyon through Shoulder Bar. View is near the mouth of Palouse River. (From Geological Society of America Bull., v. 39, p. 658)

ice-jam theory requires a very improbable dam 500 feet high and only 2 miles long in the canyon's width of about a mile, then neglects the evidence from the foreset beds that no ice masses were involved in their growth and that water was equally deep on both sides of the supposed dam. The fill-and-cut theory simply fills up Snake Canyon with gravel to the requisite depth. The Shoulder Bar thus becomes an erosional residual of a much larger deposit, all the rest of it removed by later erosion. The fill-and-cut theory neglects the point that, if it is correct, there would be other remnants similarly isolated on the south side of the canyon. There are none.

The flood theory views this extraordinary deposit as bottom gravel dragged across Snake Canyon, back into the mouth of the tributary valley and up for 500 feet *above* the canyon floor by the equally extraordinary volume of flood water it envisages.[3] The giant ripple marks can have no other explanation. The long foresets record a decrease in velocity as this rush of water was swung around westward by higher basalt hills and forced to flow back into the Snake from the south side. There is, of course, no thought by any theorist that the gravel could possibly have come off drainage slopes south of the river.[4]

[3]A 20% increase in volume can double the velocity of a stream.

[4]The above explanation may be too condensed and abbreviated to carry conviction to some readers. A detailed treatment of this and other scabland problems in this significant Palouse-Snake junction area was published in 1956. (Bretz and others.)

FIGURE 30. Gravel bar south of Snake River.

A. Pit section in northwest part of Shoulder Bar. False and true foreset bedding in lower left. One bevels the other. Note a fence (F) in each picture.

B. Pit section about halfway down bar front.

C. Pit section at toe of bar. (Photographs by H. T. U. Smith)

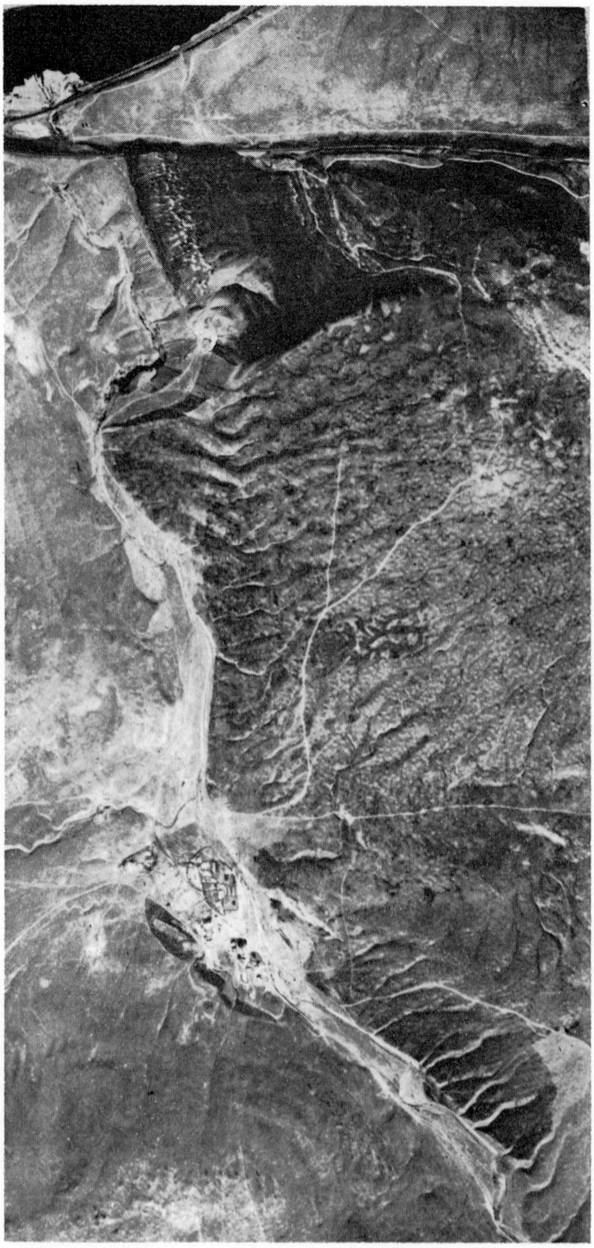

FIGURE 31. Shoulder Bar. Note parallelism of ripple marks with growing (northern) front of bar and their extension from summit down southern and western slopes. (Production and Marketing Administration photograph CCH-16-197)

PASCO BASIN

The largest downwarp on the plateau, more than 500 square miles in area, is Pasco Basin (fig. 32) at the southwest corner of the region. Here the Yakima, Snake, and Walla Walla Rivers join the Columbia. Although most of the basalt is buried under river gravel and Ringold sediments, bare basalt hills (Gable Butte and Gable Mountain) in the northern part of the Basin are island-like residuals of a basaltic upfold and stand a maximum of 700 feet above an adjacent channel in gravel.

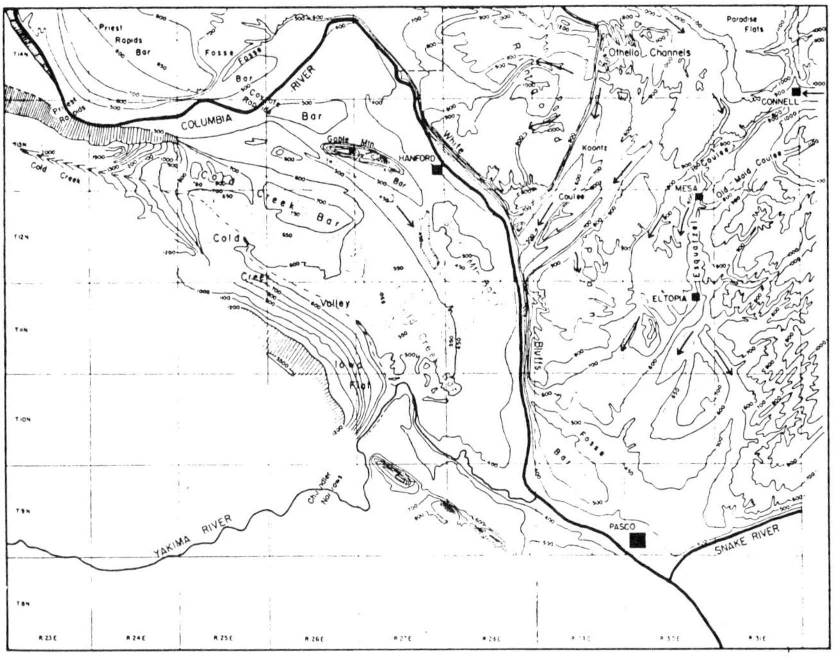

FIGURE 32. Central part of Pasco Basin. Contour interval, 100 feet; locally, 50 feet. (From Connell, Coyote Rapids, Eltopia, Hanford, Pasco, Prosser, Scooteney Lake, and Wallula topographic maps, U. S. Geological Survey)

Gravel bars are the Basin's specialty. The largest and longest bars of the entire scabland are here. One bar trails downstream from Gable Mountain for a dozen miles. Another (Cold Creek bar) is traceable for about 20 miles, although, like Gable Mountain bar, it is somewhat obscured by sand dunes. Priest Rapids bar is nearly equidimensional. It covers 20 square miles, and its summit is 400 feet above Columbia River level.

Like roads to Rome, all scabland rivers led to Pasco Basin. Had it ever been a closed depression during scabland history, it would have been filled with the huge quantities of basalt that were torn loose and broken up along the glacial stream channels. There is convincing evidence that this never occurred and therefore that the great bars are not erosional remnants as claimed in both the ice-jam and fill-and-cut theories. The best evidence for the constructional origin of these massive gravel deposits is afforded by Cold Creek bar (Bretz, 1928a, p. 678+; Allison, 1941, p. 73; Bretz and others, 1956, p. 1011+).

Figure 32 shows the relations of this broadly ridged gravel deposit to the Basin and three of its streams, the Columbia and Yakima Rivers and ephemeral Cold Creek. The higher and more obvious portion of the bar projects out into the downwarp for 13 miles, extending from the lee of the great cliffs overlooking Priest Rapids more than halfway across the broad Basin and separating the Columbia River valley portion of the Basin northeast of the bar from the 3- to 6-mile wide Cold Creek valley portion to the southwest. Yakima River, encountering the tip of the bar in the dune-infested southern part, makes an abrupt, acute-angled turn at The Horn to avoid it.

Cold Creek water rarely reaches the Yakima. It rarely flows along the length of its "valley" but is swallowed down into a growing alluvial fan of coarse debris contributed with every torrential rain from the high country immediately upstream. Gradient along the last 16 miles is 12½ feet per mile, on its fan it is 70 feet per mile, and farther upstream it increases to 150 feet per mile. Yet the rarely used, low-gradient valley is up to 6 miles wide, and the fan is from ½ mile to nearly 4 miles wide, and, farther up, the stream is in a canyon barely wide enough for a secondary road alongside it.

Clearly the lower "valley" is only an unfilled part of the Basin, partially shut off by the growth of the 100- to 200-foot broad gravel hill, the Cold Creek bar. Cold Creek never eroded this "valley" from a continuous fill up to the level of the bar top. The bar is not a terraced remnant.

Almost equally persuasive evidence is found in Priest Rapids bar. Its smooth riverward slopes are unterraced for the full 400 feet of its height, although there is no better place than along the inside of the Columbia's large symmetrical curve here for terraces to be left if the bar were really an erosional remnant. Bar structures should be found in this deposit as the Priest Rapids Dam is built.

The outlet to Pasco Basin is Wallula Gap, (figs. 33 and 34) (Bretz, 1925, p. 236+; Bretz, 1928c, p. 318+; Flint, 1938, p. 516; Allison, 1941, p. 72; Bretz and others, 1956, p. 1009), almost a squeezeway narrows where the Columbia cuts through the Horse Heaven Hills anticline. Scabland marks its cliffed walls for 800 vertical feet. All floodwater from the plateau had to escape through this constricted notch. Considering the great volumes recorded all over the scabland channel complex, the capacity of an empty Pasco Basin, the height of scab-

FIGURE 33. Wallula Gap and Horse Heaven uplift, looking southward. (Photograph by Bretz)

land on Gable Mountain and the narrowness of Wallula Gap, the flood theory suggests that the Gap was a bottleneck at any great flood's maximum crest. Using Chezy's formula with two wetted perimeters 15 miles apart and a surface gradient derived from their upper limits, it is computed (Bretz, 1925, p. 257) that 39 cubic miles of water escaped daily through this short gorge during a maximum flood. Is it any wonder that the flood theory aroused profound skepticism?

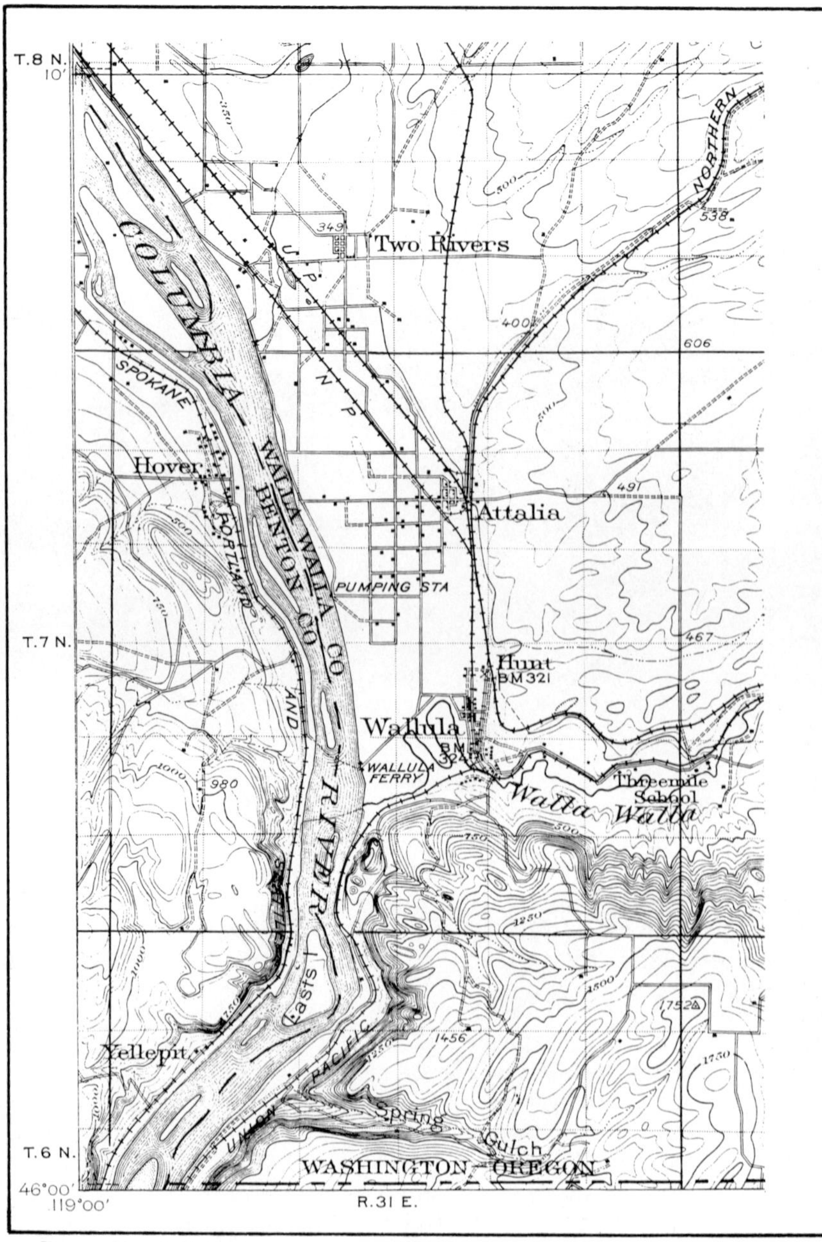

FIGURE 34. Wallula Gap. (Part of Wallula topographic quadrangle map, U. S. Geological Survey. Contour interval, 50 feet)

OTHER SIGNIFICANT FEATURES AND PLACES

(1) The records left by floods and flood-borne icebergs have been found down along the Columbia all the way to Portland and Vancouver (Bretz, 1925) and are a vital part of the epic story of the scablands. (2) Devils Canyon (Bretz, 1923b, p. 639; Bretz, 1928c, p. 206; Flint, 1938, p. 511+; Bretz and others, 1956, p. 1026) has great significance for any theory of scabland genesis. (3) Hanging mouths of the Koontz Coulee group of channels (Bretz and others, 1956, p. 995+) bear directly on this question. (4) Currents which flowed up the Snake River above Riparia, up the Yakima near Chandler, and up the Walla Walla near Reese (Bretz and others, 1956, p. 1014 footnote) are in the record. (5) Hundreds of massive icebergs that did not get caught in any jam were stranded back in Cold Creek valley (Bretz and others, 1956, p. 1012). (6) No terminal moraine now survives east of Grand Coulee. (7) A great bar at Beverly (Bretz and others, 1956, p. 993) blocks Lower Crab Creek and must be later than the last of Grand Coulee's discharges. (8) Scabland now considerably broken down by weathering occurs at high levels in channelways that functioned only during early floods (Bretz and others, 1956, p. 987+). (9) A comparison (contrast) between the non-flooded Yakima Valley's gravel terraces at Ellensburg and the flood gravel bars of scabland is illuminating (Pardee, 1910; Bretz and others, 1956, p. 979). Regretfully, these and many more items cannot be treated in this bulletin.

SOURCE OF THE WATER

Of the 4 theories for the origin of the Channeled Scabland, 3 obtain water for the channel-making from ordinary marginal and near-marginal glacial wastage. Given a forward movement in the ice sheet adequate to replace the waste, the meltwater rivers can continue flowing indefinitely. At no time, however, can they become the enormously swollen streams the flood theory asks for.

Realizing this and aware than many gravel deposits are huge river bars, the ice-jam theory endeavors to supply the necessary volume and depth by sporadic floods from abrupt failures of local dams. The flood theory asks for water enough delivered to the heads of all the main channels to initiate the entire system simultaneously. Suggestions that the glacial margin shifted sufficiently in proper places to make at least the Grand Coulee-Quincy Basin and the Cheney-Palouse routes sequential have been rejected as, of course, has been the idea that operation of smaller glacial streams through a long time could produce the great bars, rock basins, cataracts, and divergences.

What source for this unprecedented volume in a short time? The flood theory was without a plausible cause for some years, although it was constantly strengthened, in the opinion of its protagonists, by continued discoveries of more evidence for its verity. Then the addition of 2 and 2 to make 4 occurred, a simple addition that should have been made much earlier.

The source must have been a large body of water standing high enough and near enough, and with a dam of glacial ice which, failing, would catastrophically release many cubic miles of water with no way to escape except to the scabland channel heads.

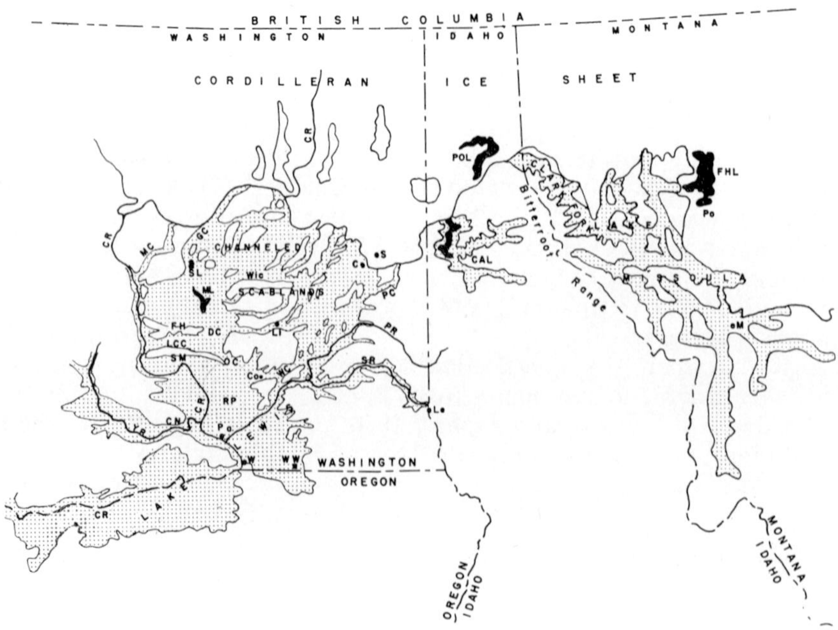

FIGURE 35. Map of Glacial Lake Missoula and Channeled Scablands. (From Geological Society of America Bull., v. 67, p. 1036)

C—Cheney
CAL—Coeur d'Alene Lake
CN—Chandler Narrows
Co—Connell
CR—Columbia River
DC—Drumheller Channels
FH—Frenchman Hills
FHL—Flathead Lake
GC—Grand Coulee
LCC—Lower Crab Creek
Le—Lewiston
M—Missoula
MC—Moses Coulee
ML—Moses Lake
OC—Othello Channels
Pa—Pasco
PC—Pine Creek
Po—Polson
POL—Pend Oreille Lake
PR—Palouse River
RP—Ringold Plain
S—Spokane
SL—Soap Lake
SM—Saddle Mountain
SR—Snake River
W—Wallula
WC—Washtucna Coulee
WiC—Wilson Creek
WW—Walla Walla
YR—Yakima River

That body of water was Glacial Lake Missoula in western Montana (fig. 35) (Pardee, 1910; Pardee, 1942; Bretz and others, 1956, p. 1034+). It was dammed by the front of the same ice sheet that rode up on the northern margin of the plateau. It had been described and mapped in a preliminary way before the Channeled Scabland problem was discovered. Altitude of its surface was

1,700 feet above the main channel heads. Its volume was estimated as 500 cubic miles. Its depth at the dam was 2,000 feet. A large deep valley led westward from the dam site to the basalt plateau's northern margin, 90 to 175 miles distant.

All the mountain valleys that held Lake Missoula led northwestward to the site of Lake Pend Oreille in the Idaho panhandle. Here the Bitterroot Range, separating Idaho and Montana in this latitude, had its northern terminus, an abrupt drop-off for thousands of feet into the large deep Pend Oreille-Spokane valley leading from the lake to the plateau. Here the rate of ice advance from the north became balanced by its rate of melting, so the glacial front pushed no farther south, and the ice made a dam that closed off these Montana valleys from their preglacial dischargeway. A lake promptly formed in front of this dam and rose as the dam thickened.

Lake Missoula's waves left a multitude of closely spaced but very faint shorelines on its enclosing mountain walls. They tell of repeated shiftings of the lake level. The highest is no more marked than any other, and no saddle in the crest of the Bitterroot Range was low enough to become an outlet. Opinion is that the only possible outlet was under or through or, most probably, over the glacial dam, whence it could flow southwestward toward Spokane.

If this Pend Oreille-Spokane valley beyond and below the site of the dam was not filled with glacial ice at the time the lake reached its highest level, and if the dam was holding back a head of possibly 2,000 feet of water, a tremendous hydrostatic pressure existed. This was a precarious situation if we assume that an almost empty valley lay immediately to the west.

Late in development of scabland theories came an illuminating discovery in one of the coalescent arms of Lake Missoula. A series of giant current ripple marks was identified on its floor (fig. 36) (Pardee, 1942). These could only indicate a tremendous current downvalley toward the site of the dam. Few geologists have doubted that this current was born of a very rapid lowering of the lake, the water escaping past a collapsed dam. These ripple marks are a record of probably the last flood to reach the plateau.

But there were several floods. The theory is elastic enough to take care of that. The margins of modern glaciers fluctuate in position with climatic change. Ancient glaciers and ice sheets have left records of marginal retreats and readvances. The history of the Pleistocene (or Glacial) epoch involves 3 very large retreats (perhaps some were complete extinctions), of the continental ice sheets, bringing on 3 long interglacial stages. The theory requires that the failure of the Lake Missoula dam occurred in part because a cycle of warmer years, centuries, or millenia had weakened it and had concomitantly filled the lake to its maximum volume. After the bursting of the dam and the cataclysmic release of hundreds of cubic miles of water, headed for the scablands, a readvance again

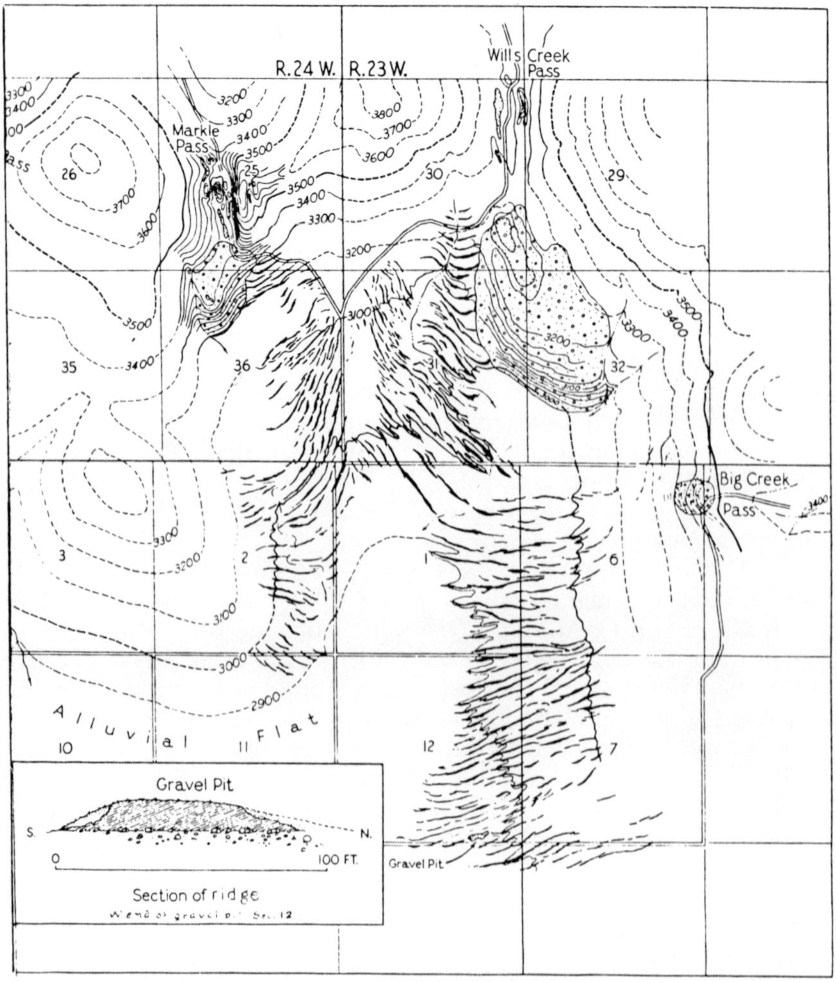

FIGURE 36. Giant ripple marks from Glacial Lake Missoula in northern part of Camas Prairie Basin. (From Geological Society of America Bull., v. 53, p. 1584)

closed the gap and initiated another Lake Missoula. As now believed, the plateau records at least 7 floods down the Pend Oreille-Spokane valley, 5 of them crossing the plateau but 2 of them finding no Okanogan lobe dam and therefore affecting only the Columbia Valley.

It was the successive return of flood water that deepened the major spillways and thus caused abandonment of many high-lying channels of earlier floodings. In this repeated experience, Grand Coulee held the winning hand. Its decent of the Coulee monocline gave it that highest of all scabland cataracts, Steamboat Falls. Retreat of this cataract needed only to reach the Columbia Valley north of the preglacial divide to provide by far the deepest channelway, nearly a thousand feet lower than the Cheney-Palouse and associated streamways.

A gravel dam in Upper Crab Creek valley at its junction with the eastern part of the Grand Coulee distributary coulee system has foresets dipping up the creek (Bretz and others, 1956, p. 982). This unmistakably records the fact that Grand Coulee outlived all its plateau contemporaries. The Beverly bar, similarly blocking the mouth of Lower Crab Creek with foreset gravel thrown back out of the Columbia Valley, tells of a post-Grand Coulee flood in the master valley alone, the last one of the series. No Okanogan lobe could have been blocking the Columbia when the last of the Lake Missoula dams broke.

CONCLUSION

Nowhere in the world is there known or suspected a story at all comparable to what we read from the scabland forms. Evidence of the magnitude of the floods has been, and may continue to be, questioned simply because the flood theory goes so far beyond any other geological explanations for past, unwitnessed events in earth history. Geologists, like most humans, tend to think in accepted patterns. The Colorado River's Grand Canyon took millenia to excavate. Why not use more time with less water and get comparable results in the Scabland?

Try that concept of moderately proportioned rivers working through a protracted time to make possible:

(1) Quincy Basin's four contemporaneous outlets;
(2) Drumheller's and Dry Falls' widths;
(3) Rock basins like Scooteney Lake;
(4) Bars like Cold Creek, Upper Crab Creek and Shoulder;
(5) Giant current ripple marks;
(6) Plucking and transportation of the boulders of Othello Channels;
(7) Back-filling as in Yakima, Walla Walla, and Snake Valleys;
(8) Divide-top scabland like that south of Hooper and Washtucna;
(9) Wallula Gap's 800 vertical feet of scabland;
(10) The anastomosing drainage pattern over an area of 15,000 square miles.

There is only one reasonable conclusion. Enormous volume alone could have been responsible. If the computed discharge rate through Wallula Gap is approximately correct, Lake Missoula would have been emptied in about 2 weeks.

Although known glacially dammed lakes have later been drained, under observation, by failure of their dams, no lake as large as Missoula has stood so high above and so near the country swept by the escaping water. No other glacial lake has been so deep against its dam. No such lake has had its dam so favorably situated for a catastrophic quarrying out by tremendous hydrostatic pressure. No second dam, like the Okanogan lobe, has diverted those waters from a major drainage lineament capable, like the Columbia around the Big Bend, of keeping floods of any conceivable magnitude within its valley walls. No area like the broad, loess-covered, basaltic Columbia Plateau has been so favorably situated for receiving and keeping the flood records.

We do not yet date the successive glacial floods, but the freshness of the Channeled Scablands' records indicates a geological recency. They await your inspection in almost their pristine forms.

CHRONOLOGICAL LIST OF REFERENCES

Pardee, J. T., 1910, The Glacial Lake Missoula, Montana: Jour. Geology, v. 18, p. 376-386.

Bretz, J H., 1923a, Glacial drainage on the Columbia Plateau: Geol. Soc. America Bull., v. 34, p. 573-608.

————, 1923b, The Channeled Scablands of the Columbia Plateau: Jour. Geology, v. 31, p. 617-649.

————, 1925, The Spokane Flood beyond the Channeled Scabland: Jour. Geology, v. 33, p. 96-115, 236-259.

————, 1928a, Bars of Channeled Scabland: Geol. Soc. America Bull., v. 39, p. 643-702.

————, 1928b, Channeled Scabland of eastern Washington: Geog. Rev., v. 18, p. 446-477.

————, 1928c, Alternative hypotheses for Channeled Scabland: Jour. Geology, v. 36, p. 193-223, 312-341.

————, 1932, The Grand Coulee: Am. Geog. Soc. Spec. Pub. 15, 89 p.

Allison, I. S., 1933, New version of the Spokane Flood: Geol. Soc. America Bull., v. 44, p. 675-722.

Flint, R. F., 1938, Origin of the Cheney-Palouse scabland tract: Geol. Soc. America Bull., v. 49, p. 461-524.

Allison, I. S., 1941, Flint's fill hypothesis for Channeled Scabland: Jour. Geology, v. 49, p. 54-73.

Pardee, J. T., 1942, Unusual currents in Glacial Lake Missoula, Montana: Geol. Soc. America Bull., v. 53, p. 1569-1600.

Hobbs, W. S., 1943, Discovery in eastern Washington of a new lobe of the Pleistocene continental glacier: Science, v. 98, p. 227-230.

————, 1947, The glacial history of the Scabland and Okanogan lobes, Cordilleran continental glacier: Ann Arbor, J. W. Edwards, 36 p.

Bretz, J H., Smith, H. T. U., and Neff, G. E., 1956, Channeled Scabland of Washington; new data and interpretations: Geol. Soc. America Bull., v. 67, p. 957-1047.

Part V

MODERN RESEARCH

Editor's Comments on Papers 12 Through 15

12 **BAKER**
Paleohydraulics and Hydrodynamics of Scabland Floods

13 **BAKER**
Large-Scale Erosional and Depositional Features of the Channeled Scabland

14 **PATTON and BAKER**
Origin of the Cheney-Palouse Scabland Tract

15 **BAKER**
A Preliminary Assessment of the Fluid Erosional Processes That Shaped the Martian Outflow Channels

In 1969 Bretz published a reflection on his fifty years of research on the catastrophic flooding of the Channeled Scabland and nearby regions. In that paper (Bretz, 1969) he presented a detailed list of the many problems that were left unanswered by his own work. The list can be divided into two categories: (1) stratigraphic problems, and (2) process problems. These comments will report on progress in answering Bretz's questions. The stratigraphic problems have been addressed by many researchers employing a variety of new stratigraphic tools unavailable to Bretz. The work on the process implications of the Bretz hypothesis began with the editor's studies (Baker 1971, 1973a, 1973b).

PALEOHYDRAULIC STUDIES

In 1973 Baker described evidence of the high-water mark left by the largest of the Lake Missoula outbursts. Using engineering hydraulic calculation procedures, he estimated the flood discharges and flow velocities for the flooding. This approach is updated in the report reproduced here as Paper 12. The key to the reconstruction of hydraulics is a "magic surface," the high-water surface of the flooding. That surface records the dynamic energy variations of the flooding. The philosophy of its analysis is similar to that employed by Gilbert (1890) in his analysis of Lake Bonneville. For Gilbert it was the lake level that served as the key element in paleohydrologic reconstruction.

To the early investigators of the Columbia Plateau, the scabland complexes were a chaos of eroded terrains. The report reproduced here as Paper 13 shows there is order in the development of the scabland features. An important discovery is that the phenomenal flow velocities of the Missoula floods are dynamically consistent with the scale and the variety of erosional landforms in the Channeled Scabland. After fifty years Bretz's hypothesis of catastrophic flood erosion is no longer considered "outrageous."

Paper 14 emphasizes this point in the Cheney-Palouse scabland tract, the region selected by Flint (1938) as an ideal place in which to disprove the flood hypothesis. Note that the references for Papers 12 and 13 are found at the end of Paper 14. This reference list also provides many other sources on the geology and regional setting of the Channeled Scabland.

THE SCABLANDS OF MARS

The final reprint in this volume (Paper 15) presents the case for the Channeled Scabland as an analog to one class of channels on Mars. Orbital photographs of Mars obtained during the Viking space mission reveal the planet to possess a remarkable variety of channeled terrains (Baker 1979). Some channels form relatively small networks up to 100 km in length that resemble dendritic valley systems of the earth. The most spectacular channels, however, comprise great anastomosing complexes up to 100 km wide that can be traced 2000 km or more across the planet's surface. These zones are called "outflow channels" because they appear full-born at localized source regions, usually collapse zones. Paper 15 suggests that immense floods of water were released from within the planet, emanating from troughs that receded headward to form the channels. The collapse zones at the heads of the channels mark the last points of fluid release.

Much of Mars is heavily cratered, similar in some respects to the Moon. Closer scrutiny reveals that the martian craters are very different from those on the Moon (Carr et al. 1977). Instead of being surrounded by rays of ballistically ejected debris, the martian craters appear to have been formed by meteors that impacted a soupy mud. They are bounded by lobes of debris that *flowed* from the impact sites. Abundant photographic evidence suggests that Mars was once underlain by a thick layer of permafrost, which probably contained a high percentage of ice (Carr and Schaber 1977). The great outflow channels probably developed from a variety of processes related to this permafrost. Local zones of volcanic heating may have built up pressure beneath the permafrost cap until the liquid water burst forth as great floods (Masursky et al. 1977). Another idea is that liquid water was trapped beneath the frozen surface layers until crustal warping caused pressure differences to develop (Carr 1978). When the pressure in a local zone reached a critical point, the water burst forth, draining the subsurface reservoir in much the same way that relatively small springs sometimes develop on the earth.

Whatever the mechanism of water release on to the martian surface, the outbursts produced an assemblage of erosional landforms that includes streamlined hills, anastomosing patterns, inner channels with recessional headcuts, bar-like forms, and grooves aligned parallel to the indicated fluid flow direction (Baker and Milton 1974; Baker and Kochel 1978, 1979). The scour features all occur below a definite "trim line," indicating that the responsible fluid had an upper flow boundary. The only terrestrial landscape that contains analogs to all the martian outflow channel landforms is the Channeled Scabland of eastern Washington.

The ages of the martian outflow channels are very difficult to estimate from photographic evidence alone. One method, looking at the number of impact craters that occur on the channeled surface, reasons that more craters per unit area will occur on older surfaces. Although this method provides relative ages, the absolute age in years requires an assumption concerning the rate of cratering. Current thinking relies on assumptions that suggest very old ages for the outflow channels, perhaps 1 to 3 billion years (Masursky et al. 1977; Neukum et al. 1979). If these ages prove correct, then the channels may have formed during an ancient episode of martian history, perhaps when the climate was warmer and the atmosphere denser than at present.

RECENT STRATIGRAPHIC RESEARCH

Bretz (1969) proposed, primarily on physiographic evidence, as many as eight separate scabland floods, of which at least four encountered an ice-blocked Columbia River and spilled over the northern margin of the Columbia Plateau. Bretz's four other floods were not diverted by ice and flowed down the Columbia River valley. Bretz (1969, p. 513) believed that the earliest plateau-crossing flood occurred during the Bull Lake glaciation, now considered correlative to the Illinoisan glaciation of the midwest United States (Pierce et al. 1976).

Baker (1973) noted that some of the physiographic relationships described by Bretz (1969) could be produced during the dynamic progression of a single flood. Only flood deposits recognized in a firm stratigraphic sequence can be considered unequivocal evidence for multiple flooding. The best stratigraphic information to date suggests that there certainly were multiple floods in the general vicinity of the Channeled Scabland. One, possibly two, floods were pre-Bull Lake and are overlain either by thick caliche or by Palouse Formation. The Palouse Formation is a sequence of loess units that mantles divide areas in the Channeled Scabland and was scoured by the Wisconsinan floods of Lake Missoula (Richmond et al. 1965). Patton and Baker (1978) describe the regional evidence for pre-Wisconsin (pre-Palouse) flooding.

Malde (1968) studied the catastrophic flood produced by the overflow and rapid lowering of Pleistocene Lake Bonneville. He traced the course of this flood through the Snake River Plain of southern Idaho to Hells Canyon. Malde interpreted the date of this event to be about 30,000 years B.P., based on a radiocarbon date for molluscan fossils associated with flood debris and on the relict soil profile developed on the flood gravel (Melon Gravel). The soil has a thick calcic horizon extending to depths greater than 2 m. The soil on the Bonneville Flood deposits is believed to have formed during and since the mid-Wisconsin (Bull Lake-Pinedale) interglaciation, or within the last 30,000 years.

Downstream from Hells Canyon at Lewiston, Idaho, probable Bonneville flood deposits are overlain by slackwater surge deposits from the last major episode of scabland flooding. Because Bonneville flooding was confined to the Snake River Canyon, it skirted to the south of the Channeled Scabland. Nevertheless, studies in the Pasco Basin may eventually recognize Bonneville Flood deposits in association with Missoula flood deposits.

The last major episode of scabland flooding, involving at least several individual floods, occurred between 18,000 and 12,000 years B.P. In the eastern Channeled Scabland the last major flood scoured channelways through the entire sequence of brown loess units that comprise the Palouse Formation. Small channels eroded into the flood deposits by postflood streams contain pale loess and loess-derived alluvium. The oldest soil profiles on postflood loess deposits show textural B horizons, with the $<2\,\mu$ size fraction increasing from 8 percent to 11 percent (Baker 1973a).

The dating of the last major scabland flooding episode is complicated by the fact that the catastrophic flood deposits occur in two facies. The main-channel facies can always be recognized as unequivocal flood deposition by its coarseness, sedimentary structures, angular boulders, broken rounds, erratics, and so forth. However, the slackwater facies consists of rhythmically bedded sands and silts deposited in the preflood tributaries to the scabland channels (Bretz 1929, 1930). An unequivocal flood origin can be established by continuous tracing between main-channel areas and slackwater areas, as in the Tucannon Valley sequence (Baker 1973a). At the Tucannon Valley one can follow a complete transition from chaotically deposited boulder and cobble gravel in the main channel (proximal facies) to rhythmites of sand and silt in slackwater areas (distal facies) 15 km up a preflood tributary of the main channel (Baker 1973a, pp. 42–47).

The slackwater facies of the Missoula flood deposits is the current subject of intense study concerning the number and timing of late-glacial floods in the Columbia River system. The upper portions of some slackwater sequences contain tephra layers that correlate to the "set S" pumice deposits of Mount St. Helens (Moody 1978; Mullineaux et al. 1978). The correlation points to a limiting upper date of about 13,000 years B.P. for the last major episode of catastrophic flooding to have affected the Columbia Basin in eastern Washington (Mullineaux et al. 1978). The regional distribution of this tephra makes it an important time-stratigraphic marker in the slackwater sequence.

Richard B. Waitt, Jr. (1979) has studied the remarkably thick late-glacial rhythmite sequences of the Walla Walla basin and the lower Yakima Valley, where he has found evidence for subaerial exposure following floodwater emplacement of some of the rhythmites. At the remarkable Burlingame Canyon section, Waitt counts at least thirty-eight well-formed distal facies rhythmites. One possible hypothesis for this new data is that each rhythmite represents a separate catastrophic flood event (Waitt 1979). Unfortunately, the proximal facies sediments for each of these inferred floods have not yeat been discovered. Thus, the possibility remains that single great floods can produce multiple *distal* rhythmites, so that the Burlingame section may represent a somewhat smaller number of late Wisconsin flood events than given by the rhythmite count (Bjornstad 1980; Patton et al. 1979).

REFERENCES

Baker, V. R. 1971. Paleohydrology of catastrophic Pleistocene flooding in eastern Washington. *Geol. Soc. America Abs. with Programs* **3**: 497.

Baker, V. R. 1973a. Paleohydrology and sedimentology of Lake Missoula flooding in eastern Washington. *Geol. Soc. America Spec. Paper 144*, 79 pp.

Baker, V. R. 1973b. Erosional forms and processes for the catastrophic Pleistocene Missoula floods in eastern Washington. In *Fluvial Geomorphology*, Morisawa, M., ed. Binghamton: N.Y. State University, pp. 123–148.

Baker, V. R. 1979. Morphology of channels on Mars. *NASA Conf. Pub.* **2072**:4–6.

Baker, V. R., and Kochel, R. C. 1978. Morphological mapping of martian outflow channels. *Lunar Planet Sci. Conf. 9th, Proc.*, pp. 3181–3192.

Baker, V. R., and Kochel, R. C. 1979. Martian channel morphology: Maja and Kasei Valles. *Jour. Geophys. Research* **84**:7961–7983.

Baker, V. R., and Milton, P. J. 1974. Erosion by catastrophic floods on Mars and Earth. *Icarus* **23**:27–41.

Bjornstad, B. N. 1980. Sedimentology and depositional environment of the Touchet Beds, Walla Walla River Basin, Washington. *Rockwell Hanford Operations Report RHO-BWI-SA-44*, 83pp.

Bretz, J H. 1929. Valley deposits immediately east of the Channeled Scabland of Washington. *Jour. Geology* **37**:393–437, 505–541.

Bretz, J H. 1930. Valley deposits immediately west of the Channeled Scabland. *Jour. Geology* **38**:385–422.

Bretz, J H. 1969. The Lake Missoula floods and the Channeled Scabland. *Jour. Geology* **77**:505–543.

Carr, M. H. 1978. Formation of martian flood features by release of water from confined aquifers. *NASA Tech. Memo* **79729**:260–262.

Carr, M. H., Crumpler, L. S., Cutts, J. A., Greeley, R., Guest, J. E., and Masursky, H. 1977. Martian impact craters and emplacement of ejecta by surface flow. *Jour. Geophys. Research* **82**:4055–4065.

Carr, M. H., and Schaber, G. G. 1977. Martian permafrost features. *Jour. Geophys. Research* **82**:4039–4054.

Flint, R. F. 1938. Origin of the Cheney-Palouse scabland tract. *Geol. Soc. America Bull.* **49**:461–524.

Gilbert, G. K. 1890. Lake Bonneville. *U. S. Geol. Survey Mon. 1*, 438 pp.

Malde, H. E. 1968. The catastrophic late Pleistocene Bonneville flood in the Snake River Plain, Idaho. *U. S. Geol. Survey Prof. Paper 596*, 52 pp.

Masursky, H., Boyce, J. M., Dial, A. L., Schaber, G. G., and Strobell, M. E. 1977. Classification and time of formation of martian channels based on Viking data. *Jour. Geophys. Research* **82**:4016–4032.

Moody, U. L. 1978. Microstratigraphy, paleoecology, and tephronchronology of the Lind Coulee site, central Washington. Ph.D. dissertation, Washington State University, 273 pp.

Mullineaux, D. R., Wilcox, R. E., Ebaugh, W. F., Fryxell, R., and Rubin, M. 1978. Age of the last major scabland flood of the Columbia Plateau in eastern Washington. *Quaternary Research* **10**:171–180.

Neukum, G., Hiller, K., Henkel, J., and Bodechtel, J. 1979. Surface ages of martian shield volcanoes and channels. *Lunar and Planetary Sci.* **10**:907–909.

Patton, P. C., and Baker, V. R. 1978. New evidence for pre-Wisconsin flooding in the Channeled Scabland of eastern Washington. *Geology* **6**:567–571.

Patton, P. C., Baker, V. R., and Kochel, R. C. 1979. Slack-water deposits: a geomorphic technique for the interpretation of fluvial paleohydrology. In *Adjustments of the Fluvial System*, Rhodes, D. D., and Williams, G., eds. Dubuque, Iowa: Kendall-Hunt Publ. Co., pp. 225–253.

Pierce, K. L., Obradovich, J. D., and Friedman, I. 1976. Obsidian hydration dating and correlation of Bull Lake and Pinedale glaciations near West Yellowstone, Montana. *Geol. Soc. America Bull.* **87**:703–710.

Richmond, G. M., Fryxell, R., Neff, G. E., and Weis, P. L. 1965. The Cordilleran ice sheet of the northern Rocky Mountains, and related Quaternary history of the Columbia Plateau. In *The Quaternary of the United States*, Wright, H. E., and Frey, D. G., eds. Princeton, N.J.: Princeton University Press, pp. 231–242.

Waitt, R. B., Jr. 1979. Forty late-Wisconsin catastrophic Lake Missoula backfloodings of Walla Walla and lower Yakima valleys, southern Washington. *Geol. Soc. America Abs. with Programs* **11**:133.

12

Reprinted from pages 59–79 of *The Channeled Scabland,* V. R. Baker and D. Nummedal, eds., NASA Office of Space Science, Planetary Geology Program, 1978, 186 pp.

Paleohydraulics and Hydrodynamics of Scabland Floods

VICTOR R. BAKER

ABSTRACT

The last major episode of scabland flooding (approx. 18,000-13,000 years B.P.) left considerable high-water mark evidence in the form of (1) eroded channel margins, (2) depositional features, (3) ice-rafted erratics, and (4) divide crossings. These can be used to reconstruct the maximum flood stages and water-surface gradients. Engineering hydraulic calculation procedures allow the estimation of flood discharges and mean velocities from these data. Despite a number of limitations on the accuracy of this reconstruction as discussed herein, the estimated paleohydraulic conditions are, nevertheless, consistent with a wide range of erosional and depositional phenomena in the Channeled Scabland.

Lithologic and structural irregularities in the Yakima Basalt were significant in localizing the plucking-type erosion of high velocity flood water. The anticlinal structures of the western Columbia Plateau resulted in numerous constricted channels. The phenomenal flow velocities achieved in these constrictions produced the most spectacular scabland features, including rock basins, potholes, and abandoned cataracts. In contrast, the eastern scabland region was characterized by more uniform flow conditions associated with more subdued erosional topography.

Secondary flow phenomena, including various forms of vortices and flow separations, are considered to have been the principal erosive processes. The intense pressure and velocity gradients of vortices along the irregular channel boundaries produced the plucking-type erosion. The great depth of the flood flows probably considerably reduced the effectiveness of cavitation as an erosional agent.

INTRODUCTION

Geomorphic features result from forces acting on resistant materials at the interface between the lithosphere and the atmosphere or hydrosphere. Until the last decade the dynamics of the forces involved in making the Channeled Scabland were generally ignored in the controversy that surrounded the origin of those forces. Baker (1973a) was the first to use quantitative procedures in relating the pattern of scour and deposition to the regimen of scabland floods. Because the Missoula floods involved the largest discharges of fresh water that can be documented in the geologic record, continued study of scabland processes establish some upper limits to our knowledge of the short-term erosive and transport capabilities of running water.

Bretz recognized that eventually the Channeled Scabland problem needed to be investigated in physical terms. In questioning his own hypothesis (Bretz, 1932a, p. 82) he states, "somewhere must lurk an unrecognized weakness. Where is it? If it exists, it probably lies in the hydraulics of the concept." Bretz asserted that the turbulence of the glacial flood and the jointing of the basalt were both important, an idea gained from his earlier studies of the Columbia River (Bretz, 1924). He added (Bretz, 1932a, p. 83), "we do not know enough about great flood mechanics to make any conclusions valid . . . Hydraulic competency must be allowed the glacial streams, however much it may differ from that of stream floods under observation."

The present paper will review some of the hydraulic and hydrodynamic principles that apply to the Channeled Scabland. Moreover, these data will be related in the next chapter to the dis-

[*Editor's Note:* References cited in this article are included in the reference list that follows Paper 14.]

tinctive erosional and depositional features of the region. The results will support an earlier conclusion (Baker and Milton, 1974) that the distinctive bed forms in scabland channels are a hydrodynamic consequence of exceedingly swift, deep flood water acting on closely jointed bedrock.

RECONSTRUCTING AN ANCIENT FLOOD

Several fortuitous circumstances have combined to allow an approximate reconstruction of maximum flood flows through the Channeled Scabland: (1) the last flooding episode was the last major event of the Pleistocene, occurring perhaps as recently as 13,000 years B.P. (but no older than about 18,000 years B.P.), (2) postflood drainage on the Columbia Plateau was isolated by the paths of major rivers around the plateau, (3) the postglacial dry climate produced only intermittent streamflow, (4) flood deposits and erosive effects contrast sharply with the loess sediments and processes immediately adjacent to scabland channels, (5) exotic lithologies transported by the flooding are easily recognized, and (6) earlier flood events were followed by the massive loess deposition over the plateau.

High-Water Marks

For the last major Missoula flood to cross the Channeled Scabland a variety of features have been preserved which serve as high-water marks. These features may be studied by geomorphic field work and by the interpretation of topographic maps and aerial photographs. The most abundant type of highwater mark is a divide crossing. Although these are best studied on vertical aerial photographs, they may also be recognized on detailed topographic maps (7.5-minute quadrangles).

Eroded Channel Margins. Eroded scarps on loess or the highest eroded rock surfaces (scabland) certainly mark a water level (Fig. 4.1). However, the various erosional features along

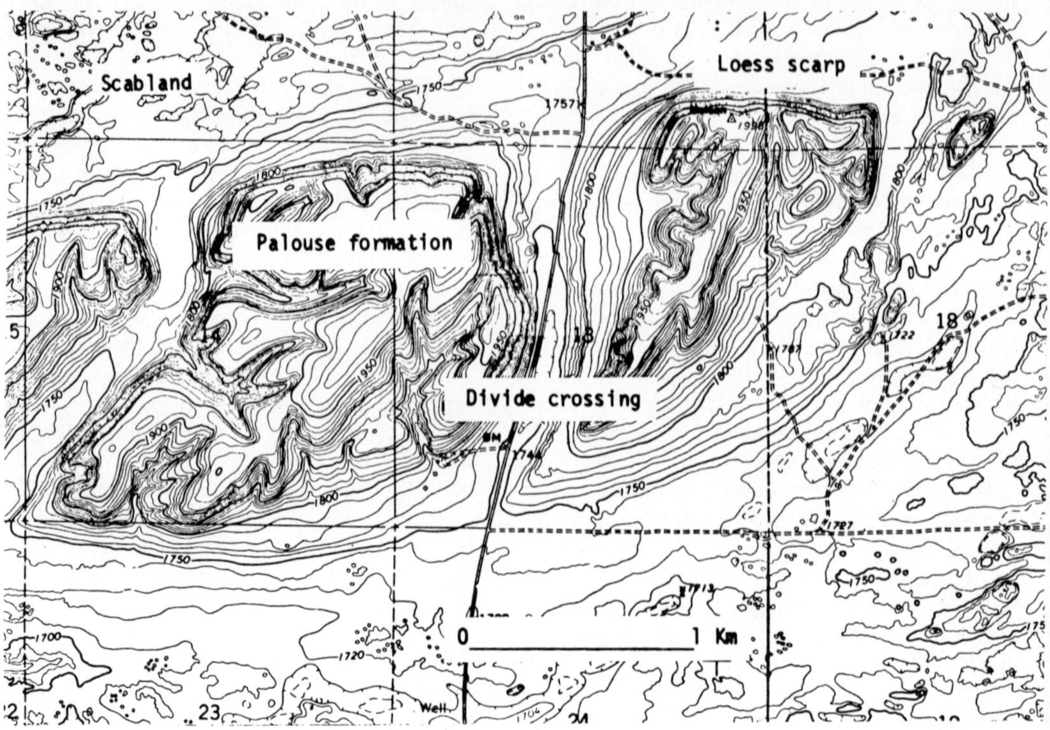

Figure 4.1. Loess scarps, scabland, and a major divide crossing in the Cheney-Palouse scabland tract. Topography is from the U.S. Geological Survey Texas Lake Quadrangle (7.5 minute, 10 feet contour interval).

scabland channel margins only provide minimum figures for high-water surface elevations.

Depositional Features. Bars of flood gravel also provide excellent evidence for flood flows, but, as with the large-scale erosive features, the flood gravel may have been under some unknown depth of water. These features must also be interpreted as minimum flow depth indicators. A more accurate procedure is to trace flood deposits laterally from scabland channels up the valleys of pre-flood tributaries. Such deposits rapidly fine away from the scabland channels and constitute "slackwater facies." Because tributaries to the Cheney-Palouse scabland drain the loess terrain of the Palouse hills, the slackwater basaltic sand can be easily recognized in these tributaries. The highest elevation of slackwater facies in the tributary valleys then provides a high-water indicator.

Ice-rafted Erratics. Exotic boulders of crystalline rocks from the Okanogan highlands or the Belt Series east of Spokane, are commonly found in mounds on the basalt or loess of the Columbia Plateau. The highest elevation of erratics that were transported by the floating ice in the Missoula Flood water is locally a useful high-water mark. The erratics are usually concentrated in areas where Missoula Flood waters were locally ponded.

Divide Crossings. Divide crossings, where flood water filled valleys and spilled over the loess-mantled interfluves, are the most common form of high-water mark evidence in the Channeled Scabland. In contrast to the dendritic drainage pattern on the loess divides, the flood-modified channels are linear or nearly linear (Fig. 4.2) and often expose bare basalt outcrops on their floors.

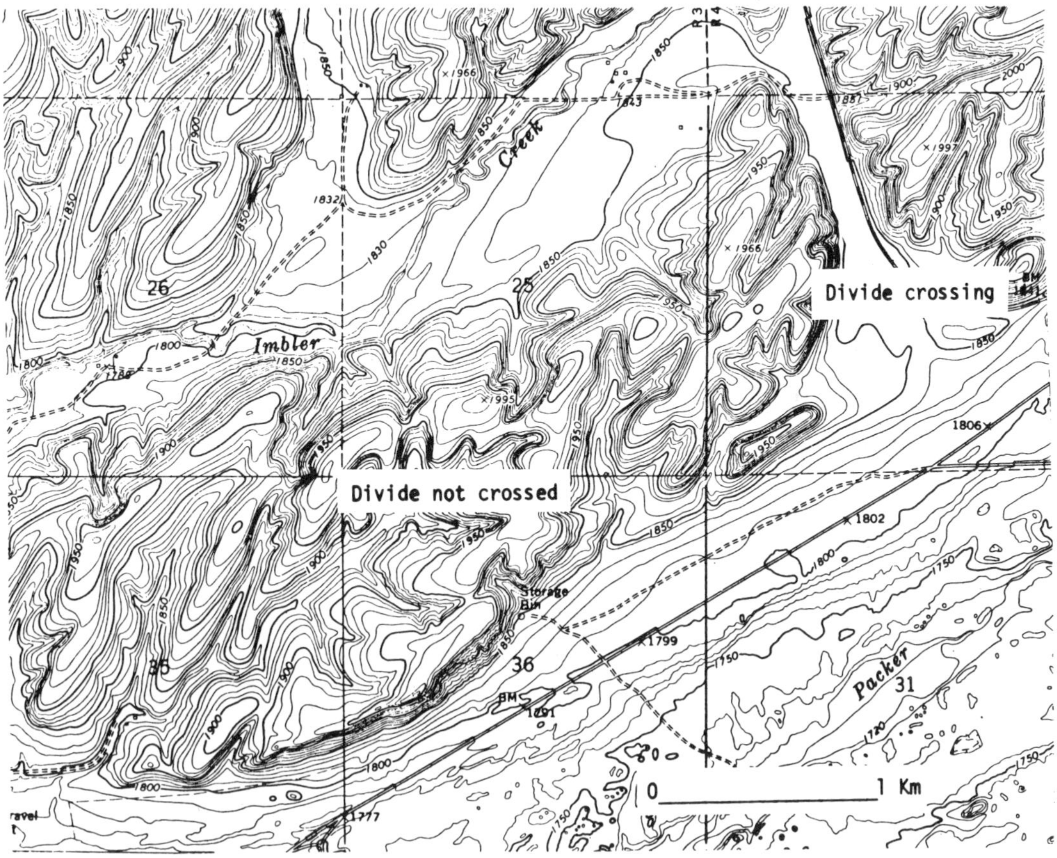

Figure 4.2. Examples of crossed and uncrossed loess divides in the Cheney-Palouse scabland tract. The divide crossing depicted is a definite example with a linear, flat-floored trough. Topography is from the U.S.G.S. Texas Lake Quadrangle.

Such divide crossings always establish lower limits for the high-water surface elevation and, in some cases, may fix upper limits as well. Obviously there is an expected range of error on these elevations. Elevation estimates are most accurate when divides were crossed by shallow water. These marginally-crossed divides are characterized by poorly-developed troughs. In areas where divides were crossed by deep flows, the approximation for the water-surface elevation is less precise. Thus, the most useful divide crossings are also the most difficult to recognize. Further refinement of the high-water surface elevation is possible through the study of aerial photographs and field observations of highest flood gravels, erosion of interfluves, and highest erratics.

The magnitude of the range of elevations obtained for the high-water surface depends on the contour interval of the topographic map and upon the nature of the crossing. Estimates are most accurate when uncrossed as well as crossed divides are present in the local study area (Fig. 4.3). A substantial portion of the topographic map coverage of the Channeled Scabland is at a 3 m (10 ft) contour interval. Thus the error range is rather small for maximum water depths of 100 m.

Uncrossed divides are generally relatively narrow and sharply defined. These divides are not dissected by flat-floored troughs and usually display no topographic irregularities with respect to surrounding terrain. The elevation of the lowest uncrossed divide may establish a maximum upper limit for the water surface. Divides that were apparently crossed are characterized by well-developed, flat-floored troughs through their crests. The incisions are commonly steep-walled and variable in length, but are usually elongated. Divides that were definitely crossed by deep flows commonly exhibit wide troughs and steep walls. These divides are found at lower elevations than those that are marginally crossed. They were probably crossed by at least 10 meters of water. Marginally crossed divides contain less well-developed troughs and are found at elevations higher than those crossed by deeper flows. These divides were in some places crossed by only a few meters of water (Fig. 4.4).

Water-surface Profiles

The plotting of the flood high-water surface begins with locating the obvious evidence, such as highest eroded scabland and major divide crossings. These data provide a rough approximation to the lower elevation limit. Refinement of the high-water surface elevation may follow through the location of the lowest divide not crossed (an upper limit) and the highest divide of marginal nature that was crossed (lower limit). It is important to select those divides just barely

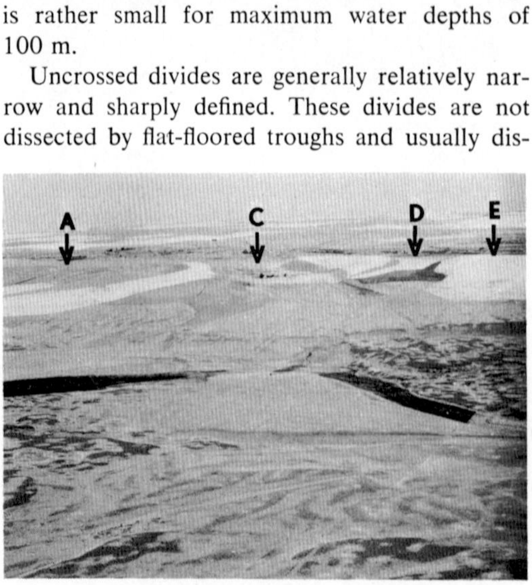

Figure 4.3. Oblique photograph of minor divide crossings (arrows) cut through a loess divide between Crab Creek and South Fork (Sections 26, 27, and 28, T.21N, R.36E.). The floors of the divide channels A, C, and D are at 600, 588, 594 meters (1970, 1930 and 1950 feet) elevation respectively. The uncrossed divide at E is at 606 meters (1990 feet).

Figure 4.4. Minor divide crossings located about 6.5 km north of Wilson Creek. The floors of the flood-eroded channels A and B are at 521 m (1710 ft) elevation. The nearby uncrossed divide at C is at 527 m (1730 ft) elevation. This type of relationship is of maximum utility in establishing the flood high-water surface.

crossed when establishing lower limits. There remains a nebulous zone between these two limits where it is uncertain whether divides have been crossed. The midpoint of this range can be taken to represent the high-water surface elevation for a local reach of the profile.

Figure 4.5 shows the high-water surface plot for the western part of the Channeled Scabland. The water surface slopes show a marked steepening at points where the flood waters encountered constricted channel sections. Water was ponded upstream from these constrictions and flowed at high velocity through the constrictions. Such constrictions were posed by the pre-flood drainage lines through the anticlinal structures of the western Columbia Plateau (High Hill, Frenchman Hills, and Saddle Mountains in Fig. 4.6). The steepening of the water surface gradient at the Saddle Mountains constriction (Othello Channels) is less pronounced than at the others. This is probably because the next constriction downstream, Wallula Gap, ponded water at 350 m (1150 ft) elevation.

Flood Flow Calculations

The high-water surface slope can be used as an approximation of the energy slope in the slope-area method of indirect discharge calculation (see Benson and Dalrymple, 1968, and Baker, 1973a). Paleoflow depths are obtained by measuring the difference in elevation between the channel bottom and the high-water surface. Channel cross sections can be derived from the large-scale topographic maps. Baker (1973a) followed standard hydraulic engineering procedures

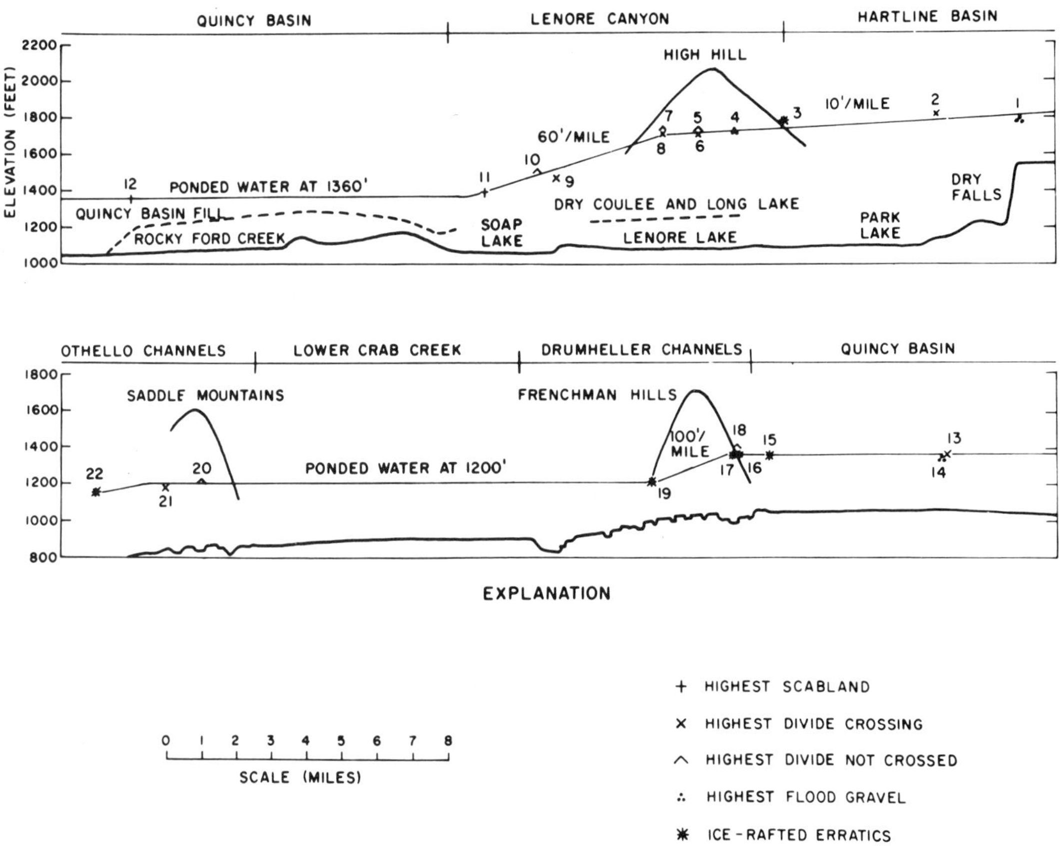

Figure 4.5. High-water surface profile of the western Channeled Scabland. The numbered high-water marks are individually described by Baker (1973a).

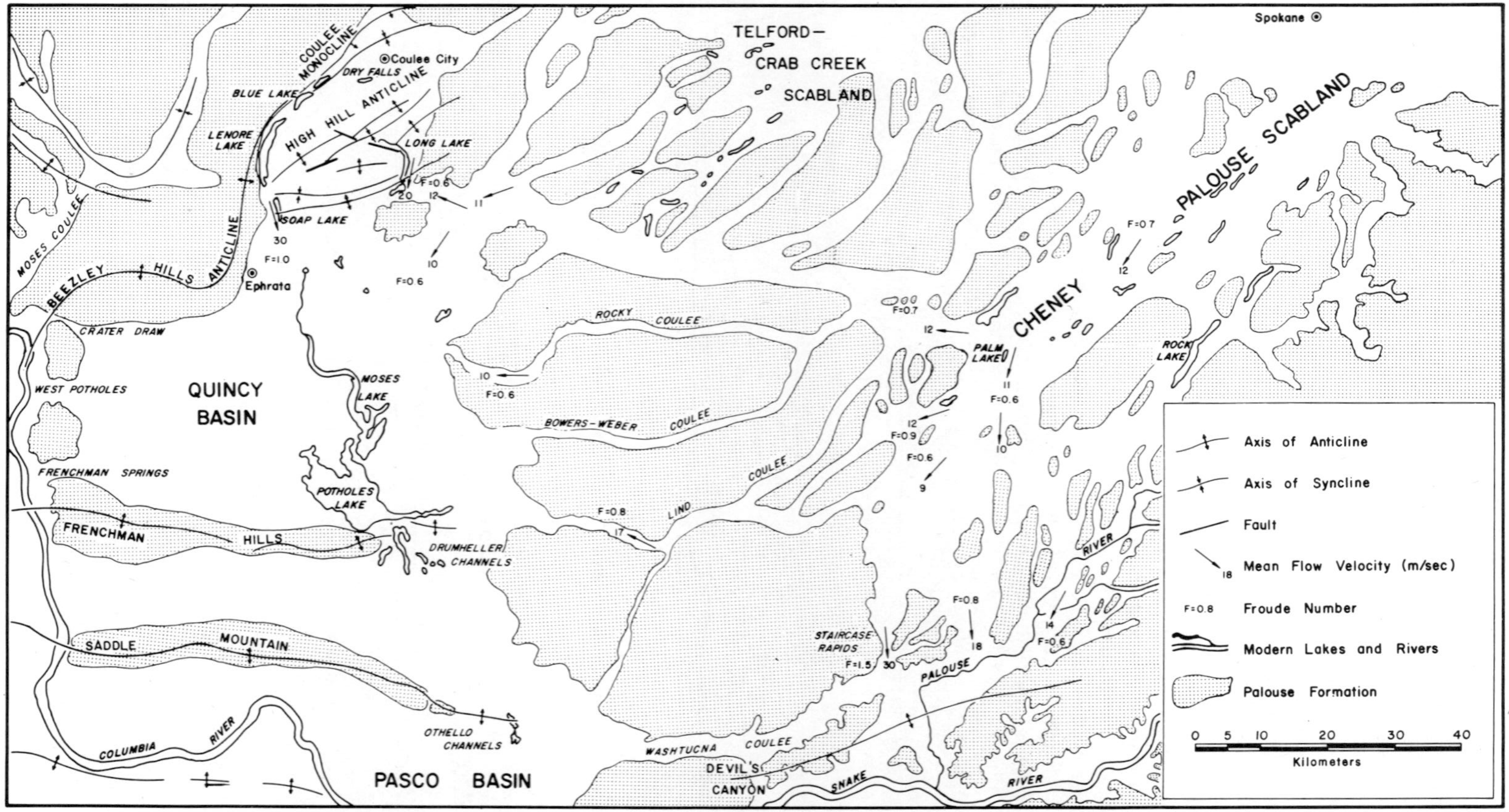

Figure 4.6. Regional paleohydraulic features of the Channeled Scabland. Mean flow velocities were determined from high-water mark evidence and channel geometry by Baker (1973a). Also shown is the regional structural pattern on the Columbia Plateau. The Palouse Formation is not shown on High Hill and Pinto Ridge (between Soap Lake and Long Lake) in order to emphasize the structural detail in that area.

to combine these data into estimates of probable flood discharges and mean flow velocities at the maximum flood stage. Discharges as great as 21.3 x 10^6 m^3/sec were conveyed through the Channeled Scabland (Baker, 1971).

The distribution of mean flow velocities in the Channeled Scabland (Fig. 4.6) is important in understanding flood erosion processes. Constricted channels in the western scablands, such as Lenore Canyon, lower Grand Coulee, Othello and Drumheller Channels, achieved flood flow velocities as high as 30 m/sec. Such high velocities were reached only because of the unique combination of great flow depth (60-120 m) and very steep water surface gradients (2-12 m/km) that characterized the Missoula Flood. The constrictions contain the best developed erosional topography. Adjacent basins such as the Hartline, Quincy, and Pasco Basins, produced ponded water and the accumulation of sediment eroded from the constrictions (see Baker, 1973a, p. 15).

Duration

Baker (1973a) made some preliminary limiting calculations on the draining of Lake Missoula and the routing of flood flows through the Channeled Scabland. The peak discharge of flood flows in the Rathdrum Prairie, close to Lake Missoula's breakout point, was 2.13 x 10^6 m^3/sec. Even at this phenomenal discharge, the lake volume of 2.0 x 10^{12} m^3 would have maintained the flow for about a day. Of course, the declining head in the lake would gradually reduce the discharge in time, spreading the water release over a broader time scale. It seems more likely that the flood peaks recorded by the high-water surface persisted no more than a few hours. Lower stages could have been maintained for a week or more.

The exact progress of the flood wave from the Rathdrum Prairie through the complex pathways of the Columbia River valley and the Channeled Scabland is not fully resolved. The initial flows filled the great valley of the Columbia upstream from the Okanogan Ice Lobe that blocked the river in the vicinity of modern Grand Coulee Dam. The water ponded to an elevation of about 770 m (2525 ft) and then spilled over the northern rim of the Columbia Plateau at the heads of the Cheney-Palouse, Telford-Crab Creek, and Grand Coulee scabland tracts. The subsequent flood events are less certain. The great cataract in the upper Grand Coulee, 250 m high, retreated northward at some time after this initial phase to the valley of the Columbia (Bretz, 1932a). The breakthrough of this cataract would then induce a great surge of water, 250 m deep, down the Grand Coulee and through the western scablands. Flood bars blocking the mouths of west-flowing scabland channels (Rocky, Bowers, Lind, and Washtucna Coulees, as well as Crab Creek) show that the Grand Coulee surge postdated flooding in the Telford-Crab Creek and Cheney-Palouse scabland tracts. The rapid drawdown of water by the upper Grand Coulee could have produced an abrupt cessation of flow in the eastern Channeled Scabland.

The stability of the Okanogan lobe during the flood event is also a problem. Waitt (1972a) has suggested that the last major scabland flood also destroyed the Okanogan lobe and released a great flood down the Columbia Valley. This flood produced a variety of catastrophic flood evidence that he has documented for the Wenatchee area, just northwest of the Channeled Scabland (Waitt, 1977b).

The ultimate control on outflow from the entire region of eastern Washington flooded by Lake Missoula was Wallula Gap, an antecedent canyon eroded by the Columbia River through the Horse Heaven Hills anticline (Bretz, 1925). Wallula Gap was inadequate to carry all the flood water supplied from upstream scabland channels. This constriction hydraulically dammed the flood, causing ponded water to reach 350 m elevation behind the "dam." Baker (1973a) calculated the maximum discharge achieved by this control point as 9.1 x 10^6 m^3/sec. Thus, the total volume of Missoula Flood water entering from upstream would have required 2½ days to pass this point, even at the maximum discharge. More likely the high flows persisted for at least a week.

The Question of Precision

The discharge calculations of Baker (1973a) can only be considered a first approximation to flood hydraulics in the Channeled Scabland. The calculations were based on hydraulic formulae which were derived for use in streams whose discharges are 3 or 4 orders of magnitude less than those of the Missoula flooding through the Channeled Scabland. The data input into these formulae was not based on a dynamic record of the flooding, but rather on the time transgressive high-water surface. Time variant aspects of the flood surges cannot be quantitatively deduced from the existing field evidence. The complex geometry of Lake Missoula itself probably exerted an unknown dynamic influence on its draining. As discussed above, the flows were routed through an extremely complex set of anastomosing channel ways.

The exact significance of the high-water surface itself is open to some question. If the flooding burst on to the Columbia Plateau at near its peak discharge, then one might expect the high loess divides to have been eroded almost immediately. A subsequent long recession of the flood hydrograph might then have greatly deepened the channels, enlarging their capacity beyond what is implied by the high-water surface. Two lines of evidence argue against this view. First, the hydrographs of jökullhlaups (Icelandic: "glacier bursts") are precisely opposite to the above hypothesis. Flood flows rise slowly to a peak and then drop abruptly (Meier, 1964). The second line of evidence involves the field relationships. The extraordinary preservation of high-stage bed forms throughout the Channeled Scabland implies that the jökullhlaup hydrograph applies. Waning flood stages lasted so short a time that bed forms such as giant current ripples were not scoured away.

A more difficult problem arises in the estimation of roughness for the scabland channels. Baker (1973a, p. 19) followed Malde (1968) in assuming a value of Manning's "n" despite the many difficulties in scaling from the empirical Manning equation (derived for "normal" rivers) to the immense flow depths of the scabland flood. The value chosen from Chow's (1959) empirical tables was $n = 0.040$ for rock-bounded channels.

Komar (1978) has recently presented a lucid discussion of the problems of roughness estimation in very deep fluid flows, such as the Missoula floods and turbidity currents. Rather than employing the empirical Manning's "n" factor, he suggests use of a dimensionless drag coefficient C_f that is related to other common resistance measures for flow in alluvial channels. By calculating the drag from the estimated flow depth and the particle diameter larger than 84% of the bed particles, one finds $C_f = 0.0026$ for the Missoula Flood. Baker (1973a) assumed a Manning's $n = 0.040$. Using Komar's equations this is equivalent to a $C_f = 0.0034$, in fair agreement with the estimate derived from the relative roughness measure. This correspondence lends added credence to the paleohydraulic reconstruction.

Despite the various limitations on the paleohydraulic reconstruction of flooding in the Channeled Scabland, a variety of sediment transport pheonmena have proven to be generally consistent with the quantitative reconstruction provided by Baker's (1973a) preliminary analysis. These include the bottom shear stresses for particle transport, the boulders carried by the flow, and especially the giant current ripples. These latter forms are almost certainly confined to the upper part of the lower flow regime of Simons and Richardson (1966). As expected, the calculated Froude numbers of the reaches containing those bed forms were in the range 0.5 to 0.9. There is no doubt that the future development of a better hydraulic theory for large floods may modify the absolute magnitudes of the events, but the relative pattern will probably remain unchanged.

RESISTANCE

Except for occasional erratics of granite, loess, and Ringold Formation, 99% of the flood bedload sediments on the Columbia Plateau were derived from the Yakima Basalt. Thus, the resistance of this material partially dictated the activity of the phenomenal flow velocities described above. The Yakima Basalt consists of flow units which average about 30 m in thickness. The individual flows can be traced scores of kilometers. Bingham and Grolier (1966), Mackin (1961), Schmincke (1967a), Waters (1961), Wright and others (1973) have described the flow-by-flow stratigraphy in terms of such criteria as size and shape of columns, vesicle types, pillow zones, mineralogy, spiracles, and chemical composition. Local interbeds of tuffaceous sediments and buried soil zones are also distinctive. Swanson and Wright (Ch. 3, this volume) give a more complete discussion of this topic.

Regional Structure of the Columbia Plateau

The Columbia Plateau is structurally a basin with the basalt surface dipping toward its center from surrounding uplifts. The Channeled Scabland begins on the northern rim of the basin, where the Columbia River has cut a deep gorge around the basalt plateau. From this northern rim the basalt surface has a general regional slope of 6-8 m/km southwestward from elevations of 850 m near Spokane to 120 m near Pasco. Bretz (1923a, 1928a) described the development of a consequent stream pattern along this dipslope prior to the late Pleistocene flooding. Deformation of the dipping plateau surface increases to the west (Fig. 4.6). A series of east-west ridges occurs in the western part of the plateau. These hils are structural anticlines that separate structural basins on the Plateau surface. The Frenchman Hills, High Hill-Pinto Ridge, Saddle Mountains, and Horse Heaven Hills all posed impediments to the passage of flood water across the Plateau. As shown by the high-water profile of the western Channeled Scabland (Fig. 4.5), water was ponded in the basins directly upstream from each of these anticlinal ridges.

The major structural basins of the western Channeled Scabland are the Hartline Basin, near Coulee City; the Quincy Basin, containing Moses Lake; lower Crab Creek, near Royal City and Othello; and the Pasco Basin, containing the Hanford nuclear works. Each of these basins filled with flood water until it spilled over the low points in the ridges that form the basin rims. The Hartline Basin contained water ponded upstream from High Hill and Pinto Ridge (Fig. 4.7). This water flowed through this southern divide at three points, Lenore Canyon, Dry Coulee, and Long Lake (Bretz, 1932a). Even greater ponding occurred in the Quincy Basin, covering over 1500 km^2. Water spilled from three divides (Fig. 4.8) in the upraised western rim of the basin, Babock Ridge and, through one divide (Drumheller Channels) on the southeastern margin of the basin. Bretz' pronouncement that all four outflow channels operated simultaneously shocked the geologic community of his day.

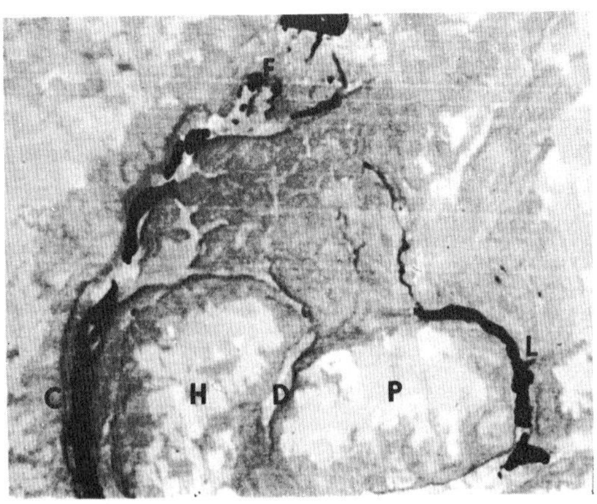

Figure 4.7. LANDSAT image of the Hartline Basin and the scabland channels leading from its southern margin: C-Lenore Canyon, D-Dry Coulee, and L-Long Lake. Other features are Dry Falls (F), High Hill (H), and Pinto Ridge (P). The scene measures 35 x 25 km. Compare to Fig. 4.6.

Regional Structure and Stream Patterns

Studies that relate regional stream patterns to regional structure are a classical activity for geomorphologists. Since the field observations of J. W. Powell and the masterful deductions of W. M. Davis, there has evolved a concept of river valley adjustment to structure over long time scales—millions of years. A genetic terminology is applied to these relationships. *Consequent* streams follow the initial structural slope of the land. *Subsequent* streams become oriented along secondary features, such as faults and the regional strike. *Discordant* streams attract special interest because they cut across structural trends.

The Neogene outpourings of Yakima Basalt forced the Columbia River to assume a huge bight around the northern and western margin of the Columbia Plateau. In the west, the basalt was folded during the Pliocene to form the anticlinal ridges discussed above. The Columbia kept pace with this uplift, carving deep gorges through the Frenchman Hills, Saddle Mountains, and Horse Heaven Hills. Thus, its relationship to these anticlines is *antecedent*. Nevertheless, there was much local ponding of drainage by the active folds, as indicated by deposition of the Pliocene Ringold Formation. Ponding behind an anticline and subsequest overflow produces a discordant structure called *cross-axial consequent*.

Figure 4.8. Orbital (Skylab) photograph of the western margin of the Quincy Basin showing the divide crossings that drained water from the basin (Q) into the gorge of the Columbia River. These cataracts include Frenchman Springs (F) and Potholes (P). Note the giant current ripples on West Bar (W). This image measures 21 x 30 km.

The above relationships were well-known in the 1920's when Bretz began his studies. However, the discordant patterns he described in the Channeled Scabland completely defied the time scale that was generally applied to the structural control of rivers. The catastrophic flood outbursts filled basins and overtopped them in a matter of hours or days. Nevertheless, regional structure still exerted a dominant control on the channel (*not* valley) erosion by the flood flows.

The Lower Grand Coulee illustrates structural control of scabland flood erosion to a remarkable degree. Flood water spilling over a pre-flood divide to the north encountered the Coulee monoclinal flexure near present-day Dry Falls (Bretz, 1932a). This created a huge cataract, 250 m high that receded northward to form the Upper Grand Coulee. Below Dry Falls the flood water prefer-

Figure 4.9. Oblique aerial photograph of the head of Lenore Canyon near Park Lake. The dipping beds of the Coulee Monocline are exposed at the top, center. Downstream from this point, the dipping portion of the flexure has been completely eroded away to form Lenore Canyon. Longitudinal grooves mark the bare basalt surface in the foreground.

entially eroded the fractured rock on the steep eastward dipping limb of the monocline. This process excavated Lenore Canyon (Fig. 4.9) from the bent and broken basalt units.

Scabland divide crossings across anticlinal ridges also show the preferential erosion of fractured zones. Plucking erosion was concentrated at the tension-jointed anticlinal crests (Fig. 4.10). Joint control of rock basins and scabland channels is perhaps most spectacular on the Palouse-

Figure 4.10. Erosion by scabland flooding at the crest of High Hill anticline in the southern Hartline Basin. Note the dip of Yakima Basalt units to the right and left of the eroded anticlinal crest (center).

Figure 4.11. Joint-controlled rock basins and channels on the Palouse-Snake divide crossing southeast of Washtucna, Washington. The Palouse River occupies the inner channel of the main canyon.

Snake divide crossing (Fig. 4.11). There it appears that the flood erosion simply etched out the regional joint pattern from the exposed basalt surface (Trimble, 1950).

Small-scale Structural Features of the Yakima Basalt

Planes of weakness within the basalt bedrock were an important influence on fluvial erosional forms produced by Missoula flooding. The individual basalt flows average 25-60 m thick and are characterized by a variety of depositional and cooling features which allowed variable resistance to flood erosion (Fig. 4.12). Most flows have a scoriaceous upper portion grading into the rubbly top typical of aa lava. Some flows exhibit the classic Tomkeieff (1940) sequence of 2-tier columnar jointing. Columns in the lower colonnade may be 1-2 m in diameter (Fig. 4.13). An entablature of long slender columns and hackly fragments is present in all flows. In some flows an upper colonnade of much smaller columns occurs above the entablature. The cooling history of the flows controls the nature of columnar jointing (Spry, 1962). Irregular cooling surfaces developed because of erosion during the intervals between basalt emplacement. This often resulted in flaring columns in the next higher Yakima basalt flow. Rapid cooling produced hackly fragmented lava (Fig. 4.14A) and the "brickbat" zones described by Mackin (1961, p. 10).

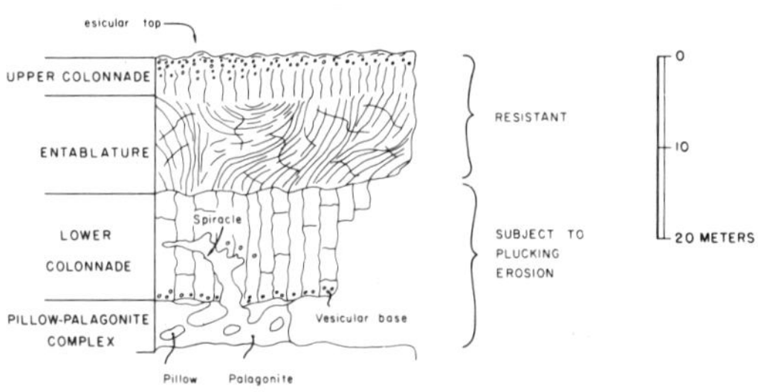

Figure 4.12. Cross section of an idealized Yakima basalt flow showing structural features important to flood erosion processes. The upper colonnade and the pillow-palagonite complex are only present in some basalt flows. Diagram is modified from Swanson (1967) and Schmincke (1967a) and uses the terminology of Tomkeieff (1940), Waters (1960) and Spry (1962).

Between the outpourings of basalt, enough time elapsed to permit weathering, growth of forest cover, and the formation of lakes. Local sedimentary intercalations in the basalts include conglomerate beds, clay layers, and freshwater diatomite. The lavas which overrode these lake beds formed pillow-palagonite complexes (Fig. 4.14B and C) and spiracles (gas chimneys) as described by Waters (1960).

Observations of basalt boulders transported by Pleistocene flooding (Baker, 1973a) revealed that the largest boulders were always portions of hackly jointed entablature (Fig. 4.15). Fragments of individual columns were common in scabland bars, but boulders greater than 1-2 m in diameter were not produced from lower colon-

Figure 4.13. Yakima Basalt (Rosa Member) showing massive columnar joints that were eroded by flood water flowing over Frenchman Springs cataract. The local basalt stratigraphy is discussed by Mackin (1961).

Figure 4.14. Structural characteristics of Yakima Basalt flow units. A. Entablature forming a resistant bench in flood-eroded scabland near Lenore Caves. B. Pillow lavas with overlying columns exposed in the wall of Moses Coulee near Appledale, Washington. C. Pillow-palagonite (Rosa Member) exposed in the Cheney-Palouse scabland tract. D. Entablature and colonnade exposure by flood erosion in Lenore Canyon.

nades because of the well-developed cooling-contraction joints in these zones.

Ring Structures

Basalt flow surfaces in the northern Columbia Plateau sometimes display unusual circular structures outlined by dikes that surround a crater-like depression or central mound (Fig. 4.16). The structures range from 75-500 m in diameter and occur either as single rings (Fig. 4.17) or as multiple concentric dike segments. The relatively resistant dikes were preferentially preserved by flood erosion of the surrounding basalt.

McKee and Stradling (1970) attribute the structures to lava escape through tension joints that surrounded sags on a cooling lava crust (Fig. 4.18). However, new studies by Hodges (1976) revealed palagonite in the central mounds of many structures. The structures may result from ground water rising into the molten interior of a very thick basalt flow. Doming results from rapidly accumulating volatiles, with associated auto-intrusion of ring dikes and possible venting. The fragile palagonite cores were then easily eroded by the scabland flooding.

Figure 4.15. Boulder of basalt entablature measuring 18 x 11 x 8 m. This boulder is located on the Ephrata Fan 2.5 km west of the Rocky Ford Creek Fish Hatchery.

Figure 4.16. Oblique aerial photograph of concentric ring structures on a flood-eroded basalt surface north of the Karakul Hills, near Keystone Siding, Washington.

HYDRODYNAMIC EROSION

Cavitation

The enormous flow velocities achieved in many scabland channels caused considerable reduction of absolute pressure, perhaps to the fluid vapor pressure. Shock waves produced by collapse of the air bubbles which form in such situations are recognized as intense erosional agents by civil engineers (Symposium, 1947). Embleton and King (1968, p. 271) suggested that this process

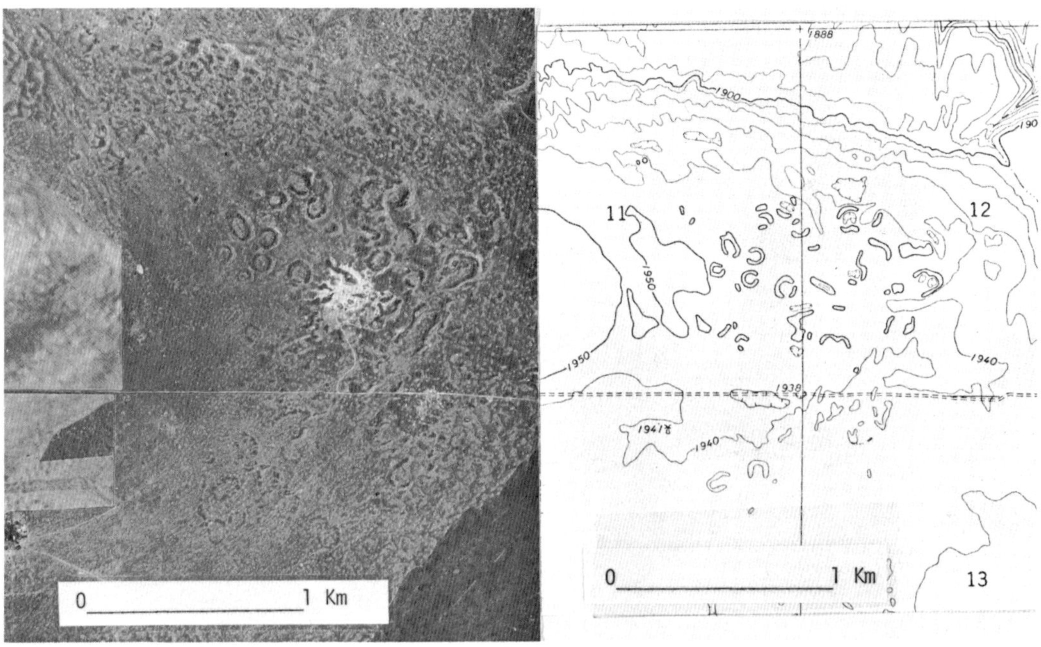

Figure 4.17. Vertical aerial photograph and topographic map (10 feet contour interval) of the ring structures shown in Fig. 4.16. These rings occur at 40 km east of the prominent rings described for the Odessa area by McKee and Stradling (1970) and by Hodges (1976).

of cavitation may have been responsible for the excavation of scabland rock basins. The bizarre "p-forms", potholes, and other erosional features which occur on rock surfaces in Scandinavia are interpreted as the result of cavitation erosion by subglacial melt-water streams flowing in concentrated routes or tunnels (Dahl, 1965). However, Schumm and Shepherd (1973) have suggested that such topography need not be subglacial.

Engineering experiments have shown that bubble collapse near materials such as steel or concrete produce hammer-like blows with local pressures as high as 30,000 atmospheres (Barnes, 1956, p. 494). These very intense local pressures are capable of shattering the surface of nearly any solid material. The fractured surface layers are quickly carried away by turbulent water and fracturing continues. Barnes (1956, p. 503) proposed that such processes in high velocity glacial streams may initiate the formation of potholes. He added, however, that cavitation forms are very short-lived. Once the potholes form, they would then trap abrasive particles of bedload and continue to enlarge by the rotary action of the water (Alexander, 1932).

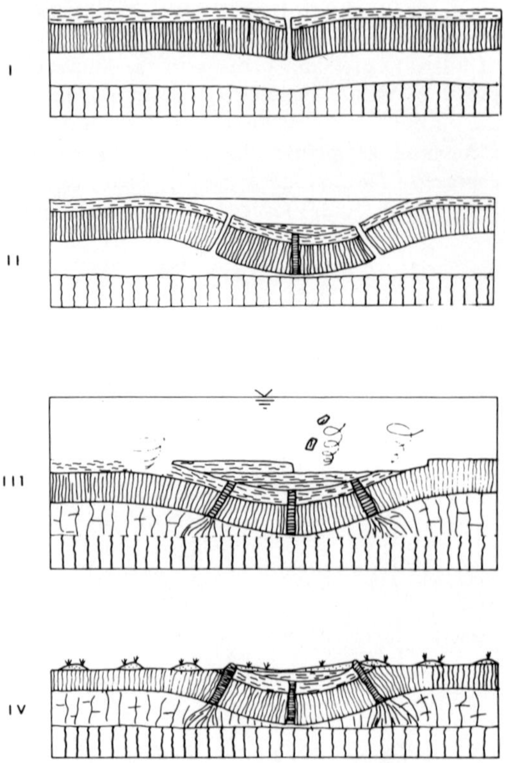

Figure 4.18. Hypothetic sequence of development for basaltic ring structures (modified from McKee and Stradling, 1970). I. Flowout of lava from the molten interior of a crusted lava flow. II. Sagging of the lava flow crust induced by loading from a point source induces circular fractures that allow additional lava escape. III. Plucking erosion by macroturbulent flood water. IV. Present relationships showing grass-covered silt mounds.

Cavitation can only occur for certain critical conditions. Hjulström (1935, p. 311) noted that the minimum velocity necessary for cavitation to occur in a river is about 12 m/sec. This figure is somewhat misleading, however, because it applies only to relatively shallow, swift streams.

The critical conditions necessary to initiate cavitation can be estimated from Bernoulli's equation (conservation of energy in fluid flow) written as follows:

$$\frac{V^2}{2g} + \frac{P_a}{\gamma} + d = \frac{V_c^2}{2g} + \frac{P_v}{\gamma} + Z, \qquad (4\text{-}1)$$

where V is the mean stream velocity, V_c is the velocity at the cavitation point, P_a is the absolute pressure (atmospheric), P_v is the fluid vapor pressure, d is the stream depth, Z is the assumed datum for position head ($Z = 0$ by this convention), γ is the specific weight of water, and g is the acceleration of gravity.

Barnes (1956) assumed that obstructions on the streambed would generally increase the local fluid velocity to about twice the mean stream velocity. If $V_c = 2V$, then

$$V = \sqrt{\frac{2g}{3}} \sqrt{\frac{P_a - P_v}{\gamma} + d}\,. \qquad (4\text{-}2)$$

For terrestrial conditions near sea level, this equation can be further reduced by assuming $g = 9.81 m/sec^2$, $P_a = 1$ atm., $P_v = 0.024$ atm. (21°C), and $\gamma = 9.8 \times 10^3$ N/m³. The equation for the critical cavitation velocity V_c (m/sec) is then a simple function of flow depth (m):

$$V_c = 2.6\sqrt{10 + d}. \qquad (4\text{-}3)$$

Equation 4-3 indicates that at very high flow depths, the velocity must rise to very high values in order to achieve cavitation (Fig. 4.19). Only in a few narrow constrictions were the Missoula flood flows of sufficient velocity to achieve cavitation. All these areas, which are named in Fig. 4.19, show intense bedrock erosion. Nevertheless, it is likely that the erosion was localized by flow separations and other macroturbulent phenomena as discussed below.

Macroturbulence and Plucking

A fundamental hydraulic characteristic of very deep, high gradient flood flows is the development of secondary circulation, flow separation, and the birth and decay of vorticity around obstacles and along irregular boundaries. Such three-dimensional flow phenomena in rivers are poorly understood even by hydraulic engineers. Indeed, Simons and Gessler (1971) have suggested that fluvial morphologic studies can make a valuable contribution to engineering river mechanics by describing examples of such phenomena in nature. Matthes (1947) has termed these phenomena collectively as "macroturbulence". The most important erosive form of macroturbulence is the "kolk", a Dutch term which Matthes used to designate intense energy disipation by upward vortex action. The intense pressure and velocity gradients of the vortex produce a phenomenal hydraulic lift force along the filament of the vortex (Fig. 4.20). The precise magnitude of this suction effect is unknown, but it certainly is greatly in excess of normal hydraulic lift forces.

The conditions necessary for the generation of kolks according to Matthes (1947) include the following:

(1) A steep energy gradient,

(2) A low ratio of actual sediment transported to potential sediment transport,

(3) An irregular, rough boundary capable of generating flow separation.

Jackson (1976) has recently presented an alternative model for kolks in which he attributes them to the oscillatory growth and breakup stages of the turbulent bursting phenomenon. Bursting, as described by Offen and Kline (1975), characterizes the turbulent structure of the outer part of the turbulent boundary layer. It is not yet clear, however, that this mechanism as observed

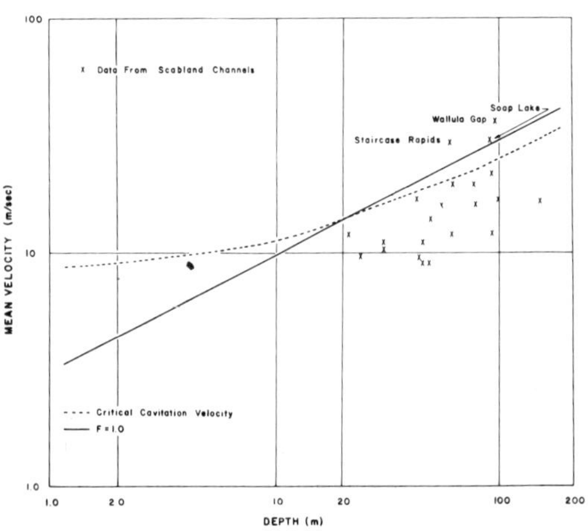

Figure 4.19. Influence of flow depth on the critical mean flow velocity necessary for cavitation to occur according to Barnes (1956, p. 499). Hydraulic data from scabland channels are plotted for comparison. Most of the flood flows were sub-critical and below the critical depth-velocity combinations necessary to produce cavitation erosion.

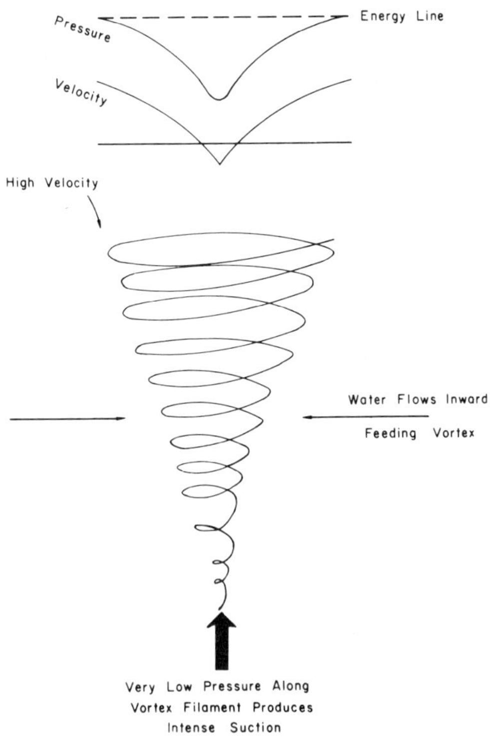

Figure 4.20. Characteristics of a hydraulic vortex ("kolk"). Such macroturbulent effects are set up by flow obstructions. The intense pressure and velocity gradients of the vortex produce a hydraulic lift. Alternatively, Jackson (1976) ascribes "kolks" to the turbulent bursting phenomenon.

in smooth-walled flumes will apply to the extremely irregular flow boundaries of scabland channels.

Turbulence as a general phenomenon is the end result of chaotic motion of various eddies that are superimposed on an average motion. When the size of the eddies corresponds to the absolute size of the flow, the scale of turbulence dictates the term "macroturbulence". The smallest eddies, "microturbulence", are almost independent of the size of the flow and tend to vary with the flow Reynolds number. The dimensionless period of macroturbulent flows, on the other hand, is generally independent of the Reynolds number. Whereas microturbulence is treated in stochastic terms, macroturbulence constitutes a more-or-less deterministic secondary flow that is superimposed on the prevailing mean flow components that result in the overall unidirectional movement of water.

The concept of macroturbulence is intuitively obvious to anyone observing fluid motions. Jackson (1976) cites Mark Twain's (Clemens, 1896, p. 44-48) description of distinctive patterns of turbulent fluid motions on the water surface of the Mississippi. Such concepts contributed to the original eddy concept of turbulence, a concept perhaps best summarized in the following rhyme, attributed to Richardson (1920):

Big whorls have little whorls,
Which feed on their velocity;
Little whorls have smaller whorls
And so on unto viscosity.

Thus eddies of a given size, or order, develop from larger eddies by borrowing energy from their "parents." This division process continues to such a small scale (microturbulence) that the eddies can no longer borrow sufficient energy to further divide.

Bretz (1924) first suggested that scabland-type erosion was a result of plucking rather than abrasion. He observed that eddies generated by the irregular channel walls at The Dalles area of the Columbia River are responsible for the plucking of basalt columns. He stated that inner channels of "The Dalles" type, potholes, and other scabland features could only be produced by a river of high discharge and steep gradient eroding close and vertically jointed rock. It was later suggested by Bretz and others (1956, p. 1028) that the hydraulic lift necessary for plucking action was provided by the kolk action of Matthes (1947). Bretz (1969) added, "plucking action thus was greatly augmented wherever submerged basalt ledges with the proper form appeared during flooded regimens."

The morphology of pothole erosion at Lenore Caves (Fig. 4.21) provides some insight into the details of the hydraulic plucking process. Like many scabland potholes, the one at Lenore Caves is asymmetrical in cross section parallel to the flood flow direction. Columns were preferentially plucked on the downstream side from beneath the resistant pothole rim of entablature. The result was a cave extending 20 m under the basalt (Fig. 4.21A). The symmetry of erosion is consistent with experimental flume studies of vortex scour. These studies indicate that at a flood peak, vertical vortices will scour at the downstream end of a hole on a stream bed and that the vortex axis will slant upstream (Fig. 4.21B).

The irregular rock steps of the Channeled Scabland created numerous points of flow separation. The lips of potholes (Fig. 4.21), canyon walls, and cataracts are but a few examples. Of course, separated flow is one of the most basic concerns of fluid mechanics. Leonardo da Vinci observed and sketched recirculating eddies. Surprisingly, however, the problem has not been fully explored in analytical terms. Nevertheless, an excellent empirical body of knowledge has evolved for the interaction of various geometries with fluid flow fields.

The dynamics of a turbulent separated flow at a downward step has been especially well studied. The temporal mean fluid pressure measured on the bed reaches a maximum at the point of reattachment and gradually decays downstream. Throughout the separated region and downstream there are large fluctuations of instantaneous pressure at any given observation point. The bed experiences rapidly varying normal forces that alternately pull and push at it. Turbulent shear stresses developed in and downstream from separated regions are much greater than in the boundary layer. Experiments by Allen (1971b) have confirmed that the greatest erosion by turbulent separated flows occurs at the points of flow reattachment. This conclusion applies to any of several erosional mechanisms, including cavitation, corrosion, fluid stressing, and solution.

Thus, secondary flow is easily induced around various kinds of obstacles or obstructions on a river bed. However, another important type of macroturbulence can be generated in straight channels free of obstruction. Prandtl (1926) showed empirically that secondary currents exist at the corners of rectangular closed conduits. Einstein and Li (1958) showed theoretically that transverse instability is often spontaneous in straight stream channels. The result of superposing transverse movements on the main flow in a stream is a helical array of vortices aligned parallel to the flow direction and showing alternating senses of rotation. These longitudinal vortices are distinctive in having their axes in the streamwise direction.

The theory initially yields two helical flows near the two banks of a relatively wide shallow channel. These two helical flows would then induce similar flows throughout the flow section until the entire fluid mass is split into secondary rotating cells (Karcz, 1967).

Karcz (1967) observed current-aligned ridges and troughs of sediment created by floods along ephemeral streams in southern Israel. These longitudinal structures have heights of 1-10 cm and spacings of 5-50 cm. They result from flows no greater than 3 m deep with velocities of .3 to

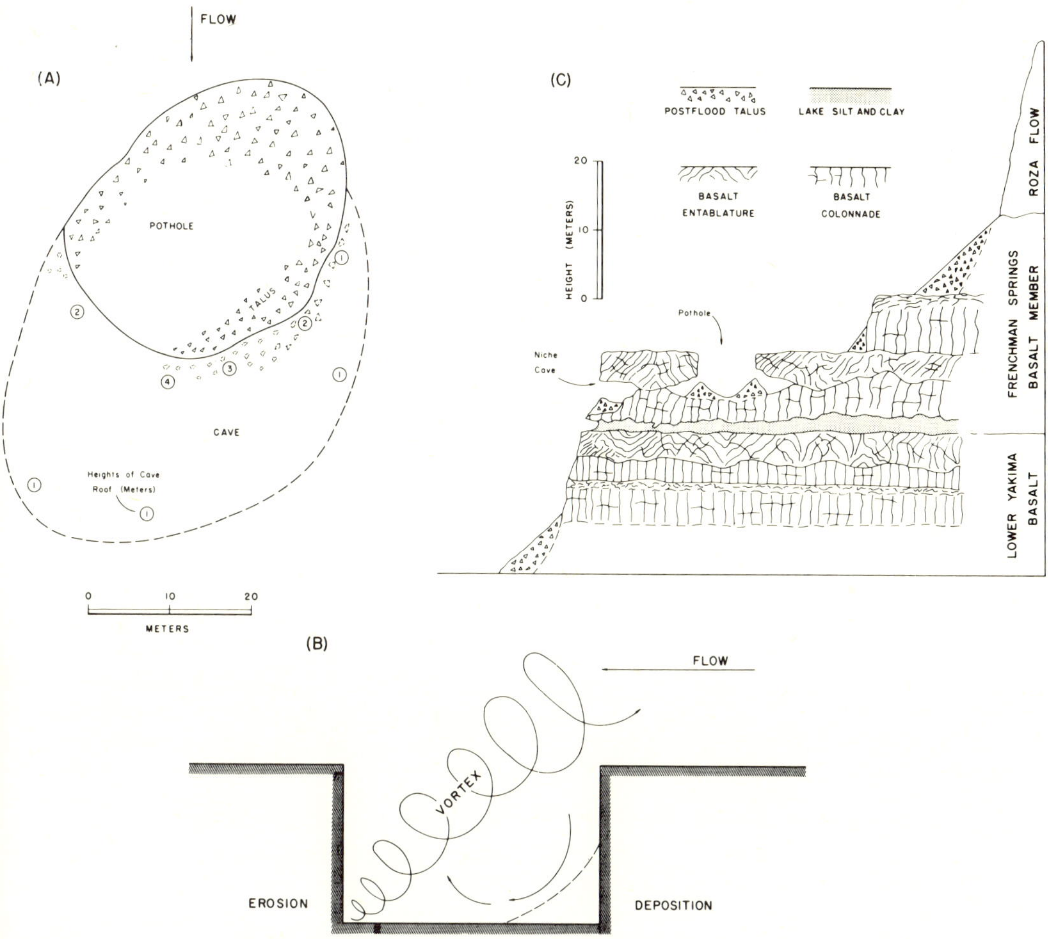

Figure 4.21. Pothole erosion at Lenore Caves, 15 km north of Soap Lake in Lenore Canyon. The sketch map (A) shows undercutting of the pothole rim on the downstream side. This is consistent with experimental observations of vortex scour (B). The relationship of pothole to the local basalt section is shown in (C).

3 m/sec. Karcz (1967) attributed these fluvial structures to secondary currents aligned in a series of longitudinal vortex tubes. Unfortunately, an adequate theory has not emerged for relating such large-scale vortex filament spacings to flow properties. For the small-scale longitudinal forms known as parting lineations in aqueous sedimentation, Allen (1970, p. 69) has shown that the responsible vortices increase in their spacing with both depth and mean flow velocity. The immense scale differences preclude the quantitative application of these results to the much larger groove phenomena in the Channeled Scabland.

The Einstein and Li (1958) analysis applied to fully turbulent flow in mathematically straight channels. Of course, any additional irregularity will create additional vorticity to interact with the longitudinal flow field. If the longitudinal vortices produce longitudinal bedforms, the bedforms will further enhance the flow field leading to stronger longitudinal erosion. This will eventually produce morphological forms that mimic certain hydraulic attributes of the responsible fluid. If defects occur in the bed, these may produce transverse flow separations that break up the longitudinal pattern of vorticity. Self-enhancement would then create transverse bed forms such as potholes and cataracts. Scabland erosion appears to derive from pronounced positive feedback.

In Chapter 5 it will be shown that a sequence of erosional forms appears to be developed on rock surfaces in the Channeled Scabland. The initial channels had relatively smooth floors. These are marked by longitudinal grooves, which mimic the longitudinal vorticity of the streaming fluid. Deeper erosion created irregular surfaces that generated flow separation and/or kolks. As erosive activity concentrated at these sites, the result was greater accentuation of the surface irregularities. A critical threshold had to be crossed to achieve this change from longitudinal forms to the production of irregular scabland.

The macroturbulent erosion mechanism has yet to be adequately evaluated in physical terms. Future research needs to focus on problems that are similar to those faced by Williams (1959) in his analysis of meteorite pitting: (1) the vortices should be able to form at the observed sites of intense erosion according to the principles of fluid physics, and (2) the presence of a vortex should set up a velocity distribution that will locally increase the erosion rate. Problem (1) seems to be satisfied by the development of flow separation at rock steps and by the development of longitudinal vortices as discussed above.

Problem (2) appears to have several ramifications. Williams (1959, p. 62) states the physical rationale for the velocity being greater in vortex tubes than in the general downstream flow field. In the idealized case of a vortex with constant circulation, the tangential velocity V at each point in the fluid is inversely proportional to the distance from the vortex center r:

$$V = \frac{A}{2\pi r}, \qquad (4\text{-}4)$$

where A is a constant of circulation. The gradient of V is inversely proportional to r^2:

$$\frac{dV}{dr} = -\frac{A}{2\pi r^2}. \qquad (4\text{-}5)$$

If a fluid is brought into a vortex with rotation A, it gradually moves down the axis of flow. While in the vortex the fluid will behave such that it develops less vorticity ($2\pi Vr$) when it leaves the flow than when it enters. By this picture, velocity from the general flow field tends to accumulate at the periphery of vortices until it is as great or greater than in the fluid entering the vortex.

SEDIMENT TRANSPORT MECHANICS

Incipient Boulder Motion

An early step in any sediment transport problem is to predict the conditions necessary to initiate the movement of bedload. One criterion for incipient motion expresses these critical conditions by the DuBoys equation for boundary shear, a function of flow depth and energy slope. Both depth and slope can be directly determined from the high-water surface evidence in the scablands, assuming that energy slope is approximated by the water surface slope. These parameters are then combined in the equation:

$$\tau = \gamma RS, \qquad (4\text{-}6)$$

where τ is the mean shear stress for initiating

particle transport, γ is the specific weight of the transporting fluid (9.8×10^3 N/m^3 for clear water), R is the hydraulic radius (cross-sectional area/wetted perimeter), and S is the energy slope. In relatively wide channels (as typical for the Channeled Scabland) the depth of flow D is a close approximation to the hydraulic radius.

Actual particle entrainment is considered to be a stochastic process, dependent on instantaneous shear values exerted on the bed rather than on the mean values. Nevertheless, equation 4-6 gives a crude guide to the physical conditions associated with incipient particle movement.

Several geological studies of boulder movement have developed correlations between the grain sizes of the largest transported boulders and tractive force. Figure 4.22 compares the results of these studies to data from the Channeled Scabland. The analysis is described in detail by Baker and Ritter (1975). The data are approximated by the trend line:

$$D = 65\tau^{0.54}, \qquad (4\text{-}7)$$

where D is the intermediate particle diameter (mm) and τ is the shear stress (kg/m^2). Only the smaller scabland boulders from giant current ripples were used in this analysis to avoid macroturbulent effects on the boulder initiation and movement processes.

Size-Distance Relationships

In rivers which have established an equilibrium between form and process over a long geologic history there are very regular downstream changes in hydraulic geometry (Leopold and Maddock, 1953), meander wavelength, and gra-

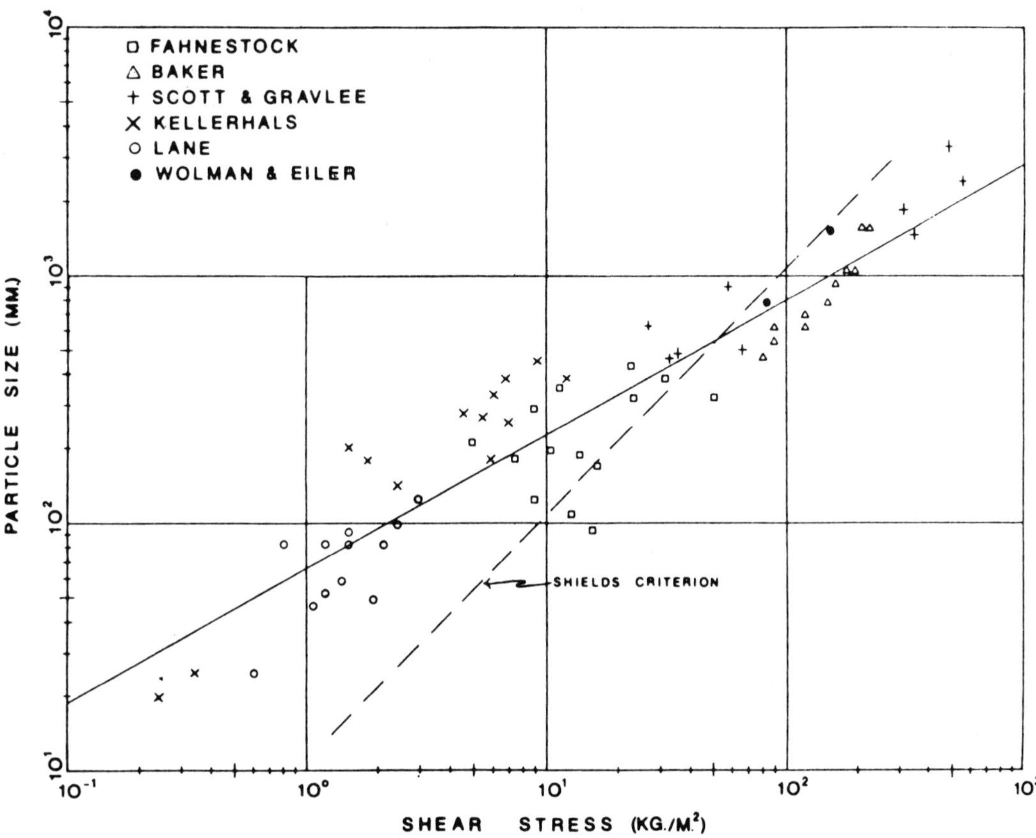

Figure 4.22. Intermediate particle diameter versus shear stress for extremely coarse bedload. The triangles indicate data for boulder sizes and shear in relatively uniform scabland reaches.

dient. Changes in the maximum sizes of transported sediment generally conform to Sternberg's Law:

$$D = ae^{-bx}, \qquad (4\text{-}8)$$

where D is the maximum particle diameter (mm) in a given reach, x is distance (km) downstream, and a and b are empirical constants. In the downstream reaches of rivers that are comparable in length to the Missoula Flood flow path (100's of kilometers), particle sizes are generally less than 50 mm. Coarser sizes may occur in steep, headwater tributaries, but these short stream segments are unsatisfactory as analogues to the large-scale Martian examples.

Alluvial fans typically have a-values (Equation 4-8) of 4 meters (Blissenbach, 1954), but size declines extremely rapidly down-fan. The b constant is often as great as 0.8. Similarly, outwash channels have coarse material in their proximal reaches, a being .3 to .4 meters (Bradley and others, 1972; Boothroyd and Ashley, 1975). The b constant indicates slower decline in size, typically .3 to .15. Even at these rates, 25 km of transport and selective sorting may reduce sizes to 10 mm in the distal portions of outwash trains. Most outwash plains have sandy (<2mm) distal facies (Boothroyd and Nummedal, 1978).

Baker (1973a) showed that debris eroded from catastrophic flood constrictions is distributed downstream according to the following equation:

$$D = ax^{-b}, \qquad (4\text{-}9)$$

in which variables are defined as in Equation 4-8. Measurements from a variety of catastrophic flood channels (Fig. 4.23) show that boulder size falls off very rapidly in the expanding reaches below constrictions. Studies in the Cheney-Palouse Scabland (Baker and Patton, 1976) indicate that very coarse debris (boulders) is localized in a narrow zone immediately below the constriction. Sizes fall off to granules and sand laterally toward channel margins. Material deposited in the lee of scabland streamlined forms (loess "islands") is 99% granules and finer sediment.

Hydraulic Jumps

Figure 4.19 shows that the flow through several scabland constrictions (Soap Lake, Wallula Gap, and Staircase Rapids; see Fig. 4.6 for locations) was supercritical. When this flow passed through the subsequent downstream expansion, if the change in flow regime was abrupt enough, a hydraulic jump may have occurred at the mouth of the constriction. Komar (1971) has analyzed this problem in general terms for the passage of a turbidity current from a submarine canyon on to the submarine plain at the mouth of the canyon. The jump is a stationary surge or shock wave in which the speed of advance of the upstream wave is balanced by the velocity of flow downstream. Kinetic energy in the constricted, high-velocity flow is transformed to potential energy in the deep, low-velocity flow downstream from the jump. This transition involves an abrupt loss of mechanical energy through the generation of intense turbulence within the jump. Whereas the transition point may be a site of erosion or nondeposition, the immediate downstream area is a site of intense deposition as revealed in the laboratory experiments of Jopling and Richardson (1966). Komar (1971) suggests that only the

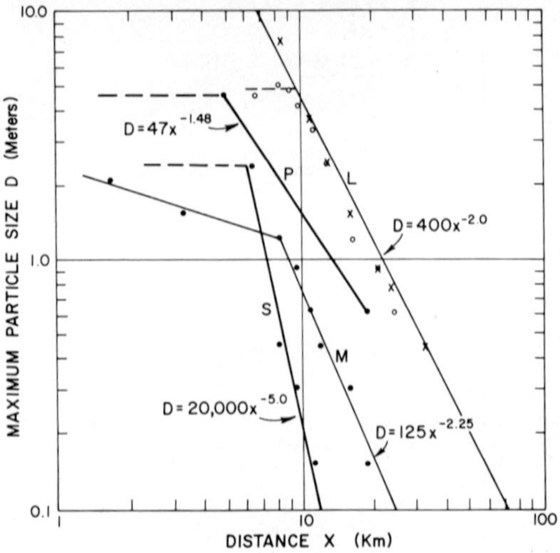

Figure 4.23. Downstream decrease in intermediate diameter of the largest boulders deposited below constrictions by catastrophic terrestrial floods. Data were obtained from the following sources: (L) Lenore Canyon, Channeled Scabland (Baker, 1973a), (P) Portland area below Columbia Gorge (Trimble, 1963), (M) Michaud Fan of the Bonneville Flood (Trimble and Carr, 1961), (S) Sentinel Gap, channeled scabland (Baker, 1973a). The constricted channels vary in width from 5 km to 16 km and water flowed between 120 m and 220 m deep at peak flood stage.

coarsest fraction will be deposited because the downstream flow will still be highly competent.

The Soap Lake constriction (Fig. 4.24) appears to show the morphologic and sedimentologic features that one would expect from a hydraulic jump. Soap Lake itself occupies a deep rock basin cut into the basalt bedrock (Bretz and others, 1956). Its location right at the mouth of the constriction is precisely at the point where one would expect the maximum flow velocities and the abrupt transition with the generation of intense turbulence. Downstream from Soap Lake is the immense gravel fan of the northern Quincy Basin. Baker (1973a) has described the unusual fall-off in maximum particle sizes that occurs on this fan. Although these factors point to the presence of a hydraulic jump at Soap Lake during the Missoula flooding, the precise physical description of that jump is hampered by the difficulty in reconstructing the flow geometry that prevailed at the precise time of the jump.

Figure 4.24. View north across the Ephrata Fan toward the Soap Lake constriction.

CONCLUSION

Recent studies of catastrophic flood phenomena have provided the beginnings of a quantitative understanding for the hydrodynamics of exceedingly swift, deep flood water acting on bedrock. Although these processes may be of limited academic interest on the earth, they may have major significance for Mars (Baker and Milton, 1974). In future applications of these ideas, however, we must remember that our new understanding of catastrophic flood hydraulics and dynamics would not have been possible were it not for the thorough qualitative studies of earlier investigators. This relationship has been explicitly stated by Mackin (1963, p. 139), as follows:

> "In general, the larger the problem, the more many-sided it is, the more complicated by secondary and tertiary feedback couples, and the more difficult it is to obtain the evidence, the more essential it is to the efficient prosecution of the study that the system first be understood in *qualitative* terms; only this can make it possible to design the most significant experiments, or otherwise to direct the search for the critical data, on which to base an eventual understanding in quantitative terms."

The "critical data" for the Channeled Scabland were discovered only because of nearly one half century (1923-1969) of qualitative research by J Harlen Bretz.

13

Reprinted from pages 81–115 of *The Channeled Scabland,* V. R. Baker and D. Nummedal, eds., NASA Office of Space Science, Planetary Geology Program, 1978, 186 pp.

Large-Scale Erosional and Depositional Features of the Channeled Scabland

VICTOR R. BAKER

ABSTRACT

The Channeled Scabland is a great anastomosing complex of highly overfit steam channels eroded into the basalt bedrock and overlying sediments of the Columbia Plateau. Both the erosional and depositional bed forms in these channels can be described according to a simple hierarchical classification. The catastrophic flood flows produced macroforms (scale controlled by channel width) through the erosion of rock and sediment and by deposition (bars). Mesoforms (scale controlled by channel depth) are also erosional and depositional. Microforms (scale controlled by the inner part of the turbulent boundary layer) are not discussed.

Large-scale bedrock macroforms are crudely developed, especially in the Cheney-Palouse scabland tract. The streamlined residual forms in loess are the most striking macroforms. Their characteristic elongation of 3 times their maximum width results from the balancing of skin resistance and pressure drag factors to create an equilibrium landform. The scabland bars are the characteristic depositional macroforms. Pendant bars, the most common type, accumulated wherever large flow separations were generated by various flow obstructions or diversions.

The bedrock mesoforms include longitudinal grooves, potholes, rock basins, inner channels, and cataracts. A sequence of bed form development probably operated for bedrock erosion in the Channeled Scabland. With time the forms changed from (1) longitudinal grooves, to (2) rock basins and potholes, (3) inner channels with recessional cataracts at their heads, and finally to deeply incised inner channels.

The depositional mesoforms for major channel flows were giant current ripples varying from 18 to 130m in chord length and from 0.5 to 7m in height and composed predominantly of gravel. The bed forms may be empirically related to depth-slope, mean velocity, and stream power. Correlation coefficients for the relationships are all greater than 0.9. Nevertheless, the prediction of these hydraulic parameters from ripple dimensions applies only to the narrow range of flow conditions which characterized the Missoula Flood through the reaches containing the bed forms.

REGIONAL CHANNEL PATTERNS

J. H. Mackin has been quoted as saying, "to understand the scabland, one must throw away textbook treatments of river work" (Bretz and others, 1956, p. 960). Certainly a failing of Bretz' critics in the Spokane Flood debate was their insistence that the Channeled Scabland conform to "established" geomorphic processes. The scale of the problem was key, and a completely new frame of reference was required. Bretz (1932a, p. 28) provided the required viewpoint: "Channeled Scabland is river bottom topography magnified to the proportion of river-valley topography."

The Spokane Flood debate might have been resolved more easily if the participants could have viewed modern orbital photography of the region (Fig. 5.1). At a glance one can appreciate Bretz'

[*Editor's Note:* References cited in this article are included in the reference list that follows Paper number 14.]

observation. Clearly the Channeled Scabland is a plexus of *channels* rather than a network of valleys.

Channel Anastomosis

The general pattern of the Channeled Scabland is large channels eroded in loess and underlying basalt. The channels form locally anastomosing complexes with individual channels that have relatively low sinuosity. The term "anastomosis" should not be confused with "braiding" (as was done in the A. G. I. *Glossary of Geology*). "Braiding" refers to branching and rejoining around alluvial islands or bars. Braided streams are part of a continuous series of fluvial forms that develop in quasi-equilibrium with external controls on the river systems. "Anastomosis" has no genetic connotation. It refers to channel morphology whether in alluvial streams ("braided") or bedrock streams. The scabland anastomosis is deeply cut into rock. Anastomosis occurs in the Channeled Scabland because preflood valleys did not have the capacity to convey the Missoula Flood discharges without spilling over preflood divides into adjacent valleys. This crossing of divides produces the effect of channels dividing and rejoining. Before the era of aerial photographs and adequate topographic maps, Bretz (1928b) used field surveys to show over 100 channnel ways in the scablands, 50 of which bifurcated in a downstream direction. Modern data sources now show that anastomosis occurs on a variety of scales. The regional scale (Fig. 5.2) is controlled by preflood topography. Individual channels may also be controlled by relatively straight geologic structures, such as High Hill anticline in the Channeled Scabland. Small-scale anastomosis includes the minor divide crossings used for reconstruction of the flood high-water surface.

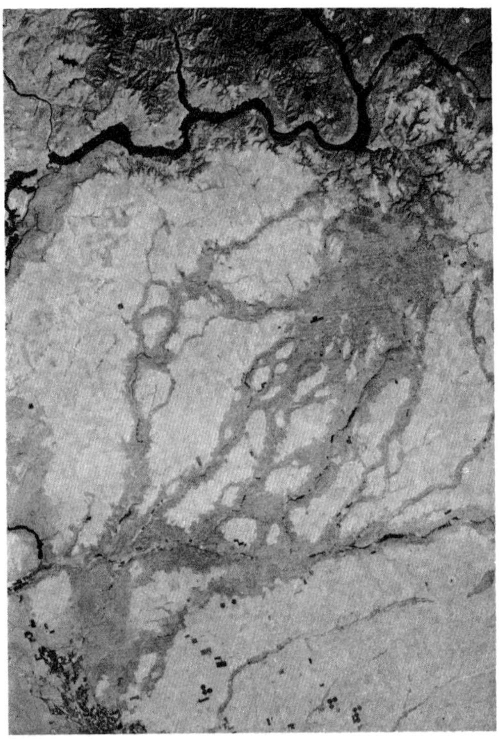

Figure 5.1. Anastomosing channel pattern in the Telford-Crab Creek scabland complex. This LANDSAT image depicts a scene 70 x 150 km (LANDSAT E-1039-18143-5, 31 August 1972).

STREAM OVERFITNESS

Misfit streams are streams that are either too small or too large for the valleys in which they flow (Dury, 1958, 1964). The underfit variety is relatively common, and such streams often show smaller channel widths and meander wavelengths than the winding valleys in which they flow. The disparity between river and valley size is explained by a reduction in stream discharge, either by capture (Davis, 1913) or by climatic change (Dury, 1965). Although there is considerable debate over the validity of the climatic explanation of underfit streams, the concept does seem to apply to alluvial valleys (Baker and Penteado-Orellana, 1977).

An overfit stream is too large for the valley in which it flows. Dury (1964) considered overfit streams in the context of sudden increases in discharge with rapid channel enlargement. Although such a condition would not persist long in an alluvial valley, the erosion of bedrock probably provides an opportunity to preserve overfit stream relationships.

The Crab Creek area, near the town of Wilson Creek in the Channeled Scabland, provides an excellent example of overfit relationships introduced

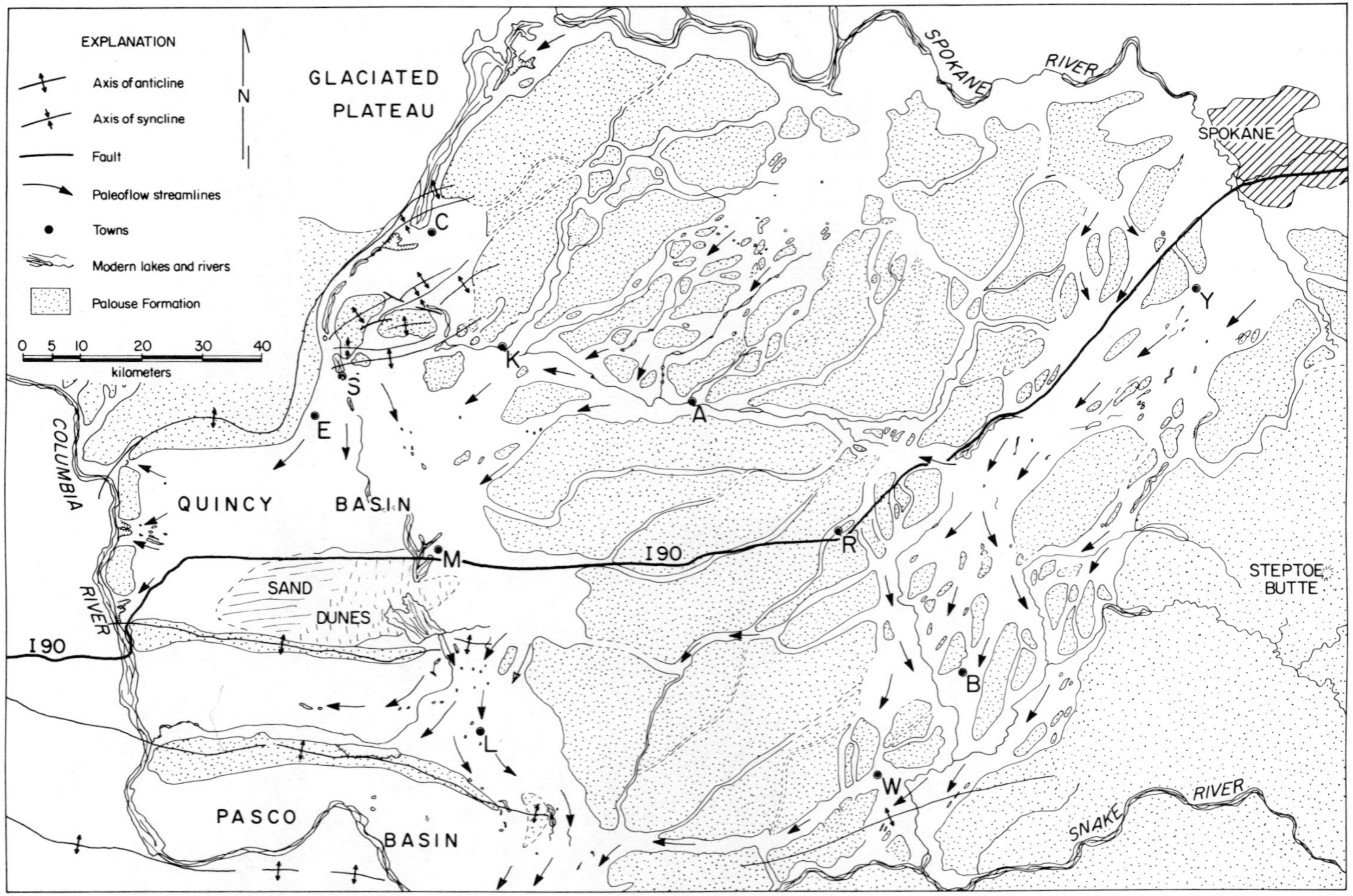

Figure 5.2. Regional pattern of the Channeled Scabland as mapped from LANDSAT imagery (E-1039-1843-5 and E-1004-18201-7). (Y) Cheny, (R) Ritzville, (B) Benge, (W) Washtucna, (M) Moses Lake, (E) Ephrata, (S) Soap

by catastrophic scabland flooding. (See Bretz and others, 1956, their Plate 8; Baker, 1973a, his Fig. 22.) Prior to the catastrophic flooding, the topography near Wilson Creek was probably very similar to that of the modern Palouse Hills region near Pullman, Washington, with the interstream divides thickly mantled by the Palouse loess. The major streams had been superimposed onto the basalt from this loessal cover. The streams flowed in relatively narrow valleys. The well-formed valley meanders had a wavelength of 2000m. Using empirical relationships that characterize most rivers, the normal bankfull discharge of Crab Creek was approximately 850 m^3/sec.

The last major scabland flood completely filled the valley of Crab Creek and adjacent streams. Water spilled over many of the divides between the stream valleys. The Missoula flood flows in this area where approximately 2,800,000 m^3/sec (Baker, 1973a, his Plate 1). Bretz (1928b) noted that the flooding could not tolerate the leisurely preflood curves of the incised stream. Slip-off slopes were vigorously attacked producing what he called "trenched spur buttes". Huge streamlined bars were deposited downstream from the former valley bends. Many of these have giant current ripples on their upper surfaces (Fig. 5.3). The preflood topography was reduced to mere bottom roughness elements by flooding 3 to 4 orders of magnitude greater than any flooding these streams had ever experienced. In the upper parts of the Wilson Creek drainage, the scabland erosion presents a striking contrast to the adjacent loess-mantled terrain (Fig. 5.4).

A HIERARCHY OF SCABLAND BED FORMS

A fluvial bed form is defined (Am. Soc. Civil Engrs., 1966) as follows: "any deviation from a plane bed that is readily detectable by eye or higher than the largest sediment size present in the parent bed material". Most of the scientific interest in bed forms has focused on small primary forms in sand, e.g. Allen (1968). However, the Channeled Scabland affords a unique opportunity for the study of large-scale forms that are either composed of gravel (Bretz, 1928b; Baker, 1973a) or eroded into rock (Baker, 1973b). Two major scales of bed forms have remarkable preservation throughout the scablands. The larger features are scaled to channel width and consist either of depositional "bars" or various kinds of erosional residuals in rock or sediment. The smaller features are scaled to the phenomenal scabland flow depths.

Jackson (1975) has developed a hierarchical classification of bed forms generated by fluid shear and composed of cohesionless granular material. The classification relates to bed form size and to the time span of existence for various bed configurations. The bed form groups each relate to different formative processes (Jackson, 1977). Macroforms for rivers include point bars, scroll bars, alternate bars, and pool-and-riffle sequences. These bedforms do not relate to local flow conditions. They rather respond to long-term hydrologic and geomorphic factors. Mesoforms include large-scale ripples (dunes), antidunes, and large-scale lineation. The spacing of mesoforms depends on the outer zone of the turbulent boundary layer as the flow varies through a dynamic event such as a flood. In rivers the boundary layer control is approximated by flow depth. Microforms include current lineation and small-scale ripples. Microforms respond to flow structure in the inner part of the turbulent boundary layer, and their lifetime is much shorter than the periodicity of dynamic events.

Jackson's classification is especially interesting because of its genetic implications. Heretofore most hydrodynamic bed form studies have focused on the unstable motion of water over a rough boundary generating variable shear stress. Flow properties are perturbed either into longitudinal vortices that produce flow separation or into transverse roller vortices that produce alternating separation and reattachment of streamlines along the boundary (Allen, 1971a, 1971b). Until recently there has been little correspondence between experiments and theory. Jackson (1976, 1977) advocates a resolution to this dilemma through the concept of flow structures.

Recent fluid mechanics work (Laufer, 1975) indicates that turbulent shear flows contain an inherent structure consisting of discrete secondary flow patterns superimposed on the prevailing unidirectional mean flow. For the mesoscale, these structures consist of the bursting phenomenon (Offen and Kline, 1975), longitudinal vortices (Karcz, 1967), in-phase waves, and possible

Figure 5.3. Oblique aerial photograph of Bar 4 in the great scabland channel that was carved from the preflood valley of Crab Creek. The bar is developed for 3 km downstream of the trenched spur buttes (T) that mark the preflood slip-off slope of Crab Creek's meandering valley. Giant current ripples (R) mark the bar surface. These ripples average 3.3 m in height with chords of 66 m. The flood flow maximum depth here was 90 m.

transverse roller vortices. Jackson (1977) speculates that fluvial dunes (a variety of large-scale asymmetrical ripple) are produced by bursting; large-scale lineation (Coleman, 1969) produced by longitudinal roller vortices; and antidunes are produced by in-phase waves. Although this scheme requires further experimental verification, it nevertheless is a useful working hypothesis for resolving the enigma of bed forms.

Although much attention has focused on depositional forms (ripples, dunes, sand waves, etc.) the Channeled Scabland requires consideration of erosional forms as well. Allen (1971b) discusses the two theories generally applied to explain erosional bed forms. The passive bed theory holds that the fluid flow imparts its properties (bursting, vortices, etc.) on a passive bed through hydraulic character of the responsible flow. The defect theory, in contrast, emphasizes defects or irregularities on the bed which disturb the flow to generate turbulent flow separations. Actually the two theories are difficult to resolve

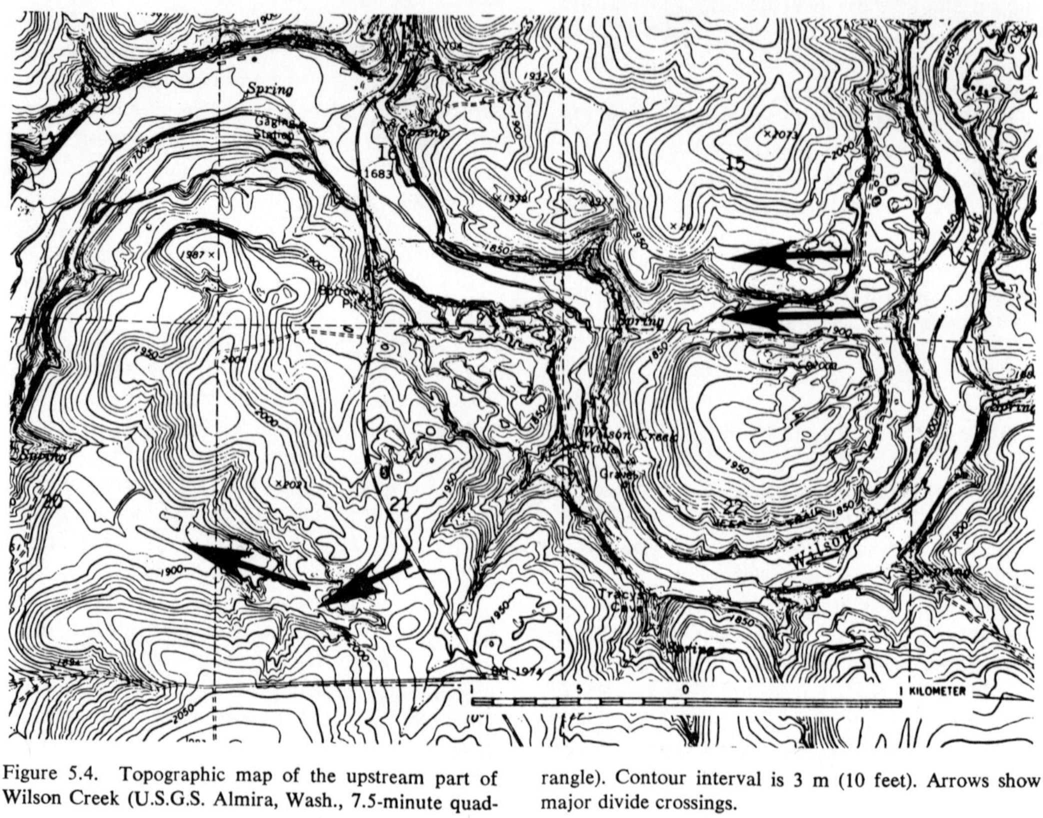

Figure 5.4. Topographic map of the upstream part of Wilson Creek (U.S.G.S. Almira, Wash., 7.5-minute quadrangle). Contour interval is 3 m (10 feet). Arrows show major divide crossings.

Table 5.1. Important Bed Forms In The Channeled Scabland

	Scoured in Rock	Scoured in Sediment	Depositional
MACROFORMS (Scale Controlled by Channel Width)	Pool-and-Riffle Sequence. Quadrilateral Residual Forms in Channel. Anastomosis.	Large-scale Streamlined Residual Forms.	Longitudinal Bars. (a) Pendant Bars. (b) Alternate Bars. (c) Expansion Bars. Eddy Bars.
MESOFORMS (Scale Controlled by Channel Depth)	Longitudinal Grooves. Potholes. Inner Channels. Cateracts.	Scour Marks.	Large-scale Transverse Ripples (Giant Current Ripples).
MICROFORMS	Scallop Pits.	Not Preserved.	Small-scale Ripple Stratification. (Restricted to Slackwater Facies.)

Figure 5.5. Excellent association of depositional macroform (bar) with superimposed mesoforms (giant current ripples). The bar occurs along the Snake River just downstream from its junction with the Palouse River (top center). The Snake River is impounded by a dam just downstream from the reach depicted in this oblique aerial photograph.

because the effects of one tend to enhance the operation of the other. In the basalt bedrock of the Channeled Scabland it is mainly structural irregularity of the rock that provides defects which then perturb the flow hydrodynamics to create distinctive bed forms.

A modification of Jackson's (1975) classification will be used in subsequent discussions of bed forms in the Channeled Scabland (Table 5.1). A rather fortuitous aspect of the depositional morphology in the Channeled Scabland is the excellent preservation of both macroforms and their superimposed mesoforms (Fig. 5.5). This association probably results from the nature of the hydrograph for flood flows in the Channeled Scabland. In most rivers the hydrograph shows a long recession. Depositional bed forms that are stable at high stage (mesoforms) are washed out, and postflood surfaces show only the highly stable macroforms such as alternate bars. In the Scablands an abrupt cessation of flood discharge resulted in preservation of many of the mesoforms, especially those located on the higher bar surfaces.

BEDROCK MACROFORMS

Quadrilateral Residual Forms

The eastern part of the Channeled Scabland is generally called the Cheney-Palouse Scabland tract (see Patton and Baker, Ch. 6, this volume). Its overall pattern is a complex of channel ways and interchannel divides. The pattern is especially pronounced on orbital photography (Figs. 5.1 and 5.2) because the loess on interchannel divides contrasts sharply with the eroded basalt scabland on the channel floors.

In overall pattern the Cheney-Palouse scabland tract resembles a braided stream. Instead of bars of sediment laid down by high discharges, however, the "islands" in the Cheney-Palouse are erosional residuals of basalt and loess. The largest of these have a crude quadrilateral shape in plan, often forming diamonds or parallelogram shapes. Morphologically these are similar in appearance

Figure 5.6. Oblique aerial photograph of the upstream and of a quadrilateral loess residual on the Palouse-Snake divide. Note the steep scarps that were eroded by flood water streaming over the divide. Washtucna Coulee is visible in the background.

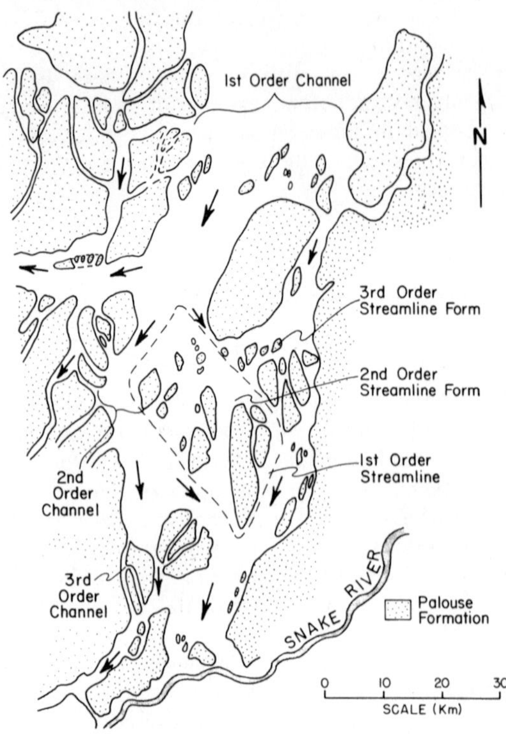

Figure 5.7. Sketch map of the Cheney-Palouse scabland showing the hierarchial relationship of channels and interchannel residual areas (mantled by Palouse Formation). An arbitrary section has been ordered to show these relationships.

to eroded bar remnants that are common in braided gravel streams. The flow split around these residual elements, eroding their upstream ends (Fig. 5.6). The flow then reconverged on the downstream ends of the residual forms. Many of the residuals were further modified by relatively shallow flood water flowing obliquely across their surfaces.

The successive division of channels and of the interchannel residual elements results in a kind of hierarchy for these erosional macroforms. Both the channels and the residuals can be ordered in the same way that Williams and Rust (1969) have ordered depositional bars in braided outwash (Fig. 5.7).

Pools and Riffles

Pools are topographically low areas produced by scour on river beds. They seem to have a very regular spacing in stream channels relative to topographically high areas called riffles. As pointed out by Leopold and others (1964, p. 203), pools and riffles tend to be spaced at 5 to 7 times the channel width. Keller and Melhorn (1973) have pointed out the significance of convergent and divergent flow for the development of pools and riffles. Pools tend to occur at locations where convergent flow increases the bottom shear stress at flood stages. In alluvial rivers, the riffles also develop at high stage when divergent flow induces deposition. In normal rivers the usual low flow condition produces a velocity reversal; the water moves more rapidly over the riffles, eroding them and transporting sediment into the relatively tranquil pools (Keller, 1971).

Because the Channeled Scabland contains some pronounced convergent and divergent flow patterns (Fig. 5.2), we would expect that flood scour might produce some incipient pool-and-riffle development. Detailed high-water profiles of the Cheney-Palouse scabland tract (Patton and Baker, Ch. 6, this volume) indicate the presence of a crude pool-and-riffle sequence. Although spacing is somewhat irregular, an average spacing of 10-15 km exists between successive riffles (steep areas on profiles). It is unclear whether this represents a pool-and-riffle sequence associated with a meandering thalweg or whether it is simply an effect caused by varying channel widths along the flood route. It is possible that both factors are interacting. The erosion of pools into bedrock requires further study.

STREAMLINED RESIDUAL FORMS

Bretz (1923b, p. 624-626) first recognized that the hundreds of isolated loess hills of the eastern scablands possessed remarkably steep, ungullied marginal hillslopes. These slopes converge to form definite prows that point up the local scabland gradient (Fig. 5.8). Although Bretz and others (1956) interpreted these hills as fluvially-eroded loess "islands", high-water mark reconstruction (Baker, 1973a) has shown that many of the hills were eroded subfluvially. There is a complete transition in scale and complexity between the small and simple streamlined hills and the large, complex zones of less modified regions of loess topography (quadrilateral forms) that separate the major anastomosing channels.

The best developed streamlined forms (Fig. 5.9) often show several or all of the following characteristics: (a) flow obstacles that localized the resistant landform, (b) upstream crescent-like scour marks, (c) downstream tapering streamlines on the adjacent channel floor, (e) oblique channels cutting through small divides at the crest of the streamlined form. The striking contrast between preserved residuals of relatively soft loess and adjacent tracts of scoured, potholed rock appears at first to be a physical paradox. Why did the harder rock erode, while the loess was preserved?

Streamlined forms are common in a variety of geomorphic phenomena. Drumlins, yardangs, and "island hills" are the characteristic forms for moving glacial ice, wind and water respectively. The water forms generally do not occur at a large scale on earth because most terrestrial rivers are too shallow. Again the deep, swift scabland flood flows account for another remarkable type of landform.

The scabland streamlined loess hills were studied by morphometric analysis of large-scale topographic maps. To simplify the analysis we considered the shapes of the hills as projections onto a plane surface parallel to the geoid (or general ground surface). Thus we only considered the shape of the forms in plan. Analysis of the streamlined shapes included three physical measurements: length, l (km) — measured parallel to suggested flow direction; width, w (km) — taken as the maximum width of the streamlined form measured perpendicular to the implied flow direction; and area, A_s (km^2) — measured with a grid emplaced over the feature. From these physical measurements a dimensionless parameter (K) can be calculated:

$$K = \frac{l^2 \pi}{4 A_s} . \qquad (5\text{-}1)$$

Chorley (1959) used this parameter to show that drumlins have a close resemblence to airfoils. Through the use of the lemniscate loop equation:

$$P = l \cos K\theta, \qquad (5\text{-}2)$$

where P and θ are polar coordinates, an equation for the streamlined form can be derived and plotted on polar coordinates (Fig. 5.10). The shapes strongly support Bretz' contention that the loess hills were streamlined by a rapidly flowing fluid. Water velocities averaged 12-15 m/sec for depths of 30-60 m in these areas.

The width (w) vs. length (l) ratio serves as a measure of the Index of Elongation (Fig. 5.11). Greater values of the ratio l/w (more gentle slopes) imply greater elongation of the form. The curve fitted to the point scatter (via regression analysis) shows that the streamlined forms exhibit nearly parallel reationships. There is little change in the l/w ratio with increasing size. The index is very close to 3.

The relationship of length to area (Fig. 5.12) measures the narrowness of streamlined forms. It is a measure of the relative percentage of the area contributed by the length component. Higher l/A_s values indicate increasing narrowness of streamlined forms.

The relationship of width to area (Fig. 5.13) measures the degree of broadness, i.e. how much of the total area is contributed by the width component. As the w/A_s increases, the tendency toward broader forms will increase.

These detailed shape analyses further document the average geometry for streamlining of the scabland loess hills. If the hills had rectangular shapes, then their total areas A_t would be

$$A_t = lw. \qquad (5\text{-}3)$$

However, from Figures 5.12 and 5.13 we have:

$$l = 1.9\sqrt{A_s}\,, \qquad (5\text{-}4)$$

$$w = 0.66\sqrt{A_s}\,, \qquad (5\text{-}5)$$

and

$$A_s = \frac{3.19}{4\,A_t}\,. \qquad (5\text{-}6)$$

That this area ratio derives from producing a streamlined shape is seen in the analogy of producing the minimum perimeter shape that has the same length and width as a square. For the latter problem, $l = w$ is the diameter of a circle with area A_c:

$$A_t = l^2.$$

$$A_c = \frac{\pi}{4}\,l^2.$$

$$A_c = \frac{\pi}{4}\,A_t\,. \qquad (5\text{-}7)$$

which is nearly identical to the result produced empirically in equation (5-6). We can further extend this theoretical exercise to calculate an ideal K factor:

$$K = \frac{l^2\,\pi}{4\,(\pi\,(l/2)^2)} = 1. \qquad (5\text{-}8)$$

By minimizing K one is essentially minimizing the skin resistance of an object with a given area. However, skin resistance is only one of two components of the drag that develops around any submerged object. The other component is pressure drag, which derives from the turbulent wake developed behind an obstacle. This factor is

Figure 5.8. Streamlined residual forms in the Cheney-Palouse scabland: A. Ground-level view showing height(s) of the flood-eroded loess scarp. B. "Upstream" end of the loess hills on the Palouse-Snake crossing. Note the prominent prows formed by convergent hillslopes pointing upstream for the flood flows. The scarps average 40 m in height. Maximum flood flow depths averaged 60 m at this location. C. "Downstream" ends of same loess hills. D. View of the channel cutting obliquely through the "snout" of a stream-lined loess hill. Note the surrounding scabland and unmodified loess topography in the far distance.

minimized by increasing the elongation of the obstacle so that less width component is presented transverse to the flow. The scabland forms are elongated to about three times their maximum width (Fig. 5.11). It is apparent that the streamlined hills are equilibrium forms, elongated sufficiently to reduce pressure drag, but not so long that they create excessive skin resistance. They owe their preservation in the high-velocity flood flows to this marvelous streamlining. Whereas the adjacent rock outcrops generated destructive macroturbulence, the loess hills fostered a smooth bending of streamlines around them.

Some slight tendency for increased streamlining (increased K factors) appears to occur with increasing Reynold's number (Fig. 5.14). The Reynold's numbers (R) used are the maximum values achieved in the flood flows and were computed from the expression:

$$R = \frac{Q\,R\,\rho}{A\,\mu}, \qquad (5\text{-}9)$$

where Q is the maximum discharge of a scabland reach (computed by slope-area procedures described by Baker, 1973a), R is the hydraulic radius of the reach, ρ is the fluid density, A is the cross-sectional area, and μ is the dynamic viscosity. These maximum flow Reynold's numbers varied from 2×10^8 to 2×10^9 in most scabland reaches.

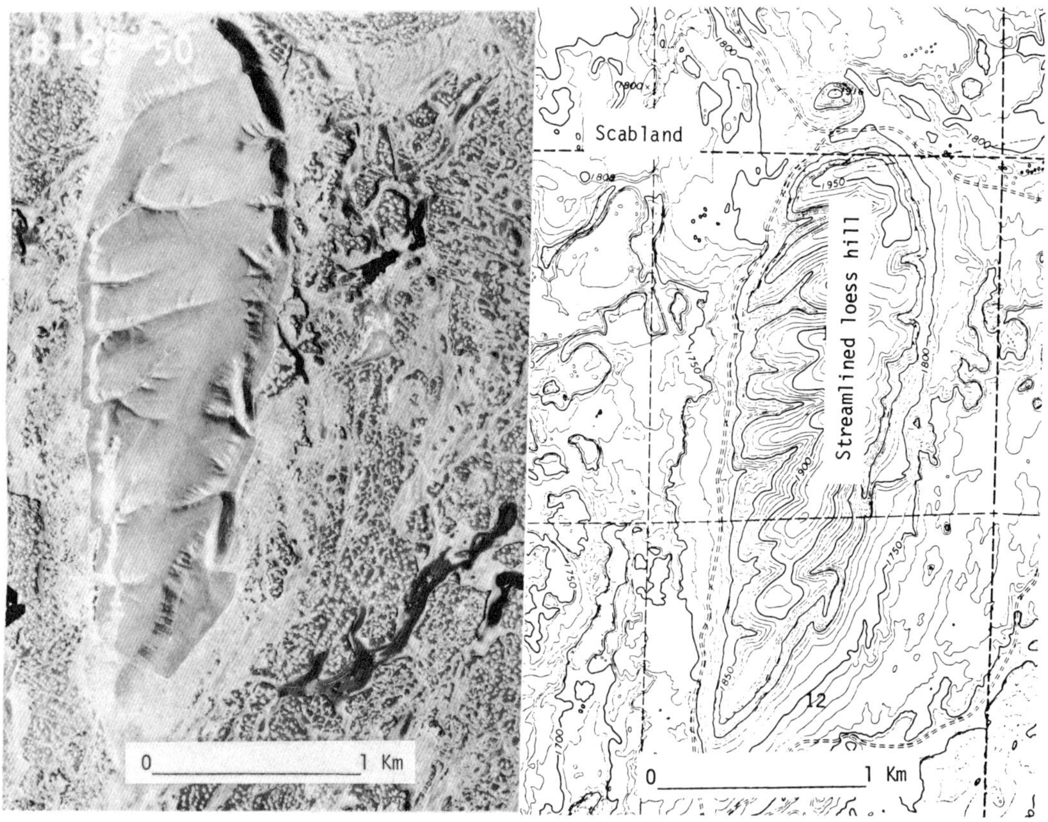

Figure 5.9. Residual loess hill streamlined by flood erosion. A small cataract, heading an inner channel, has worked its way around the blunt upstream end of the hill (on map). Longitudinal grooves and butte-and-basin topography can be seen in the marginal scablands. Water depths and velocities averaged 12 m/sec for depths of 30-40 m in this area during the flood maximum. Map contour interval is 10 feet (~ 3 m). This hill is located in the Cheney-Palouse scabland tract in sections 1 and 12, T. 18N., R. 37E. The same hill is viewed obliquely in Fig. 5.8D.

BARS OF THE CHANNELED SCABLAND

The term "bar" is used for all large-scale depositional forms in streams and rivers. Bars originally denoted impediments to navigation. Here the term is applied to all depositional macroforms in scabland channels. Unfortunately there is no generally accepted classification of fluvial bars. Indeed, a hard and fast classification of bars is probably impossible for the following reasons: (1) no classification can satisfy all three major purposes of bar studies, morphologic, hydrodynamic, and paleohydraulic; and (2) many bar forms are ephemeral members of evolutionary sequences, complexly related to initial conditions and transitory flow conditions.

The classification used here for scabland bars is limited to the Channeled Scabland. It is developed from the relationship of bars to the large-scale flow pattern in a local scabland reach.

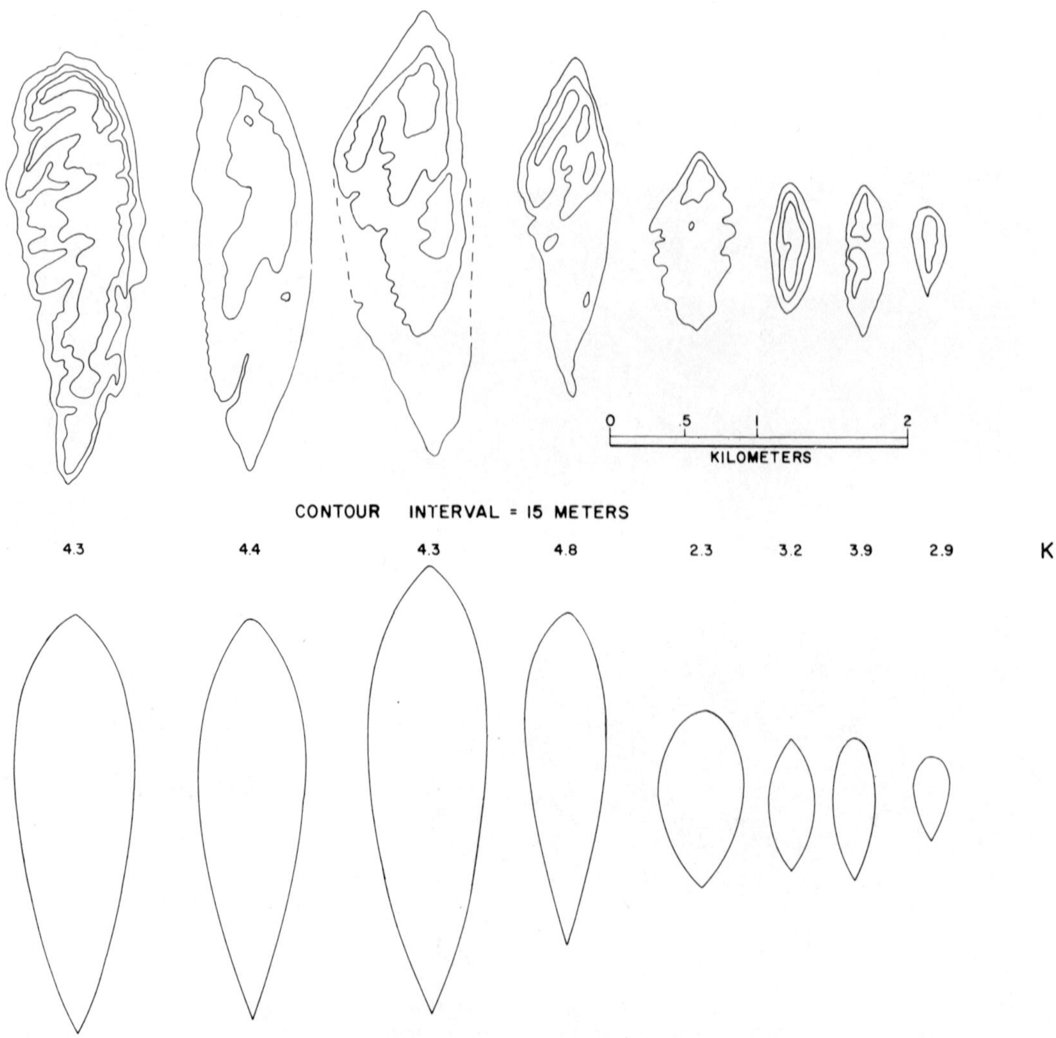

Figure 5.10. Comparison of typical streamlined scabland loess hills to equivalent lemniscate loops as calculated according to Chorley's (1959) analysis of drumlin shapes. Water flowing over and around the easily scoured loess caused the streamlined shapes, permitting less resistance to flow. Subsequent gulleying of the soft loess has somewhat modified the streamlined shapes.

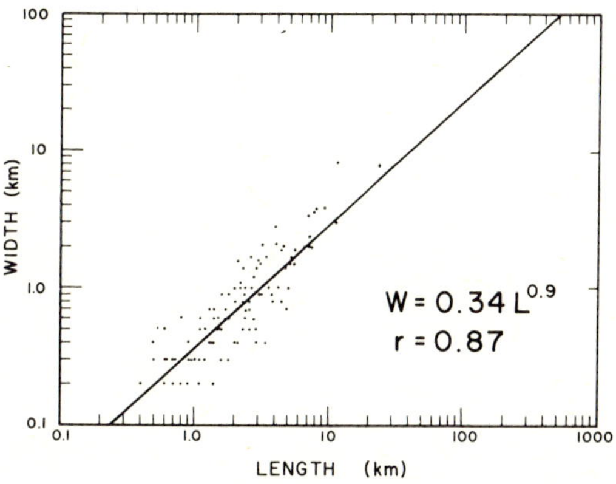

Figure 5.11. Maximum width of scabland streamlined forms versus the length of the form.

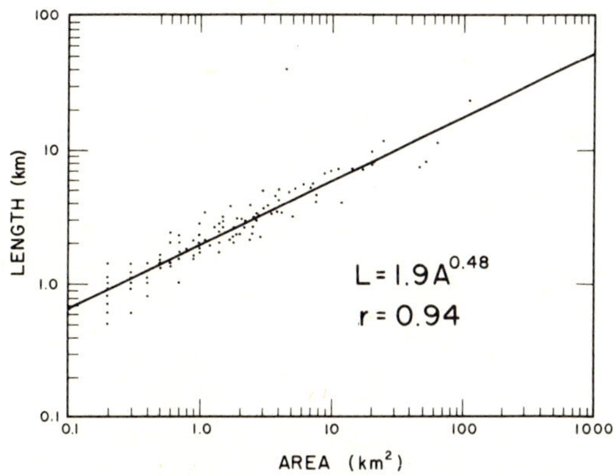

Figure 5.12. Length of scabland streamlined forms versus form area.

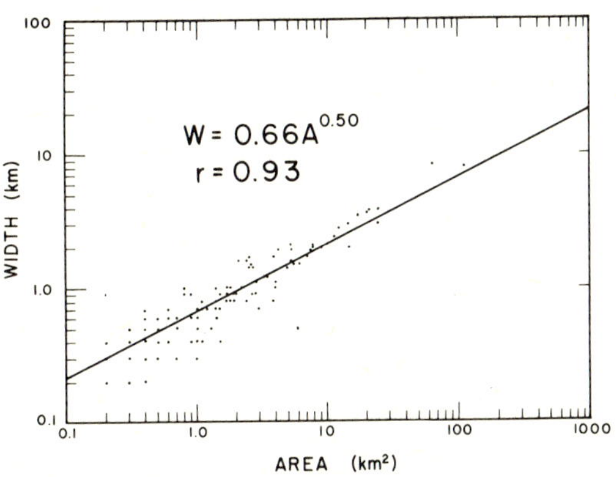

Figure 5.13. Maximum widths of scabland streamlined forms versus form area.

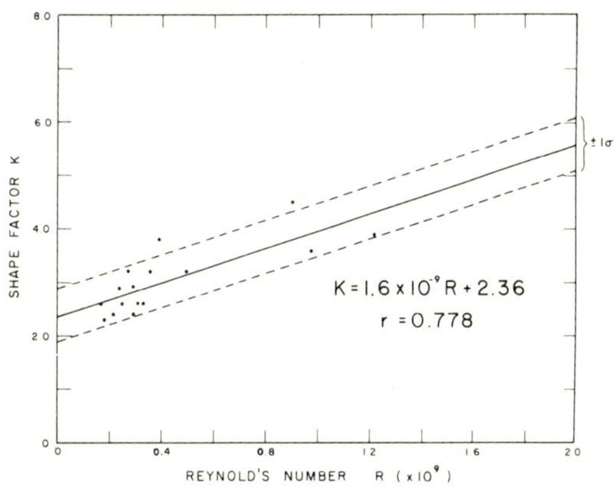

Figure 5.14. Streamlined shape factor K versus the Reynold's number for maximum scabland flood flows associated with scabland streamlined forms. The dashed lines indicate one standard error.

Longitudinal Bars

These bars are elongated parallel to the flow direction. They are characteristic of relatively uniform scabland reaches, lacking abrupt expansions and constrictions. In braided gravel rivers, longitudinal bars tend to be broad, low forms with massive bedding or crude horizontal structure within. In the Channeled Scabland, however, the longitudinal bars are mounded, streamlined forms tens of meters thick. Stratification is dominated by foreset beds that were accreted to avalanche faces on the downstream margin of the bar.

Malde (1968) introduced the term "pendant bar" to refer to streamlined mounds of Bonneville Flood gravel that occur downstream from bedrock projections on scabland channel floors. Baker (1973a) found that this was the most common type of bar in the Channeled Scabland. The locus for bar initiation may be a knob of basalt (Fig. 5.15) or the bend of a pre-flood meandering valley (Fig. 5.16). Bar deposition was apparently initiated by gravel deposition in flow separations that developed downstream from a variety of flow obstructions (Fig. 5.17). Additional material was then added as huge foresets on the downstream margins of the bars. By this downstream accretion, scabland bars maintained a zone of flow separation that induced deposition from flows that otherwise would be competent to transport even coarse boulders.

Rather than being purely depositional forms, pendant bars are best viewed as the consequences

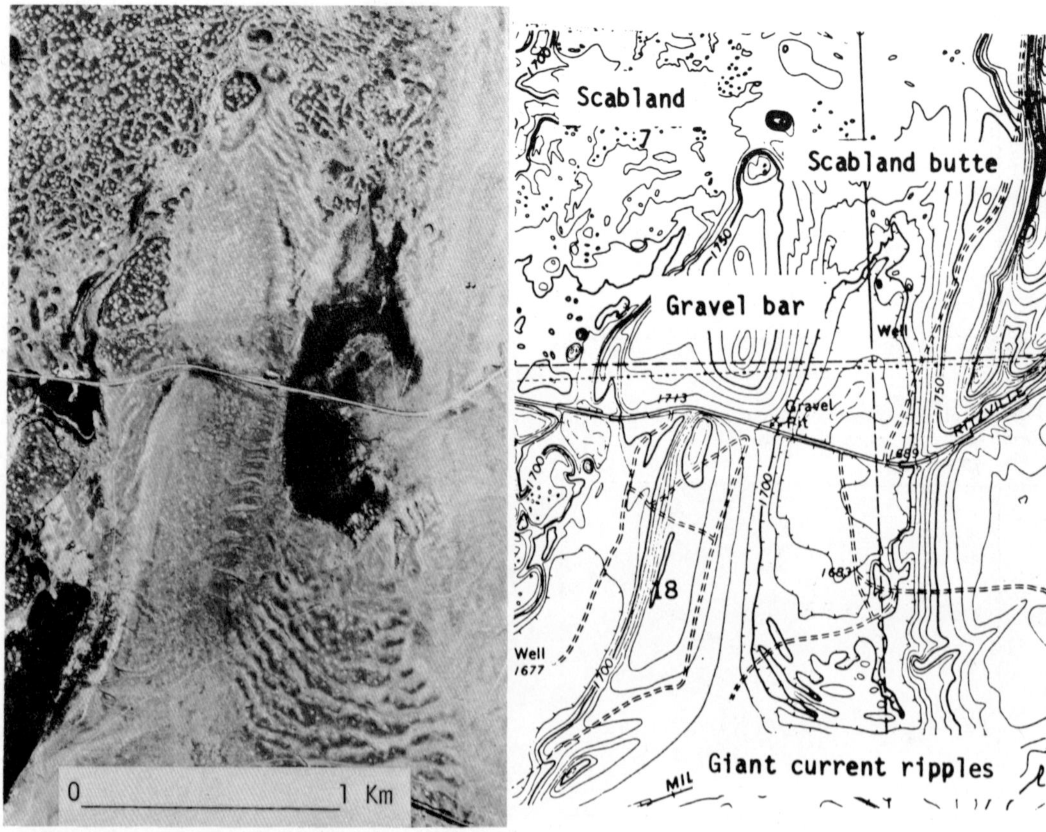

Figure 5.15. Vertical aerial photograph and topographic map of a small pendant bar in the Cheney-Palouse scabland near Macall, Washington. Contour interval is approximately 3 m (10 feet). The bar accumulated downstream from a residual butte of basalt. The bar occurs in sections 7 and 18, T. 18N., R. 38E.

Figure 5.16. Oblique aerial photograph of Bar 2 near Wilson Creek, Washington. This is a relatively small bar about 1 km in length.

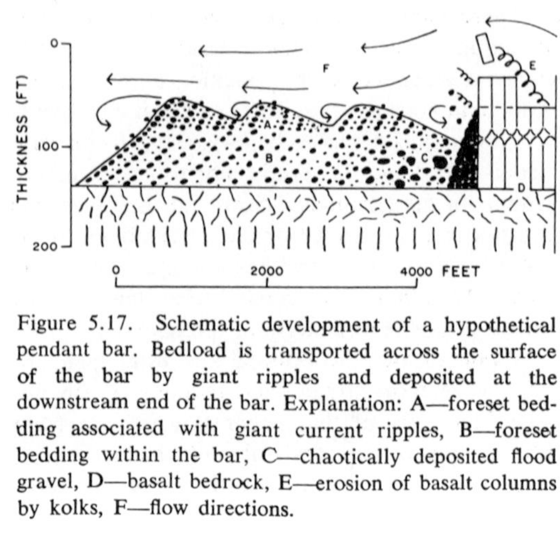

Figure 5.17. Schematic development of a hypothetical pendant bar. Bedload is transported across the surface of the bar by giant ripples and deposited at the downstream end of the bar. Explanation: A—foreset bedding associated with giant current ripples, B—foreset bedding within the bar, C—chaotically deposited flood gravel, D—basalt bedrock, E—erosion of basalt columns by kolks, F—flow directions.

Figure 5.18. Pendant bar deposits exposed near Coulee City, Washington. Note the well-sorted foresets of open-work, cobble gravel.

Figure 5.19. Large column of basalt in flood gravel comprising Bar number 3 near Wilson Creek, Washington (see description in Bretz and others, 1956, p. 977-979). Note percussion marks to right of rock hammer.

of special flow conditions that locally reduced the fantastic competency of the peak flood discharges. The transport rate of flood gravel into the separation zone downstream from the bar was simply greater than the transport rate out of that zone. The relationship of the bars to the flow conditions is evinced by the fact that they are streamlined to present minimum resistance to the flood water. Moreover, in curving reaches pendant bars never abut on the channel walls, but are separated from the walls by depression which Bretz and others (1956) termed "fosses."

Measurements of scabland bars show that, like the streamlined residual hills, they are elongated to no more than about three times their width. Thus, the pendant bars are also equilibrium forms, balancing the shear drag from skin resistance (which increases with the length of a streamlined form) and the form drag (which decreases with the width of a streamlined form). The bars grew downstream no more than three times their maximum width because the increased skin resistance produced by greater elongation would have resulted in greater transport rates away from the bar.

Pendant bars probably accumulated a rather good sample of the sediment load carried by the highly turbulent flood flows. Their internal stratification is characterized by foreset bedding (Fig. 5.18). Rapid deposition of recently eroded boulders is indicated by the inclusion of basalt columns in the bar sediments. These columns are sometimes marked only by a few percussion impacts (Fig. 5.19).

In some scabland channel reaches, pendant bars occur in groups. Near Odessa, the resistant basaltic ring structures form resistant knobs. Many of these knobs have pendant bars trailing from their downcurrent ends (Fig. 5.20).

Other Bar Forms

The absence of linguoid and transverse bars from the Channeled Scabland has been discussed by Rust (1975, p. 246). Unlike most braided rivers, in which channels are exceptionally wide and shallow, the Missoula and Bonneville Flood channels were relatively constricted. The exceptionally deep flood water allowed bars to develop prominent slip-faces. These were the stable bedforms under extreme flood conditions. The major form of modification during falling stage has been the concentration of large boulders on the bar surface by winnowing. Most braid bars, in contrast, form by deposition on subhorizontal surfaces. Foresets are rare in gravel braided streams.

Rust (1975) believes that gravel braid bars are like the scabland bars in that they form initially as primary bedforms. They are stable under the flood flows in which all bed material is in motion. In contrast, however, many sand bed and

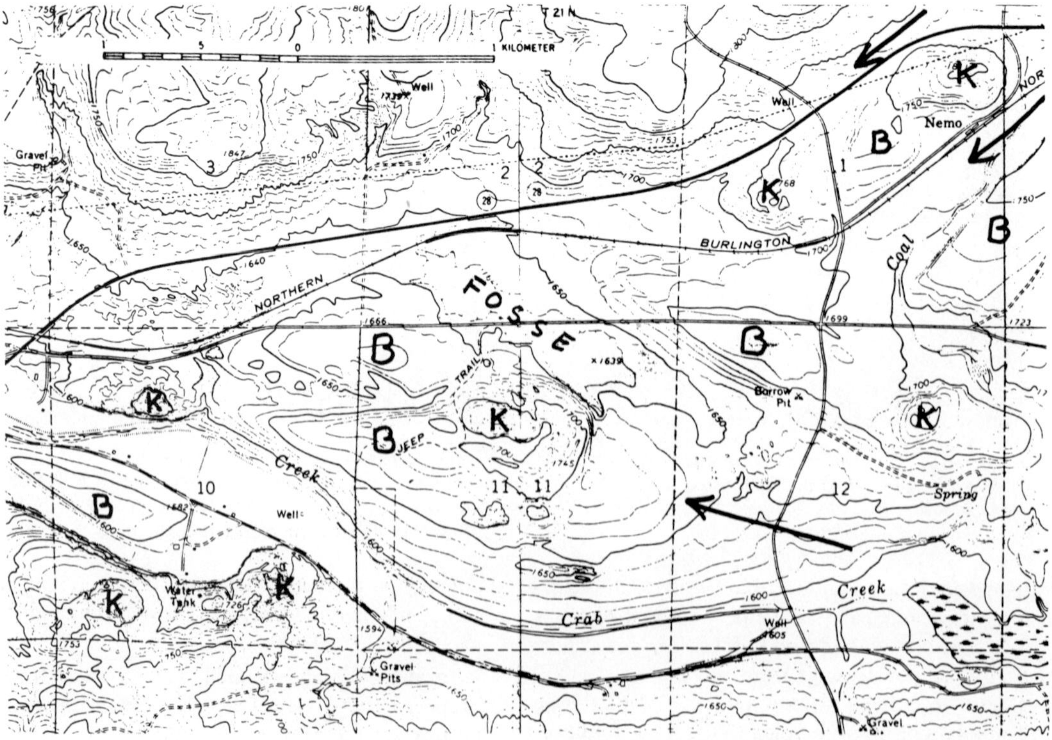

Figure 5.20. Topographic map of pendant bars (B) near Odessa, Washington. The prominent basalt knobs (K) result from the resistance of circular dikes intruded into the jointed lava flows. Contour interval is 3 m (10 feet). Topography is from the Odessa and Sylvan Lake, Washington, 7.5-minute quadrangles.

some gravel bed rivers show extensive bar modification during changing river stages. These complex, modified bar forms may characterize the great expansion bar complex of the northern Quincy Basin (Baker, 1973a, p. 39-42).

Eddy Bars

These bars occur at the mouths of alcoves or valleys that were tributary to valleys invaded by catastrophic flood flows. They are particularly well-developed along the eastern margins of the Cheney-Palouse scabland tract. Nearly every drainage entering the scabland tract is blocked at its mouth by an eddy bar. Excellent examples occur at Willow Creek (described by Bretz and others, 1956, p. 1031-1033) and at the mouth of the Tucannon River.

The contrast between the internal structures of pendant and eddy bars can be seen by comparing Figures 5.18 and 5.21. Pendant bars (Fig. 5.18)

Figure 5.21. Eddy bar deposits at the mouth of the Tucannon River. The lowest layer of horizontally bedded sand is overlain by about 60 cm of laminated silt. This is overlain by poorly sorted, angular to subangular flood gravel containing percussion flaked boulders and cobbles. The prominent boulders of silt are identical to the underlying laminated silt layer. They may be "rip-off" clasts such as commonly found in turbidites.

show well developed foreset bedding. Individual foresets contain well sorted subrounded boulders and cobbles with an open-work structure. Other foresets are entirely made up of granules and pebbles. Eddy bars (Fig. 5.21) contain a variety of grain sizes and structures. Interfingering occurs between poorly sorted boulder gravel, laminated silts, cross-bedded granule gravel, and graded sand-silt layers. The boulders and cobbles are marked by percussion flake scars. Many are "broken rounds," as described by Bretz (1929).

Foreset bedding in the pendant bars almost always dips downstream. Bedding in the eddy bars indicates varying directions of sediment transport. Crude foresets in the boulder gravels usually dip away from the main scabland channel. Frequently, however, the smaller foresets in the granule gravels dip back toward the main scabland channel. This pattern may be the result of the swirling eddys which deposited the bars. The stronger currents carried the coarsest flood debris up the tributary valley. Weaker back flow currents then deposited the finer granule gravels.

The swirling currents which resulted in eddy bars are another manifestation of macroturbulent phenomena as described by Matthes (1947). As suggested by Krumbein (1942), extremely high turbulence may prevent the equilibrium conditions that serve to distinguish bedload and suspended load. This may be the origin of the extremely poor sorting of eddy bar deposits when compared to pendant bar deposits. Another result may be the fact that giant current ripples are never associated with eddy bars. Because macroturbulent phenomena are not well understood a precise origin of eddy bars cannot be described.

The sedimentologic featuers noted for eddy bars may be seen for many kilometers upstream along scabland tributaries. However, the abundance of coarse materials decreases in an upstream direction relative to sand and silt. These deposits upstream from the eddy bar blockade are termed "slackwater deposits." Slackwater deposits are distinct from the main channel deposits, but it must be remembered that there is a complete lateral gradation between the two.

BEDROCK MESOFORMS

The bed forms eroded by the macroturbulent flood flows on bare basalt surfaces are of immense variety and unique size in the Channeled Scabland. Indeed these are the features most characteristic of "scabland." Bretz' studies of these bizarre erosional forms were largely descriptive. It is now apparent, however, that the rock bedforms exist in an evolutionary sequence that is related both to the flood flow hydrodynamics and to the resistant characteristics of the jointed basalts. Here the forms will first be described, and then their evolutionary sequence will be discussed.

Cataracts and Inner Channels

The most impressive erosional forms created by the Missoula floods are probably the abandoned cataracts (Fig. 5.22). Most famous of these is Dry Falls (Fig. 5.23), 5.5 km wide and 120 m high. Bretz and others (1956, p. 1029) recognized that many of the cataracts were formed subfluvially rather than by the plunge-pool undercutting classically illustrated by Niagara Falls. This contention is supported by the high-water mark evidence left by the early Pinedale (late Wisconsin) flood (Baker, 1973a; his Fig. 7). Bretz (1932a) described in detail the initiation of a 250 m cataract near Coulee City, Washington, and its 32 km upstream recession to create the upper Grand Coulee (Fig. 5.24). The unique capacity of vertically jointed basalt to maintain the lip of a recessional cataract was held to be the primary consideration in this type of erosion.

All scabland cataracts show multiple horseshoe-shaped headcuts. Potholes and Frenchman Springs have two parallel headcuts or alcoves (Fig. 5.25). At the base of each alcove is a large closed depression. Even with post flood modification, the closure in these depressions is as much as 35 m. Probably the cataract headcuts acted as efficient funnels during the maximum flood flows. The water surface was sharply drawn down over the cataract producing intense macroturbulent scour beneath the locally steep water-surface gradient. Plucking erosion was concentrated in the columnar-jointed zones, and large blocks of entablature were undermined at the cataract lip.

The headward recession of scabland cataracts produced distinct inner channels. Field mapping shows that the margins of these channels and the cataract lips are held up by relatively resistant basalt entablature (Fig. 5.26 A and B). Bretz and others (1956) envisioned powerful kolks at the plunge pool locations, undermining the cataract lip. Sediment, usually basalt columns and large blocks of entablature, were transported away from the cataract lip in a state of quasi-suspension, buoyed by the intense macroturbulence (Baker, 1973a, p. 26-29). The competency of the kolks is difficult to estimate theoretically. Boulders up to 30 m in diameter (estimated from aerial photography) were transported from the lip of West Potholes cataract. Certainly the vertical vortices (kolks) were exceedingly powerful agents of lift and transport.

In the upper Grand Coulee (Fig. 5.24) cataract recession was initiated at the structural step provided by the Coulee Monocline. The undermining and recession process proceeded rapidly enough for the cataract to recede 32 km to the gorge of the Columbia River on the northern margin of the Columbia Plateau. On the western rim of the Quincy Basin (Fig. 5.25) cataract recession was initiated at the canyon walls carved by the Columbia River.

Many scabland channels were excavated from preflood stream valleys. At Lenore Canyon and Moses Coulee (Fig. 5.27) the preflood tributaries now form hanging valleys that drain Palouse loess topography unmodified by catastrophic flooding. The floors of these valleys were deepened by flooding, often with the production of distinct inner channels (Fig. 5.28).

Potholes and Buttes

Perhaps the most prevalent topographic form in eroded rock of the scabland tracts is butte-and-basin topography. The usual development is small anastomosing channels and rock basins surrounding buttes and mesas, with a total relief of 30-100 m (Fig. 5.29). The rock basins range in size from shallow saucers to the scale of Rock Lake,

Figure 5.22. Oblique aerial photographs of scabland cataracts and inner channels. A. West Potholes cataract on the western rim of the Quincy Basin. B. Frenchman Springs cataract also on the western rim of the Quincy Basin. C. Hudson cataract at the head of Hudson Coulee in the Hartline Basin. D. Palouse Falls, a small cataract in the inner channel eroded through the Palouse-Snake divide crossing (the Palouse River now occupies this flood channel).

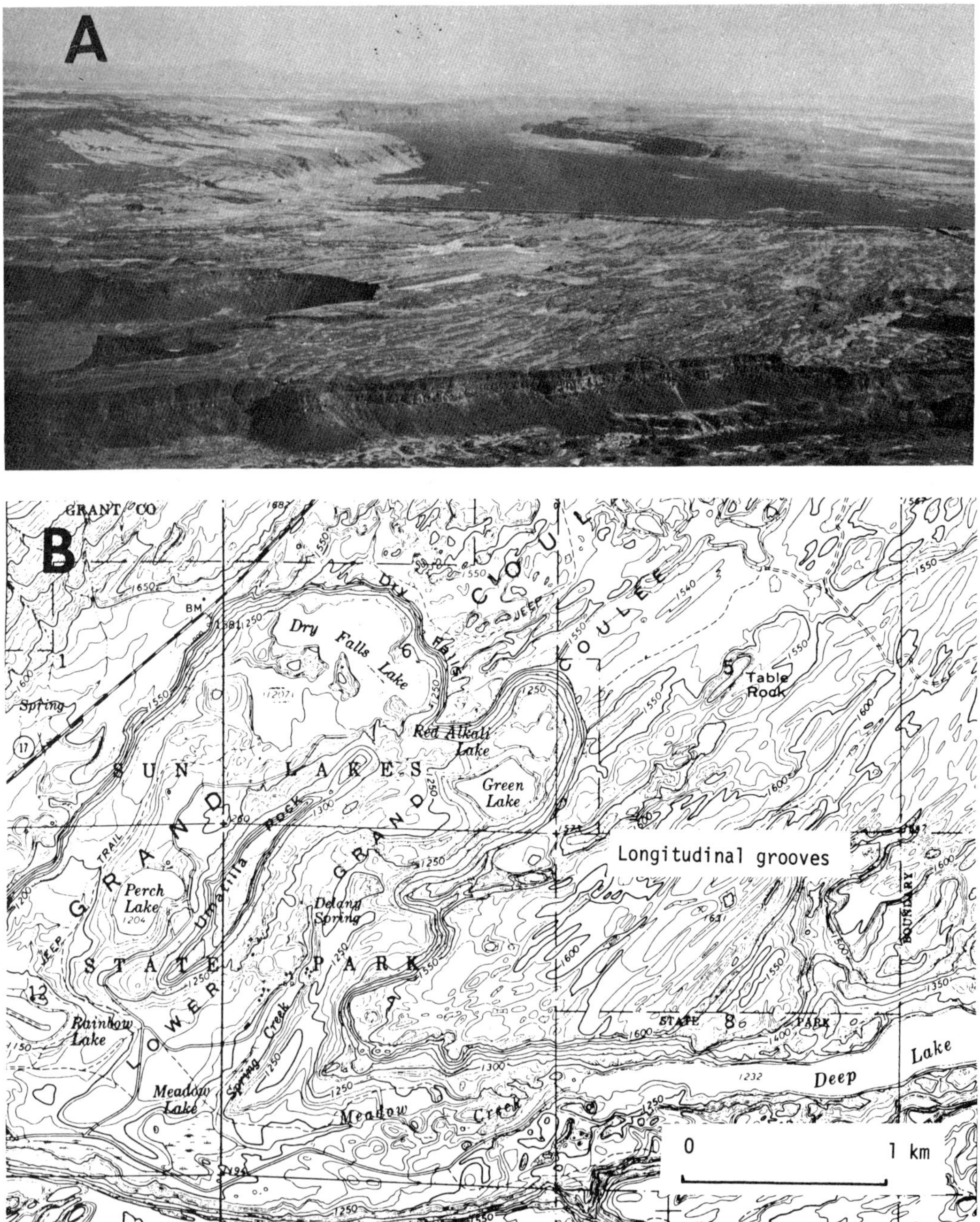

Figure 5.23. A. Oblique aerial photograph of the Dry Falls cataract group. The cataract is 120 m high and 5.5 km wide. Longitudinal grooves are visible just upstream from the cataract head, and the upper Grand Coulee (containing Banks Lake) extends to the horizon. B. Topographic detail of the Dry Falls area showing the inner channel development (contour interval is 10 feet; elevations in feet). Note topographic expression of longitudinal grooves.

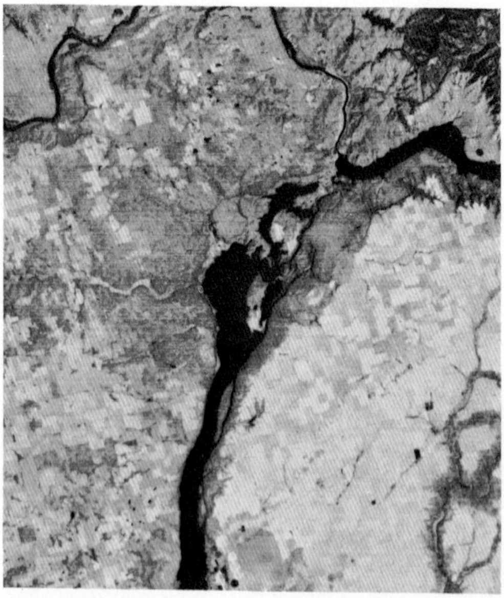

Figure 5.24. Orbital photograph of the upper Grand Coulee showing the immense inner channel left by headward retreat of the 250 m deep cataract. LANDSAT image E-2 936-17451-5.

11 km long and 30 m deep. Bretz (1932a, p. 26-28) described this combination of features as follows: "The channels run uphill and downhill, they unite and they divide, they deepen and they shallow, they cross the summit, they head on the back-slopes and cut through the summit; they could not be more erratically and impossibly designed."

Some typical scabland potholes are shown in Figure 5.30.

Longitudinal Grooves

Where broad expanses of a single basalt surface were eroded by catastrophic flood flows, the characteristic form of the erosion is a series of elongate grooves. These features generally have their long axes parallel to the prevailing flood flow streamlines. The grooves are common on basalt entablature surfaces in the Hartline Basin (Fig. 5.23). They average 5 m in depth and 50 m in width. Relatively large grooves are developed near Palm Lake in the Cheney-Palouse scabland tract (Fig. 5.31).

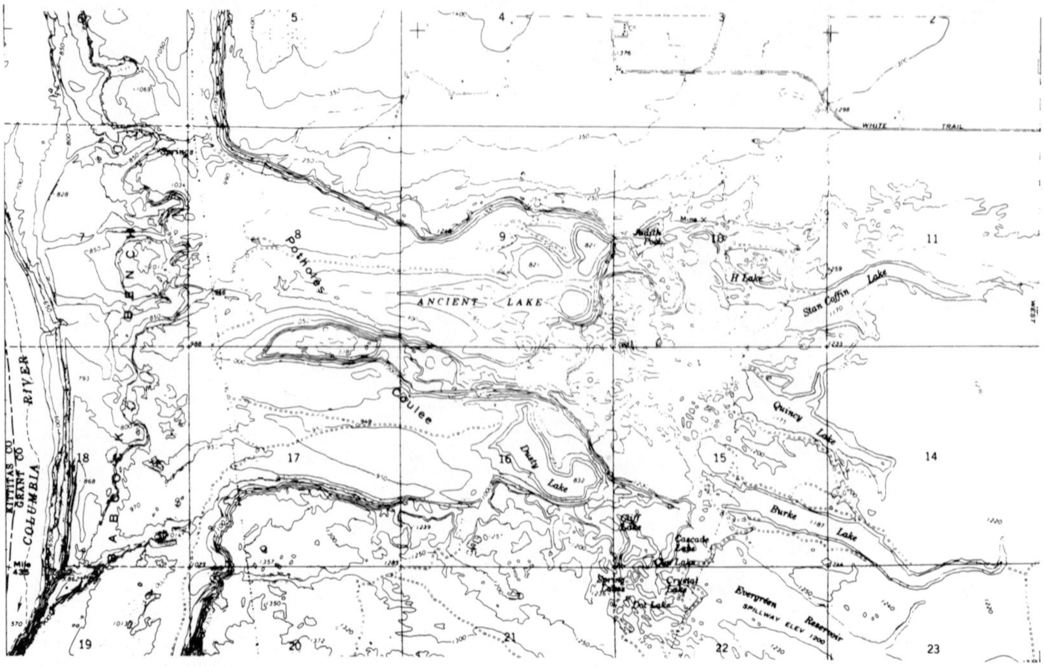

Figure 5.25. Topographic map of the Potholes Cataract. Topography is from the Babcock Ridge 7.5-minute quadrangle.

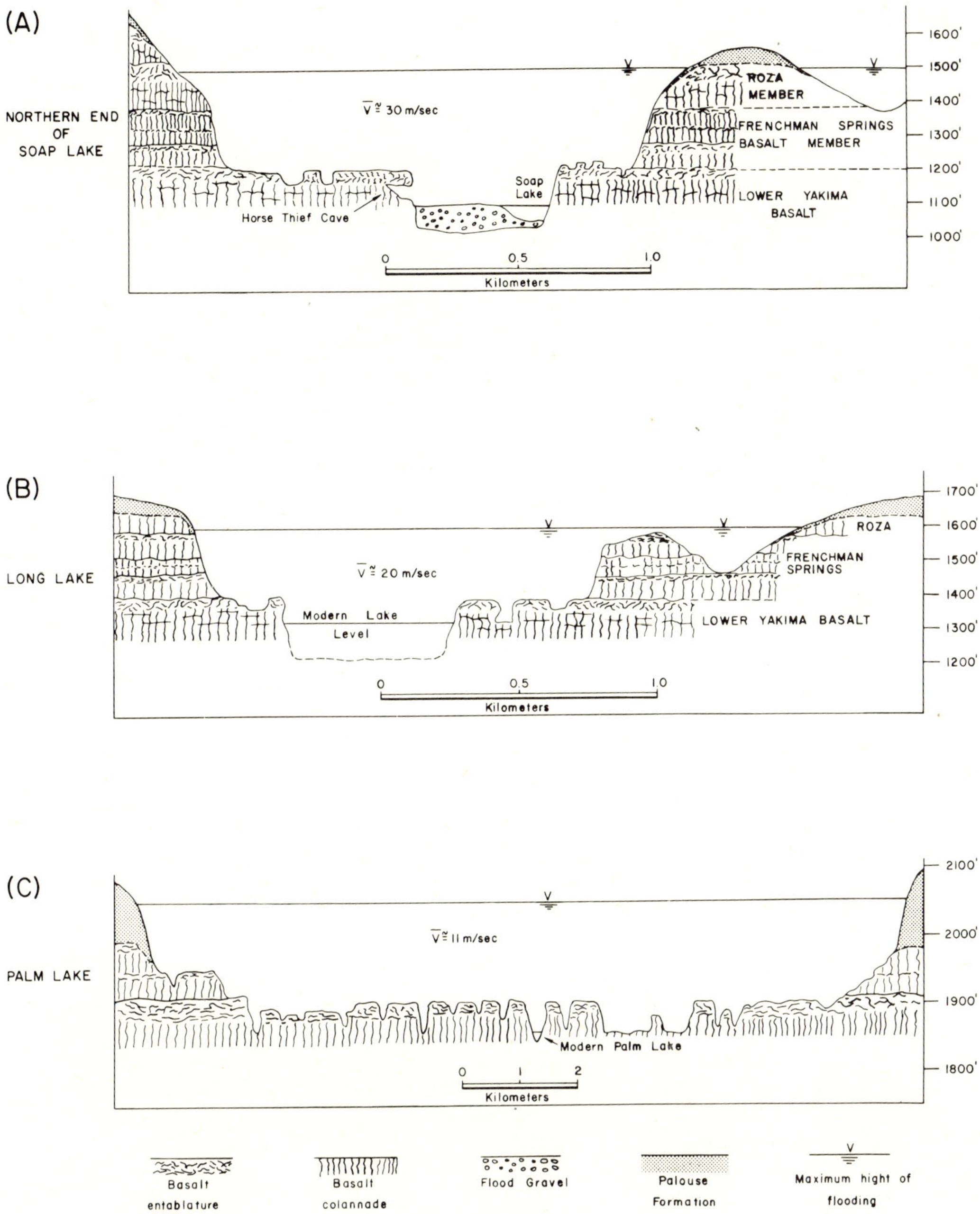

Figure 5.26. Representative cross sections of scabland channels showing structural characteristics of basalt, erosional features, and flood stage plus maximum mean flow velocity ($\overline{V}$). The sections at Soap Lake (A) and Long Lake (B) show inner channels with flanking butte-and-basin scabland topography. The section at Palm Lake (C) shows longitudinal groove topography.

Sequence of Erosional Forms

Recent experimental studies of fluvial erosion utilizing simulated bedrock (Shepherd, 1972; Shepherd and Schumm, 1974) indicate that a sequence of erosional bed forms may develop in bedrock as a function of time. First to appear in these experiments were the faint streaks of longitudinal lineations associated with potholes and transverse erosional ripples. The lineations then became enlarged into prominent longitudinal grooves. Eventually the grooves decreased in number, and finally one narrow, deep inner channel formed. In the experiments the inner channels were incised below base level, and nickpoints migrated headward upstream (see Schumm and Shepherd, 1973, p. 7, for a longitudinal profile). Deposition of bedload occured downstream from the headcut, culmination in a low-gradient inner channel with a sand bed and simulated bedrock banks. The lower part of this channel was actually an elongate basin, incised below base level.

Shepherd (1972) observed that some of the scabland erosional topography described by Baker (1973a) was analogous to features produced in the flume experiments. The most important differences derive from the differential erosional resistance provided by the jointed basalt. As shown by Ippen and others (1962) this effect is a predictable consequence of the high shear stress gradient that occurs on the inside of curves in an experimental trapezoidal channel.

The progressive erosion of a typical scabland divide crossing (Fig. 5.32) is envisioned as follows. The first flood water to overtop a divide encountered soft Palouse loess and Ringold Formation (Phase I). The high velocity water quickly exposed the underlying basalt, leaving an occasional streamlined loess hill as a remnant of the former cover (Phase II). The entablature of the uppermost basalt flow was then encountered. This probably yielded to groove development, possibly associated with longitudinal roller vortices. The first exposure of well-developed columnar jointing, perhaps at the top of a flaring colonnade along the irregular cooling surface, introduced a very different style of erosion (Phase III). Large sections of columns could now be removed at this site with the simultaneous development of vertical

Figure 5.27. Inner channel development in Moses Coulee. This oblique aerial photograph shows a distinct inner channel (I) on the coulee floor with marginal butte-and-basin scabland (S). Giant current ripples (R) occur on the surface of a flood gravel bar at the mouth of the inner channel. Preflood tributaries to the drainage were truncated by flood erosion to form hanging valleys (V) separated by truncated spurs.

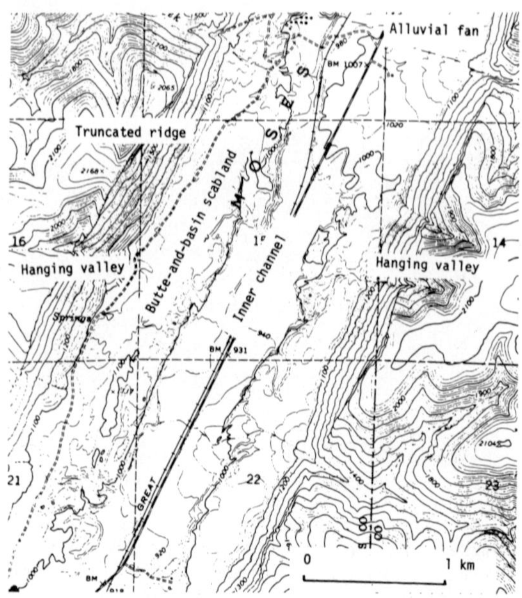

Figure 5.28. Topographic map of lower Moses Coulee showing the relationships discussed in Fig. 5.27. The contour interval is 10 feet (elevations in feet).

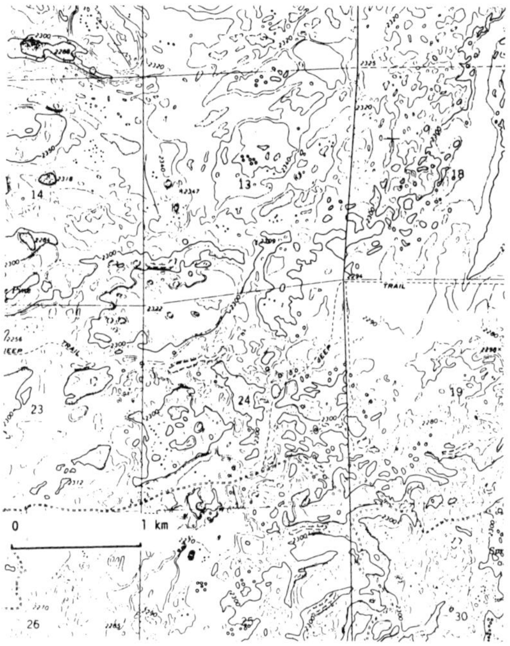

Figure 5.29. Butte-and-basin scabland developed 12 km south of Creston in the Telford-Crab Creek scabland tract. Topography (contour interval is 10 feet) is from Telford, Washington, 7.5-minute quadrangle.

vortices (kolks). With the enlargement and coalescence of the resultant potholes, the surface assumed the bizzare butte and basin topography that characterizes much of the Channeled Scabland (Phase IV). The eventual topographic form was the development of a prominent inner channel (Phase V). Such inner channels may have been initiated at downstream structural steps in the basalt, and then migrated headward by cataract recession. The lateral enlargement of inner channels probably proceeded by the undercutting of resistant entablature as columns were plucked out by kolks. Horsethief Cave, north of Soap Lake, is an excellent example of this type of erosion.

SCOUR MARKS

Large boulders on scabland bars and bedrock projections along scabland channels produced a deformation of flood flow streamlines that resulted in distinctive erosional scour marks. Engineers have studied the hydrodynamics of such scour in order to protect bridge piers during floods (Laursen, 1960). The scour is generated by two basic systems of vortices (Shen, 1971): the horseshoe-vortex system and the wake-vortex (Fig. 5.33A).

The prominent crescentic scour hole on the northwest (up-stream) side of the boulder in Figure 5.33B was probably caused by the hydrodynamic stretching and accumulation of vortex filaments in the front of the boulder. Richardson (1968) described this process as a characteristic effect of a blunt-nosed obstacle on an approaching two-dimensional velocity field. The strong pressure field produced by the blunt obstacle causes a separation of the boundary layer which then rolls up ahead of the obstacle to form a horseshoe vortex. Karcz (1968) suggested that this mechanism is responsible for the current crescents that commonly occur upstream from obstacles.

The large elliptical scour hole that formed on the downstream side of the boulder in Figure 5.33B formed as a result of a wake-vortex system generated by flow separation in the rear of the boulder. Engineering experiments have shown that the wake vortex system is a function of Reynold's number (velocity of approach times obstacle diameter times fluid density, divided by dynamic viscosity of the fluid). With other variables in the Reynold's number held constant, increasing velocity results first in a pair of vertical vortices. When the flow changes from laminar to turbulent, vortices form and migrate downstream. Karcz (1968) notes that the average velocity in the wake region is quite low at this stage and deposition is likely in the lee of the obstacle. Indeed the scablands contain many examples of pendant bars that were deposited as elliptical forms in the lee of obstacles. Malde (1968) described similar features that occurred during the Bonneville Flood in the Snake River Plain of Idaho.

At very high flow velocities, vortex intensity grows and the sucking action of kolks dominates. Shen (1971, p. 23-25) observed that the wake-vortex system then acts like a vacuum cleaner in removing bed material at this stage. Thus, the scour hole in the lee of the boulder may indicate high flow velocities. Baker (1973a, p. 41-42)

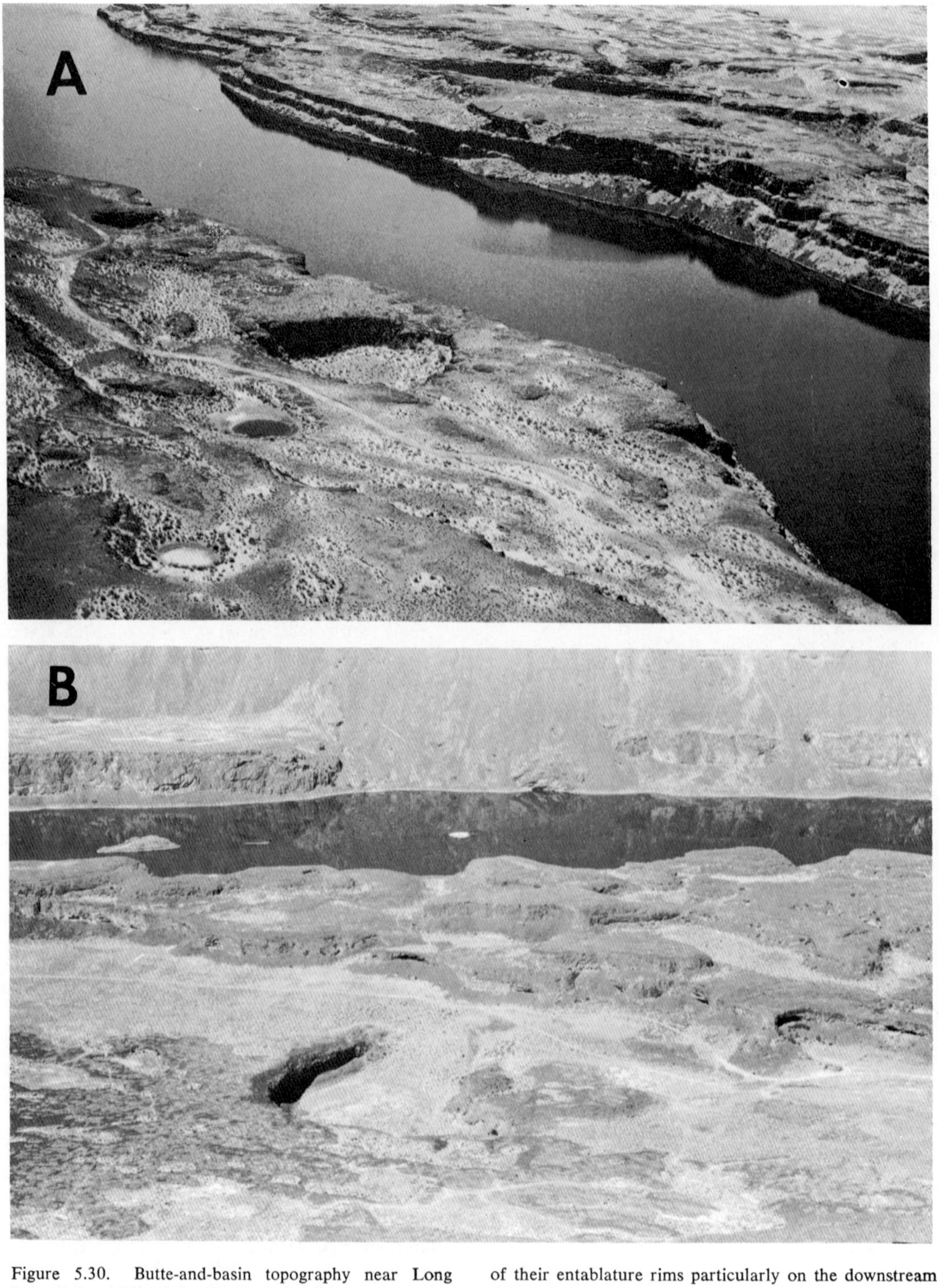

Figure 5.30. Butte-and-basin topography near Long Lake (A) and Blue Lake (B). Potholes show undercutting of their entablature rims particularly on the downstream side.

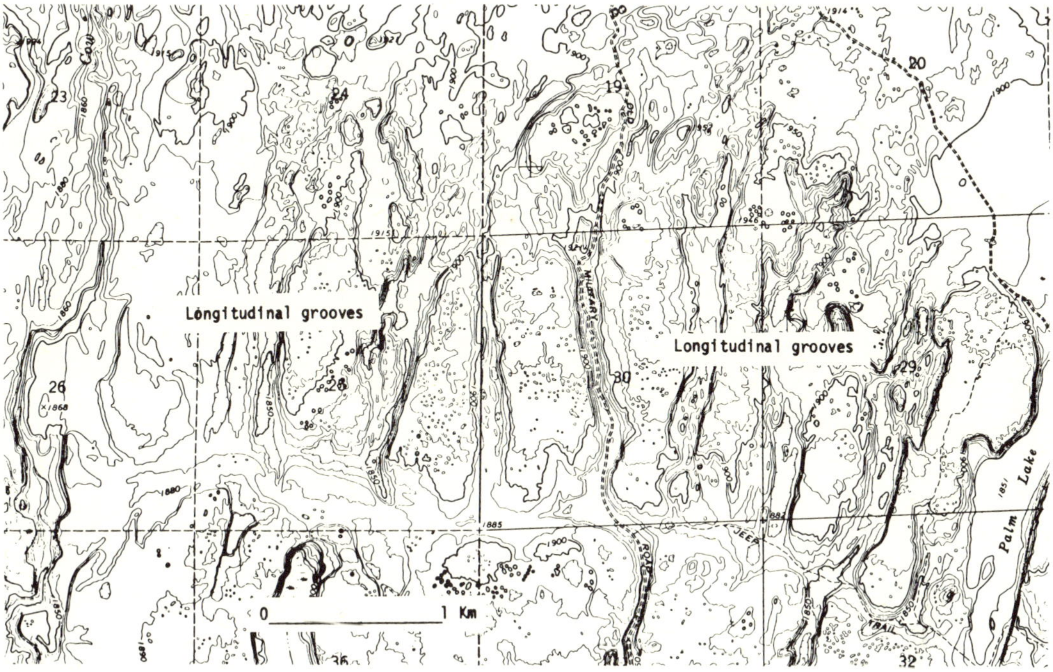

Figure 5.31. Longitudinal grooves developed on the scabland surface near Palm Lake. The small closed contours are mostly mounds of post-flood silt. The larger closed depressions are scabland basins. Topography is from the Palm Lake, Washington, 7.5-minute quadrangle.

suggested that the high flow velocities were generated during the draining of the Quincy Basin in the waning stages of the flood. This explanation is also consistent with the prominent armoring that characterizes the Quincy Basin fill.

GIANT CURRENT RIPPLES

Aerial photographs of some scabland gravel bars reveal patterns of parallel ridges and swales which Bretz and others (1956) identified as "giant current ripples." These constitute the most important bed forms used in the paleohydraulic reconstruction of the last major scabland flood (Baker, 1973a). Over 100 sets of these bed forms have been identified in various Missoula Flood channel ways. Figure 5.34 illustrates some of the variety that exists in ripple morphology.

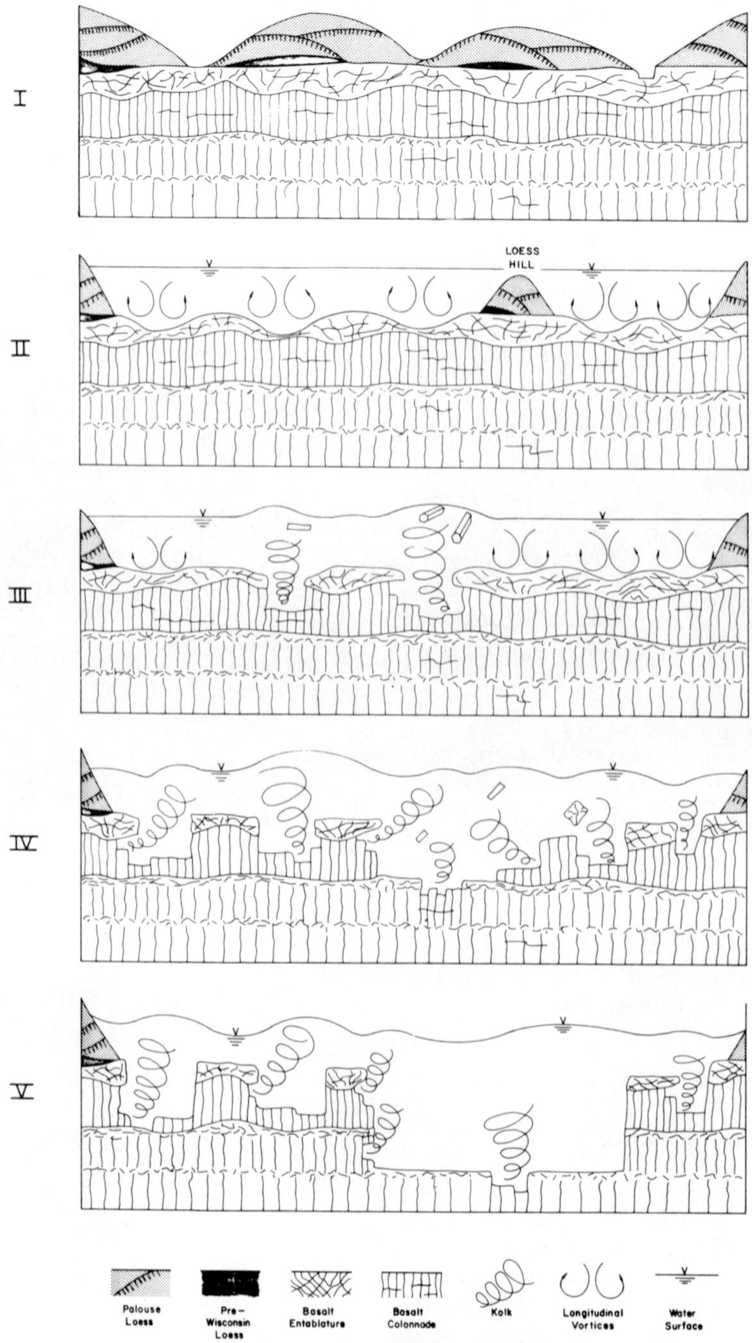

Figure 5.32. Hypothetical sequence of flood erosion for a typical scabland divide crossing. See text for discussion.

(A)

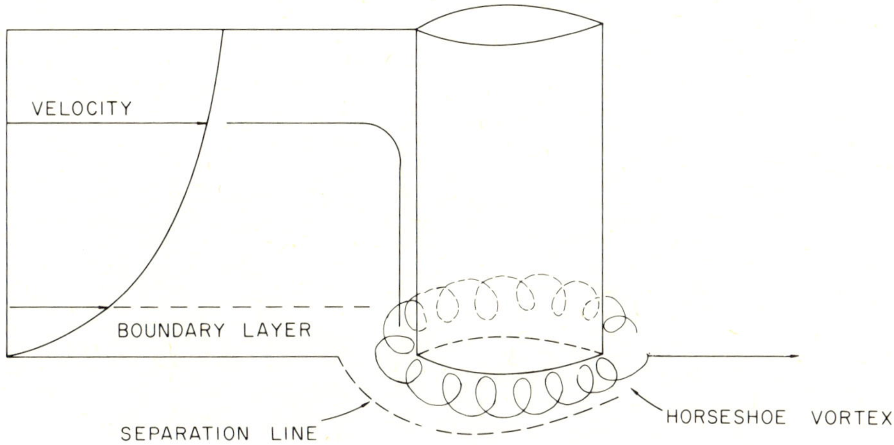

(B)

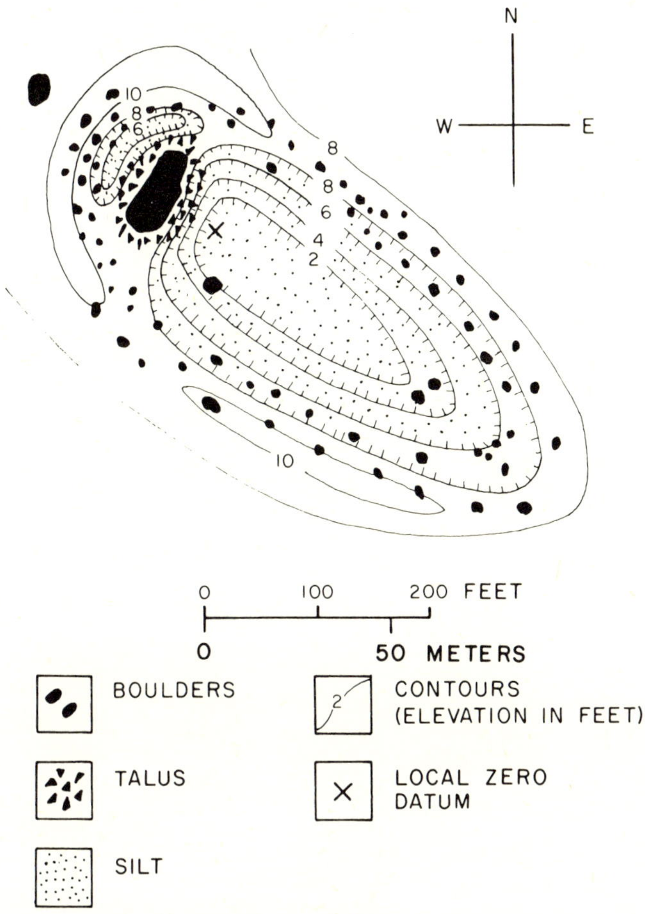

Figure 5.33. A. Formation of a horseshoe-vortex system at the front of a vertical cylinder mounted on an experimental flume bed (after Moore and Masch, 1963). B. Scour hole development near an 18 x 11 x 8 m boulder (Fig. 5.15) 2.5 km west of Rocky Ford Fish Hatchery, Ephrata, Washington (Baker, 1973a).

Ripple Morphology

Measured heights and chords of the scabland ripple fields show a remarkable symmetry of form. Mean ripple heights for a ripple field are closely related to mean ripple chords (Fig. 5.35). The chords generally range from 20 to 200 m. Ripple heights, ranging from 1 to 15 m, have probably been somewhat reduced by waning flood stages and by the modification of post-flood processes.

The giant current ripples present an asymmetric appearance in profile (Fig. 5.36). The downstream-facing slopes of the ripples (lee sides) average about 18 to 20°. Upstream-facing slopes (stoss sides) average about 6 to 8°. Undoubtedly these slopes have been somewhat modified from their original depositional form.

In plan view the giant ripples show a form (Fig. 5.37) that is classified according to Allen (1968, p. 65) as transverse catenary and out of phase. The lee slopes of the ripples occur in a series of cuspate troughs on the down-current side, morphologically analogous to the troughs that form on the leeward sides of some transverse wind dunes (Fig. 5.38). The ripple chords also tend to decrease away from deeper (higher velocity) portions of the channel.

The giant current ripples are very difficult to recognize on the ground unless the observer is paying close attention to minor ridges and swales on scabland depositional surfaces. This fact prevented the recognition of these bed forms until extensive aerial photography of the region was undertaken for the Columbia Basin irrigation project (Bretz and others, 1956). The ripple patterns on aerial photographs derive from a combination of localized eolian silt deposition and vegetation patterns (Fig. 5.39). Ripples near Spirit Lake, Idaho have a cover of pine (*Pinus*

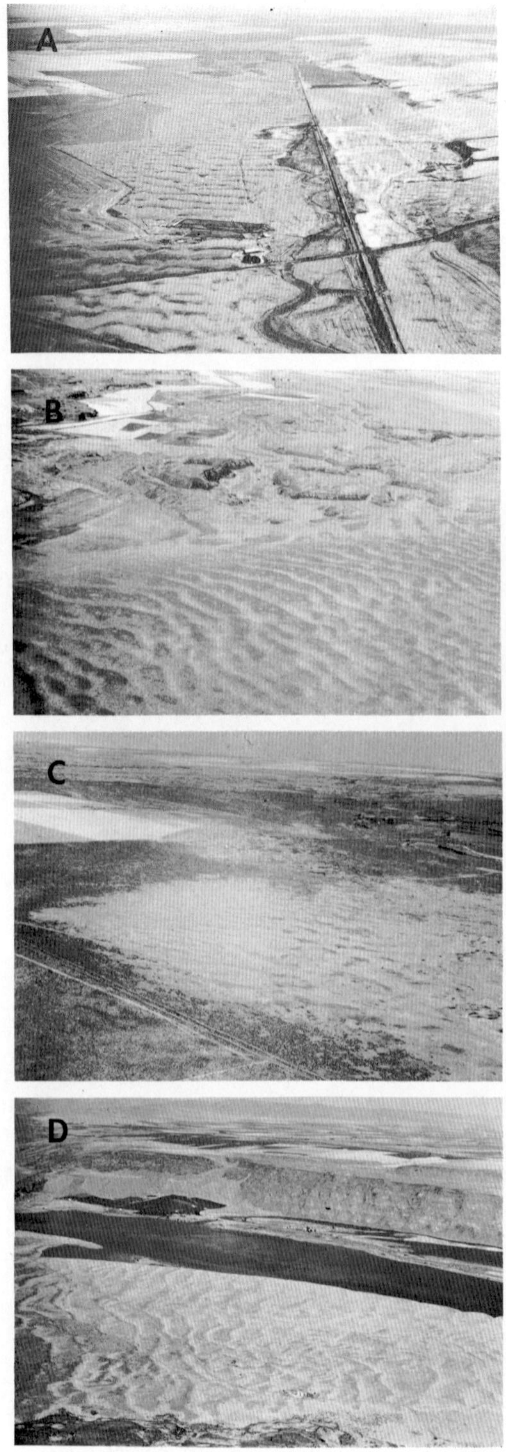

Figure 5.34. Typical sets of giant current ripples in the Channeled Scabland: A. Lind Coulee, B. Marlin, C. Artesian Lake, D. West Bar. Locations for these ripple trains are given in Baker (1973a).

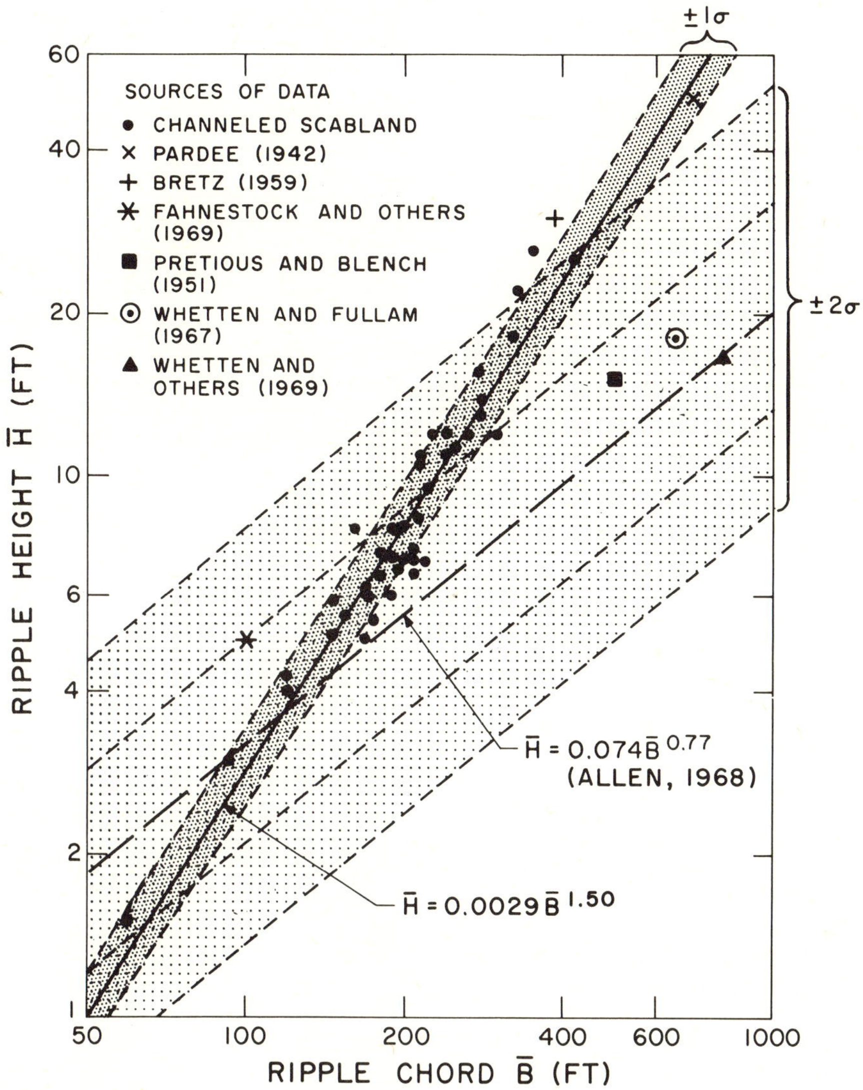

Figure 5.35. Logarithmic relationship of height as a function of chord for 40 sets of giant current ripples. The general relationship for large-scale current ripples determined by Allen (1968) is shown by a dashed line. Five additional measurements by various authors are also plotted. Standard errors on the regressions are indicated by the letter σ.

ponderosa) that reflects the higher rainfall of that region (Fig. 5.39A). In drier parts of the Columbia basin eolian silt fills the swales between ripple crests. This silt supports grass while the adjacent ripple crest supports sagebrush (*Artemisia tridentata*) (Fig. 5.39B). The contrast between these two plants results in especially striking patterns.

In some local areas, the postflood loess deposition has completely buried giant current ripples. At TSCR ripples near the Tokio grain elevator the silt has built up to form ridges in the former swales between the ripple crests. These ridges mimic the ripple pattern that is buried beneath (Fig. 5.39C), but the image is reversed. The modern swales lie over the former ripple crests, and the modern ridges (of silt) occur along the former ripple troughs.

Over the last 10 years high-quality large-scale topographic maps have become available for most of the Channeled Scabland. These maps clearly show the superposition of giant current ripples on scabland bars (Fig. 5.40).

Internal Structure and Sediments

The sediment comprising the giant current ripples is some of the coarsest known to occur in large-scale depositional bed mesoforms (Fig. 5.41). The largest particles may be 1.5 m or greater in diameter, and the median size generally occurs in the pebble fraction. In all observed examples less than ten percent of the sediment is finer than granule gravel.

Internally the ripples consist of foreset-bedded gravel deposited at an angle of about 27°. Exposures (Fig. 5.42) show that individual foresets are remarkably well sorted. Layers of cobbles alternate with discrete layers of granules or pebbles. The sorting gives the gravel a distinctive open-work texture.

Trough-filling cross stratification (Fig. 5.43) is relatively rare in scabland bars of flood gravel. Where present, this stratification type probably represents a filling of scour holes or the migration of sinuous crested giant current ripples.

The coarsest fraction of the ripple sediments, boulders and cobbles, generally form an armor on the ripple stoss slopes. The armor is in the form of an imbricate pavement that probably acted to decrease flow resistance on the ripple surface during the waning stages of flood flow. This smoothed surface may be partially responsible for the preservation of the ripples. In most rivers the depositional mesoforms (usually composed of sand) are washed out during waning flood flow stages (Jackson, 1975).

Hydraulic Significance

Bretz and others (1956, p. 980) suggested "An interesting sidelight on the hydraulics of these glacial rivers will appear when the giant current ripples are given careful detailed study." Baker (1973a) analyzed 43 sets of giant current ripples in Missoula Flood reaches. Statistical correlation and regression analyses were used, treating the values of mean ripple height (H) and chord (B) as the dependent variables. The independent variables were the various hydraulic parameters calculated for each reach. These results will now be summarized.

Depth

Depth is defined as the difference in elevation between the ripple field and the high-water surface above that field. Therefore, this is the maximum depth achieved during the passage of the flood through the reach.

Figure 5.36. Oblique aerial photograph of giant current ripples 3 km west of Odessa, Washington. The current flowed from right to left in the photo, creating gravel bed forms with a mean height of 2.6 m and a mean chord 66 m. The west-facing lee sides of these ripples have an average slope of 18.2 degrees. The east-facing stoss sides of the ripples have an average slope of 6.5 degrees.

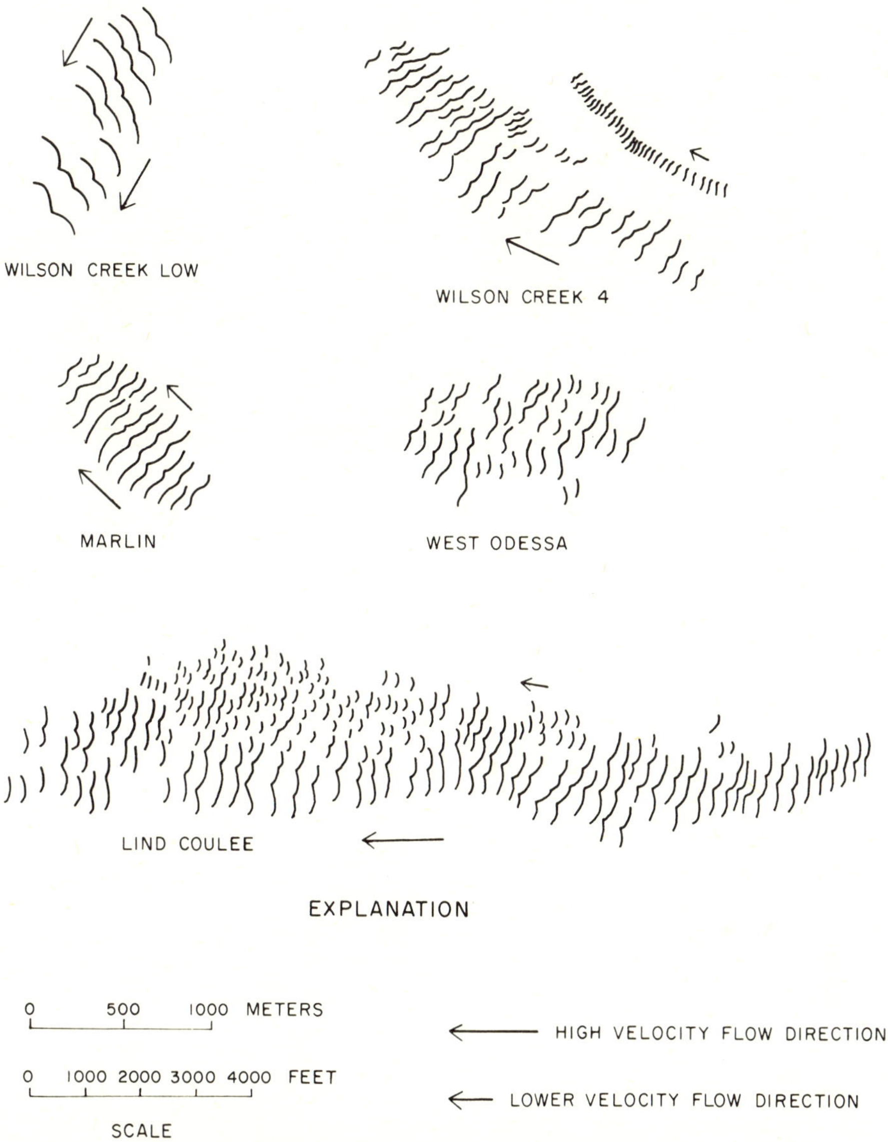

Figure 5.37. Giant current ripple trains viewed in plan. These patterns of ripple crests, traced off vertical aerial photographs, show a decrease in bed form size away from the thread of maximum flow velocity.

Figure 5.38. Oblique aerial photograph of the giant current ripples at Malaga, Washington (Sec. 28, T. 22N., R. 21E). Note the sinuous crests and cuspate troughs that are accentuated in this low sunangle photograph. These ripples were created by catastrophic flooding down the Columbia approximately 13,000 years B.P. (photograph by David A. Rahm).

In Figure 5.44, mean ripple height for a given ripple train is plotted as a function of maximum depth. In Figure 5.45, mean ripple chord is plotted as a function of depth. Both relationships show relatively low correlation coefficients (0.648 and 0.758 respectively). The broad bands defined by one standard error to either side of the regression lines are a further measure of the scatter in the data. The somewhat greater correlation of ripple chord to depth, in contrast to height versus depth, probably arises from the fact that original depositional ripple chords are less modified by waning flows than are ripple heights.

Depth-Slope Product

The product of the slope of the high-water surface over a ripple field and the depth defines a second hydraulic parameter:

$$DS = \tau/\gamma ,$$

where D is the depth, S is the slope, τ is the shear stress, and γ is the specific weight of the fluid (9.8×10^3 N/m^3 for clear water). "Depth-slope" represents a maximum shear stress achieved by the passage of the flood through the reach containing the giant current ripples.

In Figure 5.46 mean ripple height is plotted as function of depth-slope. Figure 5.47 shows mean ripple chord versus depth-slope. The correlation coefficient is greater for chord than for the height

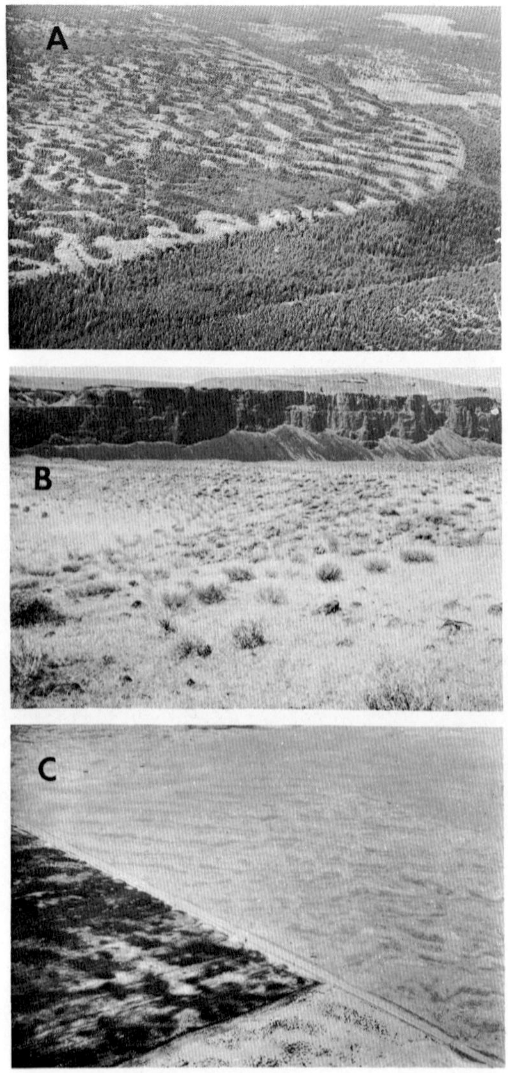

Figure 5.39. Vegetation patterns associated with giant current ripples: A. Spirit Lake, Idaho, B. West Bar, showing appearance of ripples to an observer standing on ripple crest, C. TSCR ripples.

(0.945 and 0.931 respectively). In contrast to the depth correlation, however, 89% of the variation in ripple chord is explained by the depth-slope product. Figure 5.46 shows that the equation:

$$\overline{H} = 24.6\ (DS)^{1.17} \qquad (5\text{-}10)$$

may be used within the standard error indicated

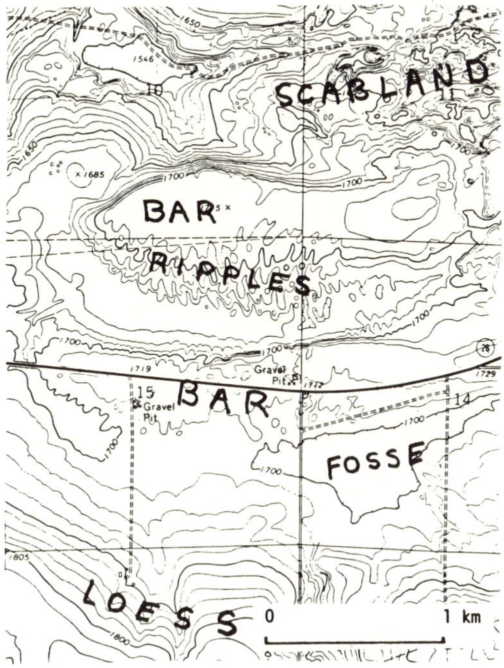

Figure 5.40. Topographic map of giant current ripples shown in Fig. 5.36.

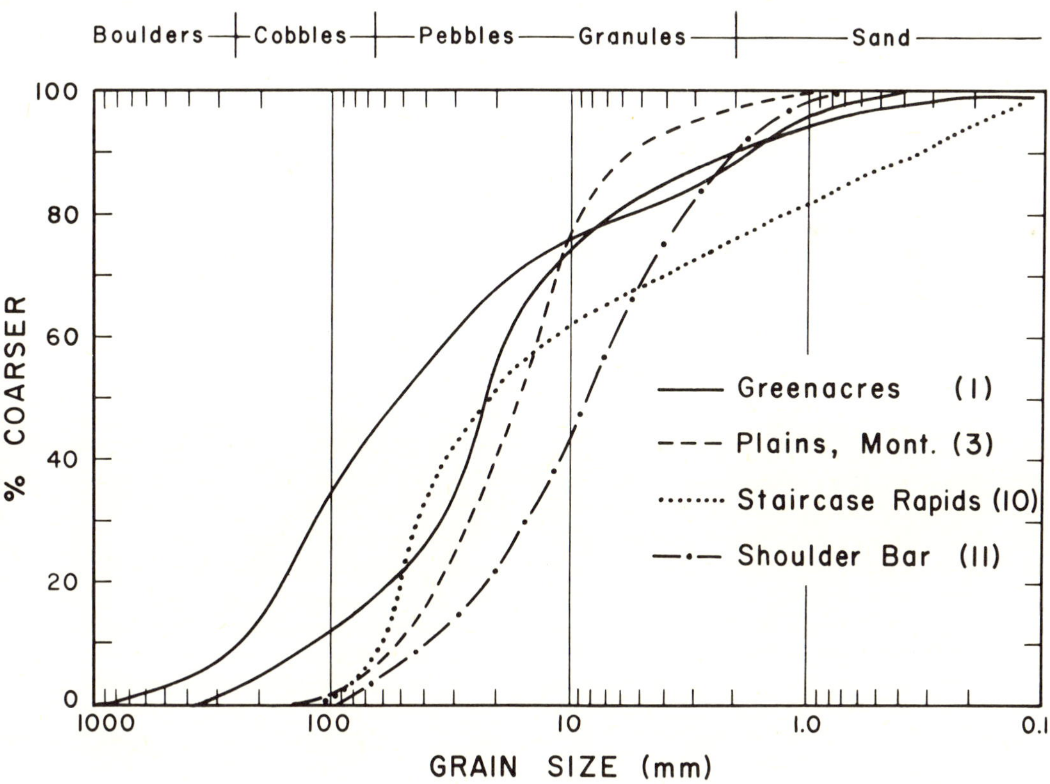

Figure 5.41. Grain-size distribution for sediment comprising giant current ripples in the Channeled Scabland. The numbers for the ripple sets are keyed to locations and descriptions in Baker (1973a, Appendix II).

Figure 5.42. Foreset bedding exposed in a gravel pit cut through a Lind Coulee ripple oblique to the probable current direction. Apparent dips of the bedding vary from 10 degrees on the left to 25 degrees on the right portion of the photograph.

Figure 5.43. Trough-filling cross stratification in flood gravel exposed near Coulee City, Washington.

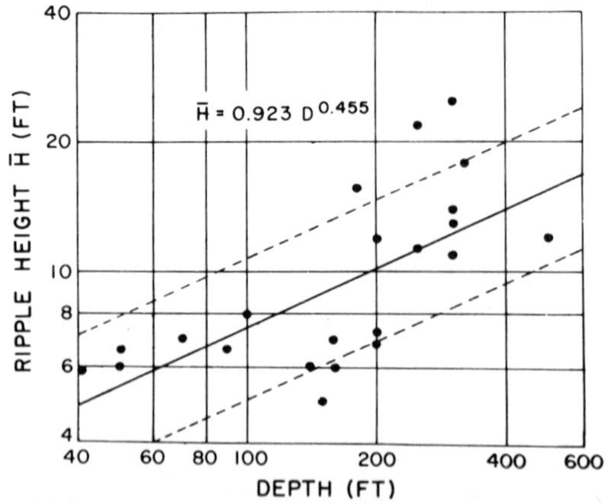

Figure 5.44. Logarithmic regression of ripple height as a function of depth. The dashed lines represent one standard error.

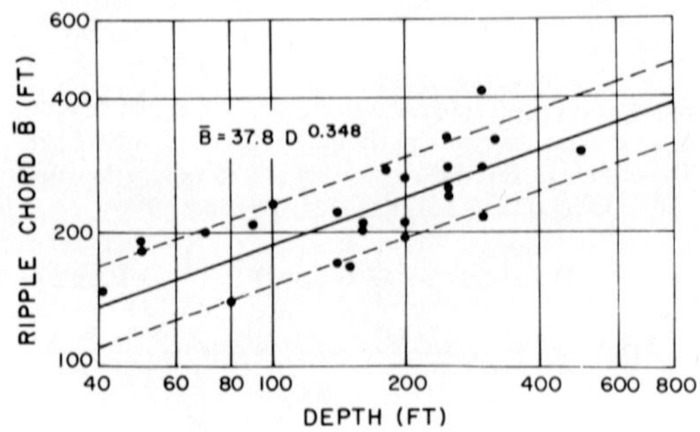

Figure 5.45. Logarithmic regression of ripple chord as a function of depth. The dashed lines represent one standard error.

to predict mean ripple heights as a function of depth and water-surface slope. Similarly,

$$\overline{B} = 393.5 \ (DS)^{0.66} \qquad (5\text{-}11)$$

may be used to predict mean ripple chords.

Because critical shear stress is related to the maximum grain size moved in traction through a given reach, ripple chord should also be related to the maximum grain size moved through the reach. Figure 5.48 shows that this is indeed the case. The correlation coefficient between ripple chord and maximum grain size, 0.993, is only slightly less than that between ripple chord and depth-slope.

Mean Flow Velocity

Figure 5.49 shows the mean ripple chord plotted as a function of mean flow velocity. The correlation coefficient for this regression is 0.81, which may be interpreted as a 65% explanation of ripple chord variation in terms of mean velocity. This relatively low correlation may arise from the fact that the mean velocity in deep flows lies considerably above the mobile bed. The bed forms, like the boulder movement discussed earlier, actually respond to velocities close to the bed.

The mean velocity ($\overline{V}$) and the mean depth ($\overline{D}$) for a subsection containing current ripples may be used to estimate the Froude number (F)

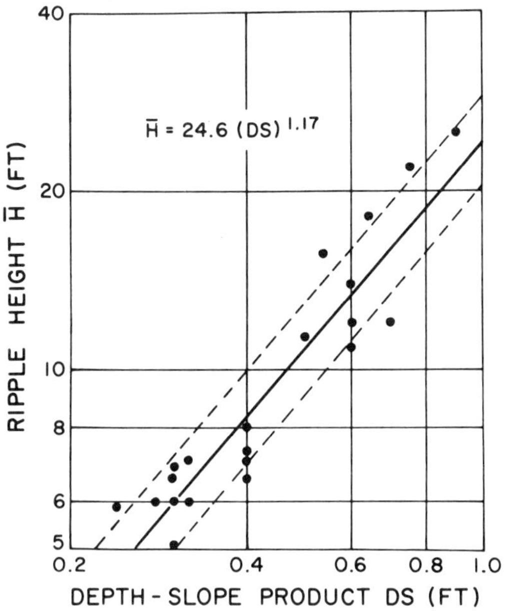

Figure 5.46. Logarithmic regression of mean ripple height as a function of the product of depth times slope. The dashed lines represent one standard error.

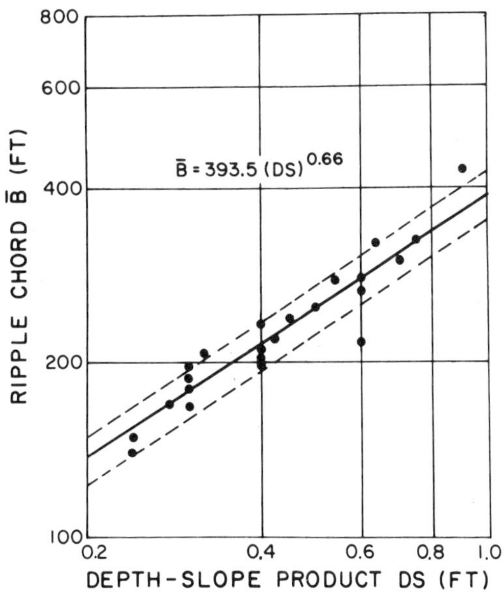

Figure 5.47. Logarithmic regression of mean ripple chord as a function of the product of depth times slope. The dashed lines represent one standard error.

through that subsection according to the formula

$$F = \frac{\overline{V}}{\sqrt{gD}}, \qquad (5\text{-}12)$$

where g is the acceleration of gravity. Froude numbers for the scabland ripples generally vary from 0.5 to 0.9. The bed forms are formed within the tranquil-turbulent range that characterizes the lower flow regime of Simons and others (1965). The Froude number calculations reinforce the concept that scabland giant current ripples are the large-scale, coarse-grained analogs of bed forms known as "dunes" (Simons and others, 1965).

Stream Power. The product of mean flow velocity ($\overline{V}$) and bed shear (τ) gives a measure of stream power (Bagnold, 1966). Stream power is commonly used to predict certain bed forms (Simons and others, 1965, their Fig. 21; Harms, 1969, his Fig. 9C). Stream power (ω) was calculated for the various scabland ripples from the expression:

$$\omega = \tau \overline{V} = \gamma \, (DS) \, \overline{V} \qquad (5\text{-}13)$$

where $\overline{V}$ is the mean flow velocity, DS is the depth-slope product, obtained from the high-water surface, and γ is the specific weight of the fluid. In Figure 5.50 ripple chord is plotted as a function of this measure of stream power. The correlation coefficient of 0.977 is the highest obtained in this analysis. The best predictive equation is:

$$\overline{B} = 8.65 \, \omega^{0.458} \qquad (5\text{-}14)$$

in which the units are as shown in Figure 5.50.

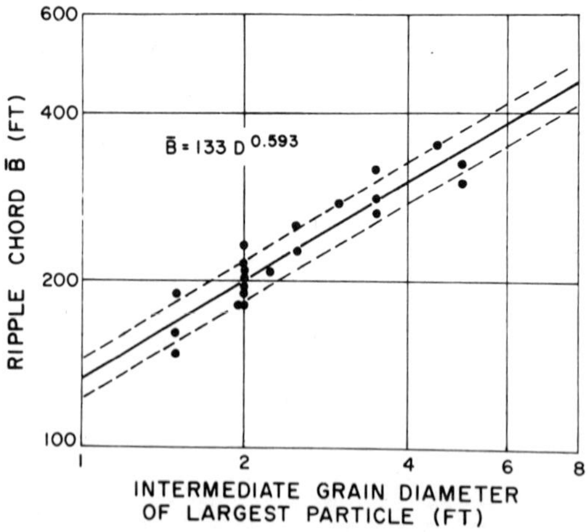

Figure 5.48. Mean ripple chord as a function of the maximum grain size found on the surface armor of the ripples. The dashed lines represent one standard error.

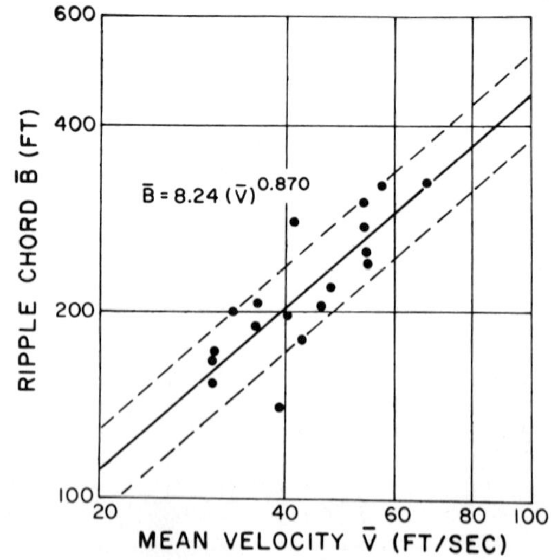

Figure 5.49. Ripple chord as a function of mean flow velocity (discharge velocity) as calculated by the slope-area method. The dashed lines represent one standard error.

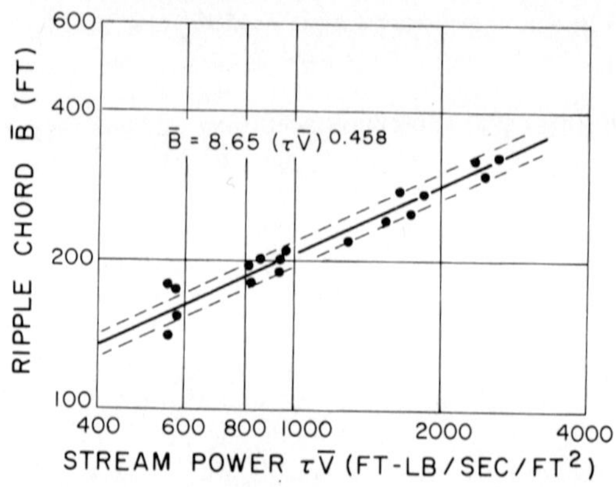

Figure 5.50. Mean ripple chord as a function of stream power. The dashed lines represent one standard error.

CONCLUSION

The bed forms of the Channeled Scabland show a remarkable consistency in scale and genesis to the hydraulics of the scabland flooding (see previous chapter). This consistency can be demonstrated in a quantitative fashion, at least as a first approximation. Nevertheless, one cannot help but look back on the remarkable insights of fifty years ago (Bretz, 1928c, p. 475-476):

> "All scabland channels possess discontinuous mounds, hillocks or hills of stream gravel. . . . They are unlike any other detrital accumulations except the much smaller features of river channels commonly called bars. With these there is exact parallelism except for size. When considered in their setting in the scabland system, with all its other evidence for great volume and great erosion, they are seen to be an integral part. They should exist! And if they are bars, the great scoured channel ways should exist! Again this assemblage of unique land forms in the Pacific Northwest is seen to be a genetic group. A lively imagination is required for the acceptance of the hypothesis, but a scientific imagination withal."

14

Reprinted from pages 117–130 and 179–186 of *The Channeled Scabland,* V. R. Baker and D. Nummedal, eds., NASA Office of Space Science, Planetary Geology Program, 1978, 186 pp.

Origin of The Cheney-Palouse Scabland Tract

PETER C. PATTON

VICTOR R. BAKER

ABSTRACT

The Cheney-Palouse tract of the Channeled Scabland is the largest continuous tract of scabland in eastern Washington. The tract is composed of a varied assortment of bedrock erosional forms, loess islands and gravel bars. Prominent bedrock longitudinal grooves and inner channels formed by macroturbulent plucking erosion of the jointed rock. Loess island forms vary as a function of their position within the flow. The three major types (submerged, partially submerged, and subaerially exposed) created sedimentologic conditions and resulting bar forms distinct from one another. Other bar forms, notably expansion bars, account for most of the sedimentation in the tract.

In form and process, much of the Cheney-Palouse is analogous to a braided stream. First, the geometry of even the most complex reaches of the tract can be classified in the same manner as braided streams are classified. Second, the degree of development of smaller and topographically higher elements of the system is similar to observations made on the hierarchy of form and process in modern braided streams. Finally, the loess islands, although erosional, appear to have acted in the same manner as longitudinal bars in braided streams during the passage of a flood wave. They diverted the flood flows toward the banks and created zones of deposition in their lee.

INTRODUCTION

This paper is being published 55 years after the first interpretation of the Cheney-Palouse scabland tract by J Harlen Bretz (1923a, 1923b) and 40 years after a paper of the same title (Flint, 1938b) rejected Bretz' catastrophic flood origin of the region's landforms. The purpose of the present paper is twofold: (1) to demonstrate the general verity of Bretz' original interpretation, and (2) to show that the details of landform genesis in the tract are consistent with the dynamics of catastrophic flooding (Baker, 1973a, 1973b).

The Cheney-Palouse scabland tract is the easternmost system of flood-scarred channel ways in the Channeled Scabland (Fig. 6.1). The tract was first denoted as a single element of the Channeled Scabland by Bretz (1923b). It begins where Missoula Flood water spilled over the upturned northeastern margin of the Columbia Plateau near Spokane, Washington. The tract terminates 135 km to the southwest where the floods crossed the pre-glacial divide between the Snake and Palouse Rivers. This longest continuous scabland tract (Bretz, 1923b) is bordered on the east by undisturbed "Palouse hills" topography and on the west by loess topography interrupted locally by the major westward-flowing distributary coulees which carried Missoula Flood waters into the depositional basins of the western scablands.

Viewed from a LANDSAT image, the Cheney-Palouse appears to be an enormous braided stream with broad anastomosing channels separated by mid-channel islands. However, the anastomosing channels are scoured into the loess and underlying basalt bedrock. The midchannel islands are not depositional bars as one finds in braided streams. Rather, they are the eroded remnants of the once continuous loess cover. Nevertheless, we will show that the hierarchical arrangement of the channel elements is analogous to the hierarchy of forms described for braided streams by Williams and Rust (1969).

The Cheney-Palouse scabland tract is a rare terrestrial example of large-scale bedrock erosion by floods whose flow was confined neither by resistant channel walls nor by major geologic structures. We have previously cited this region as a major terrestrial analog to the Martian outflow channels (Baker and Patton, 1976).

PREVIOUS INVESTIGATIONS

Bretz (1923a) described the following distinctive features of the Cheney-Palouse tract: (1) the scarped loess islands with sharp upstream prows separated by scabland channels, (2) the elongation of the loess islands parallel to the overall trend of the tract, (3) gravel terraces which were almost everywhere associated with the lee side of the loess islands, and (4) large enclosed elongated rock basins which were created subfluvially by hydraulic plucking of the basalt bedrock. He also noted that 30 - 60 m of basalt must have been eroded from the level of the pre-glacial topography of the Cheney-Palouse. He concluded that only an enormous deluge could have created these landforms.

In the controversy that arose over the origin of the Channeled Scabland (see Baker, Ch. 1, this volume), the Cheney-Palouse tract played a significant role. The probable reason is that much of the tract exhibits relatively small-scale erosional and depositional features compared to those of the western scabland. Therefore, those investigators who felt that Bretz was teetering on the brink of catastrophism with his flood hypothesis sought to explain at least this segment of the scablands in terms of normal stream erosion.

Allison (1933) and Flint (1938b) produced the most detailed hypotheses to explain the origin of the Cheney-Palouse in terms of normal stream processes. Allison (1933) proposed that an enormous dam of floating ice bergs in the Columbia River Gorge was responsible for ponding of water supplied by moderate flooding of pro-glacial streams. This ice dam supposedly ponded water to over 300 m above sea level forming glacial Lake Lewis. As the ice dam grew headward, it locally diverted streams to create the unusual scabland relationships. Allison (1933, p. 683) also hinted that aggradation of proglacial streams eventually caused divide crossings and scabland erosion as far east as the Cheney-Palouse tract. Although Allison believed that there was a "Spokane Flood," he felt that it was of moderate size and that his combined mechanism of ice damming and flooding removed the interpretation of the Channeled Scabland from the realm of Bretz' "impossible" catastrophic flood.

Flint (1938b) using Allison's ice dam hypothesis as an impetus for more detailed study,

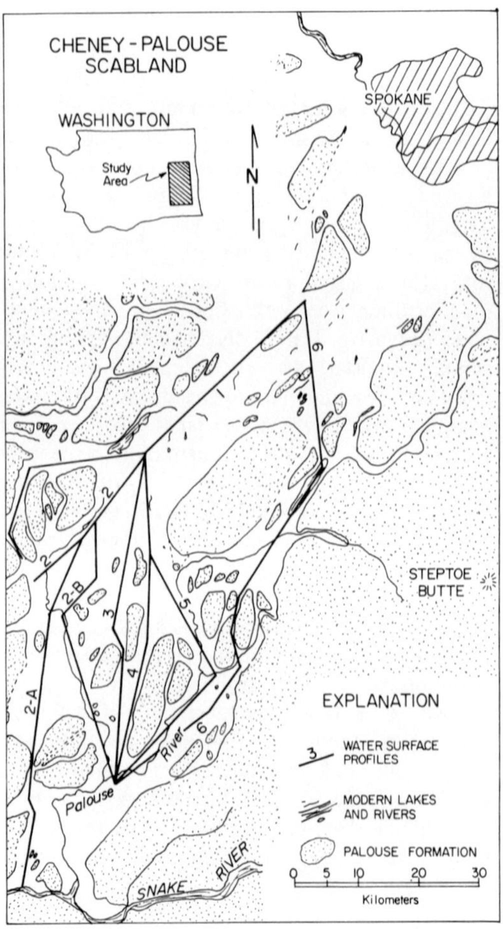

Figure 6.1. Highly generalized sketch map of the Cheney-Palouse scabland tract showing the location of various water-surface profiles that have been determined for the last major Pleistocene flood. Many of the small residual hills of Palouse Formation ("loess islands") have been omitted for simplicity.

thoroughly examined the Cheney-Palouse. Flint's hypothesis was that proglacial streams of normal discharge draining into Lake Lewis gradually aggraded their valleys as the lake level rose. Raised on their valley fills, the streams eroded the loess scarps and even spilled over preglacial divides. When the ice dam in the Columbia was finally breached, the relatively small proglacial streams re-excavated the fill leaving loess islands and various unpaired terraces. The streams then eroded into the underlying basalt, producing the scabland topography visible today.

Flint made detailed descriptions of the scarped loess islands which he interpreted as erosional remnants left by laterally planating streams. He was particularly intrigued by those loess islands that had several terraces attached to their upstream prows and downstream tails. The terraces do not occur along the loess island flanks. Flint also noted that other gravel "terraces" were only preserved at those localities where they were protected by upstream basalt knobs. Finally, several terraces had long gravel ridges on their surfaces oriented transverse to the flow which he could not explain.

Flint's (1938b) major addition to Allison's ice dam hypothesis was the creation of Lake Riparia. Flint thought that the aggradation of Washtucna Coulee caused the Palouse River to top its divide and establish a new course through Palouse Canyon to the Snake River. The sediment eroded during this process created a large fan delta and dammed the Snake River forming Lake Riparia upstream. The remnants of this fill were Mid-Canyon and Shoulder Bars. In short, Flint did not see any need to invoke a catastrophic flood because he believed that the bedrock erosion and sediment transport required to create the Cheney-Palouse tract was not extraordinary for proglacial streams operating for long periods of time.

The reinvestigation by Bretz and others (1956) refuted these alternative theories. They demonstrated that the erosional terraces were in fact constructional bars, many with enormous foreset beds that could not be rationalized in terms of slowly aggrading streams of normal discharge. More significant was the recognition that the gravel ridges originally mentioned by Flint were giant current ripples similar to those described by Pardee (1942) in the basin of Lake Missoula itself. Bretz and others (1956) reiterated the original conclusion that the erosional bedrock forms, loess islands, depositional bars, and divide crossings of the Cheney-Palouse tract were created by catastrophic flooding.

BEDROCK EROSIONAL FORMS

A variety of erosional forms characterize the scabland topography on the exposed Yakima Basalt. These include the numerous potholes, large elongate scour holes, longitudinal grooves, cataracts, and deep narrow inner channels winding the entire length of the Cheney-Palouse.

The longitudinal grooves are particularly well developed in the converging channel south of Sprague Lake between the Karakul Hills loess islands and the town of Lamont (Fig. 6.2, top, center). The widest of these grooves, now occupied by Palm Lake, is just under 300 m across. The grooves can be up to 15 to 25 m deep. The Palm Lake grooves occur as two bands which extend nearly across the channel. Both the upstream and the downstream bands are about 1.5 to 2.5 km long parallel to the flow direction. The upstream grooves are cut into the entablature surface of a basalt flow about 10 to 15 m above the level of the underlying basalt flow in which the downstream grooves are incised. The upstream grooves are spaced approximately 500 to 700 m apart, and many have been eroded headward through the basalt flow which forms the intervening plateau surface. These grooves now appear as dry canyons separated by isolated basalt mesas. The downstream grooves are spaced about twice as far apart and usually terminate upstream at a small cataract which may be up to 20 m high. These grooves do not line up with the upstream set and probably developed independently. Although some of the grooves can be traced downstream, the majority of these longitudinal forms disintegrate into a maze of butte-and-basin scabland topography.

As pointed out by Baker (1973b), the scabland grooves are analogous to grooves found in other bedrock stream channels and in experimental flume studies of simulated bedrock erosion (Shepherd, 1972; Shepherd and Schumm, 1974). Shepherd (1972) noted that in an essentially straight

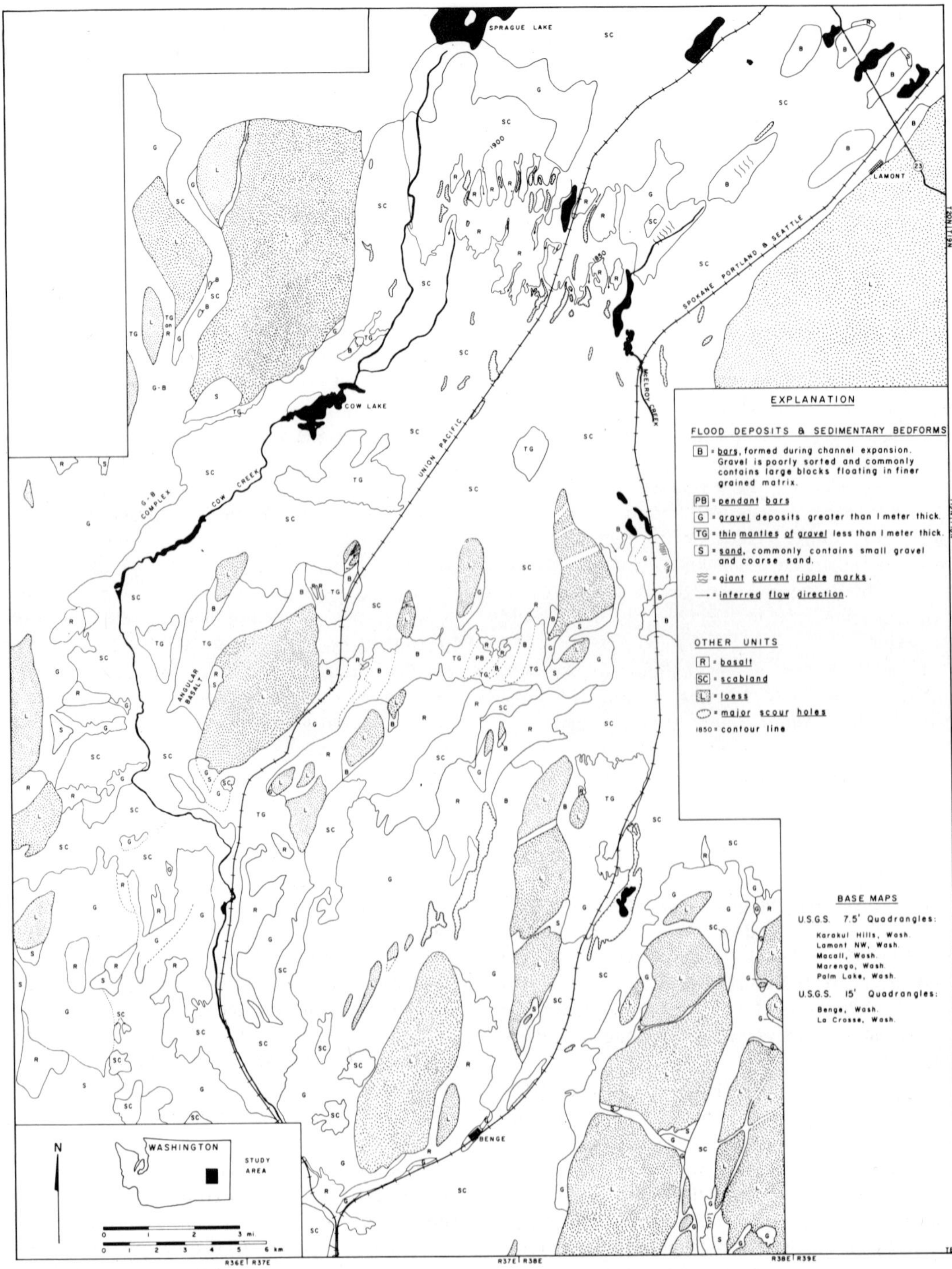

Figure 6.2. Geomorphic map of the central portion of the Cheney-Palouse scabland tract between Sprague Lake (north) and Benge (south). The map illustrates typical relationships between the various channel elements.

channel, a sequence of erosional bed forms developed on simulated bedrock. This sequence began with longitudinal lineations which became enlarged into prominent longitudinal grooves. Although potholes and erosional ripples developed with the lineations, with time, the grooves became the dominant bed form. Shepherd (1972) hypothesized that secondary circulation cells in the flow were responsible for groove erosion. Furrows were produced where the vortices attached to the bottom, and ridges were left where separation zones occurred. Shepherd (1972) varied the flow characteristics in the flume to determine which factors were most important in groove formation. He found that increases in slope and water discharge had little effect on accelerating the erosion process, but that the grooves rapidly grew when the sediment discharge rate was increased. This is because the experimental bedrock was a dense clay-sand mixture which was extremely cohesive (Shepherd, 1972). Therefore, abraison, not plucking, was the most important mechanism for eroding the experimental grooves.

Several differences should be noted between Shepherd's experimental grooves and the Palm Lake grooves. First, it is probable that the scabland floods, given their enormous discharges and their hypothesized macroturbulent nature, were underloaded with sediment. Second, the jointed basalt bedrock lends itself to plucking and quarrying, a fact that has been noted by all scabland investigators since Bretz (1924). Third and last, the experimental grooves evolved more or less simultaneously along their length while there is ample evidence that the scabland grooves evolved by cataract recession.

Shepherd and Schumm (1974) noted that with time the experimental bedrock grooves coalesced, and a dominant bedrock inner channel was formed leaving remnant paired bedrock benches as evidence of the old channel floor. Cow Creek (Fig. 6.2, west side of map) presently occupies such an inner channel. It is the deepest channel way in the Cheney-Palouse and is bordered by erosional scabland to its juncture with the Palouse River at Hooper. Other prominent inner channels in the Cheney-Palouse tract include Bonnie and Rock Lakes and the Palouse Canyon where the Palouse River crosses the former Palouse-Snake divide. These inner channels are significant in localizing the bedrock scour and causing the greatest degree of scabland development.

The long profiles of these inner channels are highly irregular. The Rock Lake system (profile 6, Fig. 6.1) has several lakes along its profile, indicating the presence of several enclosed rock basins. The reconstructed high-water surface profile (Fig. 6.3) shows that many of the larger basins coincide with steep energy gradients. Pronounced constriction of the flow induces greater erosion, as discussed by Baker (1973a, p. 15-16).

One of the most abrupt constrictions in the Cheney-Palouse tract occurred at Staircase Rapids, just north of Washtucna (see Bretz and others, 1956, p. 1000-1003). The water-surface profile (Profile 2A, Fig. 6.1) shows the pronounced ponding of water in the Rattlesnake Flats area (Fig. 6.4). The relatively subdued topography of Rattlesnake Flats contrasts sharply with the scabland and cataracts of Staircase Rapids. The flood water-surface gradient through the rapids averaged about 12 m/km (Fig. 6.4).

The influence of bedrock structure on scabland erosional forms is especially evident on the Palouse-Snake divide crossing. The major drainage lines in the unaltered Palouse Hills trend northeast to southwest in this area of the Cheney-Palouse (Lewis, 1960), and it appears that initially the flood followed the major valleys as it cut across the divide. There are several lines of evidence to support this. First, the gross orientation of the divide crossing is from the northeast to the southwest, an orientation that is also reflected in the smaller ancillary divide crossings. For example, a divide crossing northeast of Nunamaker farm (Sec. 24, T.14N., R.35E.) perfectly parallels the unaltered preflood drainage. Second, all of the remnant loess islands within the divide crossing are oriented in the same direction. This includes not only those loess islands along the far eastern margin of the crossing, but also two small loess islands directly east of the H U Ranch cataract (Sec. 27, 28, 33, 34, T.14N., R.36E.). There is, however, only one major bedrock cataract oriented in this direction and that is the cataract containing Devil's Lake (Sec. 9, 16, 17, T.14N., R.37E.). The major cataracts including H U cataract, Palouse Canyon, and the cataracts containing Wind Lake and Deep Lake all trend approximately east-southeast. A second set of smaller cataracts trends southeast.

We suggest that the initial flood flows across the divide followed the loessial topography which, in turn, was oriented according to the prevailing wind direction (Lewis, 1960). Subsequent bedrock scour was then localized by the fracture set oriented nearly perpendicular to the initial flow direction (Trimble, 1950). The weaker rock in the fracture zones were preferentially quarried during the flood. A feedback mechanism can be envisioned in which these zones of bed relief perpendicular to the flow added to the turbulence and accelerated the erosive processes.

LOESS REMNANTS

The most conspicuous macroforms of the Cheney-Palouse scabland tract are the erosional remnant loess islands. Originally described by Bretz (1923b), these streamlined hills generally have sharp upstream prows, steep faceted flanks, and long tapering tails. In the Cheney-Palouse, there are three distinct varieties of loess islands. The detailed reconstruction of flood high-water surface profiles allows us to distinguish forms that were (1) submerged beneath the flow, (2) partially submerged by several major divide crossings, or (3) unsubmerged and exposed above the

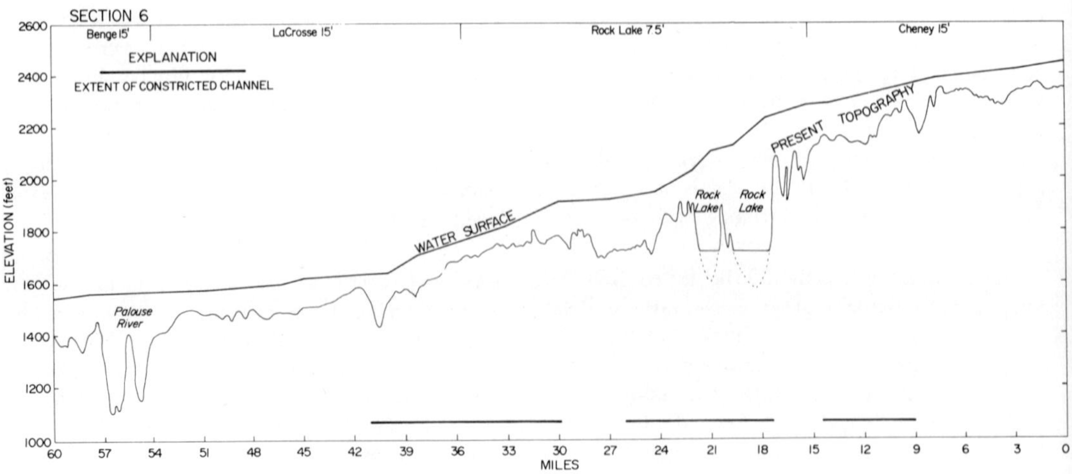

Figure 6.3. Profile 6 through the eastern part of the Cheney-Palouse tract (see Fig. 6.1 for location). Note the steepening of the water-surface gradient through constricted reaches.

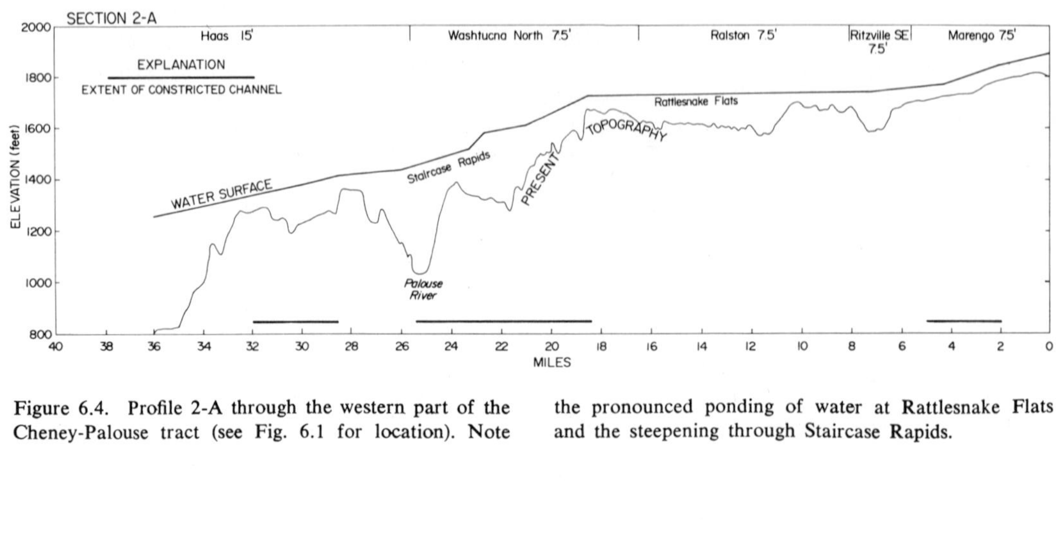

Figure 6.4. Profile 2-A through the western part of the Cheney-Palouse tract (see Fig. 6.1 for location). Note the pronounced ponding of water at Rattlesnake Flats and the steepening through Staircase Rapids.

flow. These major differences in position of the loess islands within the flow also caused distinct sedimentologic variations in the style of deposition and position of the gravel bars attached to the islands.

Submerged Remnants

Subfluvially eroded loess islands have streamlined shapes similar to airfoils (Baker, 1973b). They are the smallest of the loess forms present in the Cheney-Palouse (Fig. 6.5A). Because these forms were completely covered by the flood, the flanking scarps are not as well developed, and all preflood drainage topography on the tops of the islands was obliterated. These islands have gravel bars attached to their tails in much the same manner that wind-shadow dunes form in the lee of flow obstructions. In this case, the gravel bars drape over the tail of the eroded loess form, and the shape of the resulting streamlined hill is partly influenced by the loess island and partly by the gravel bar. This relationship can be seen in several roadcuts. The gravel thinly veneers the

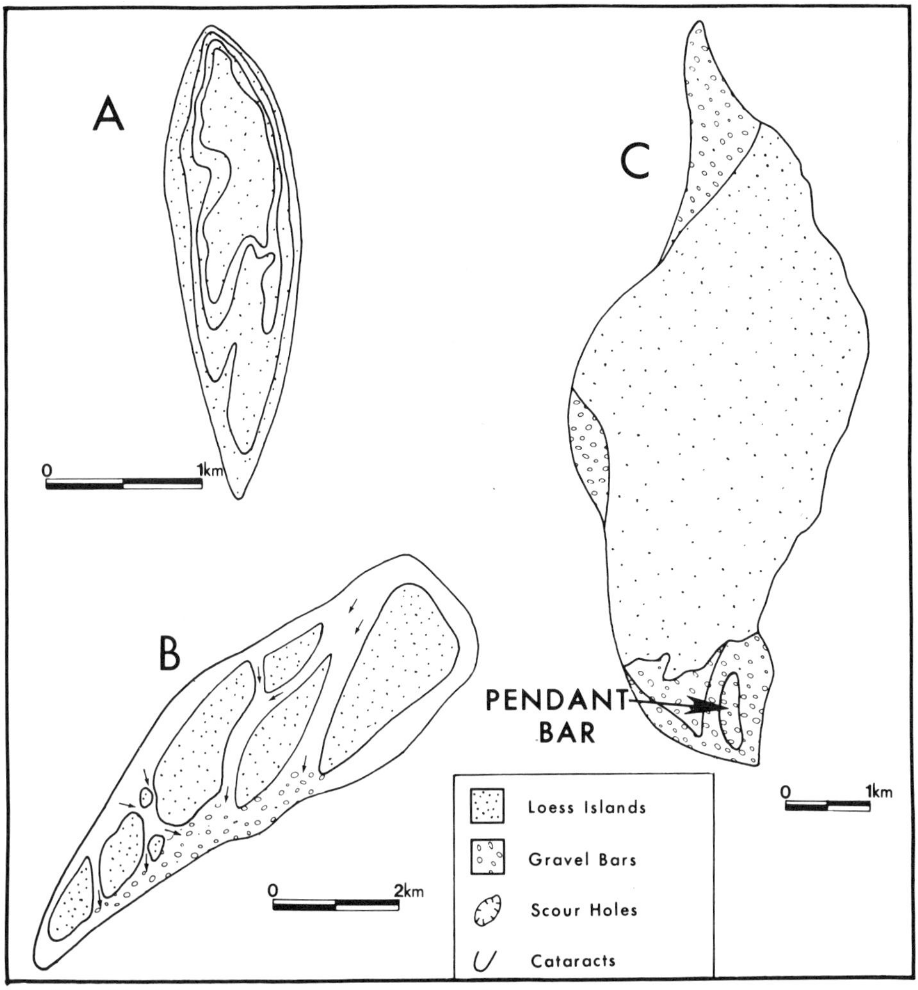

Figure 6.5. Loess island forms in the Cheney-Palouse scabland tract illustrating the three major morphologic types. A. Subfluvially eroded loess island located in the Karakul Hills (Secs. 2, 11, and 14; T. 19N., R. 36E.). B. Partially submerged loess island near Amber Lake (Secs. 1, 2, 11, 12, 15; T. 21N., R. 40E.). Arrows denote the major divide crossings and the gravel pattern indicates a bar. C. unsubmerged loess island immediately northwest of Marengo (left center of Fig. 6.2).

upstream portions of the residual hill and thickens downstream toward the "tail" of the streamlined form which is composed entirely of transported sediment. Because of the high degre of streamlining, it is often difficult to determine where the bar attaches to the loess island. The problem is made more difficult by the cover of late Pleistocene-Holocene loess draped over the entire feature.

The gravel "tails" on these loess islands are composed predominantly of cobble gravel in a matrix of granule gravel and coarse sand. The deposits are generally finer-grained than most main-channel facies in the Channeled Scabland. We suggest that deposition probably took place during waning flood stages after the loess island had become a significant obstruction inducing a zone of flow separation in which sediment could accumulate.

Partly Submerged Remnants

A second variety of loess island includes those transected by one or more major channels still shallow enough not to have eroded into the underlying basalt (Fig. 6.5B). Many such channels eroded large re-entrants on the downstream margins of the islands which were later partially filled with sediment. Consequently, these loess islands have the thickest accumulation of flood sediment. Their deposits exhibit large-scale avalanche cross-bedding, cut-and-fill sequences, soft sediment deformation structures, and a rapid vertical variation in grain size. It is not uncommon for a layer of open-work gravel to be succeeded by a layer of coarse sand followed by another layer of gravel. Light colored beds made up of reworked loess are also common. This mixture of fine and coarse sediment indicates that an extremely wide range of grain sizes were in transport and that perhaps pulses in the flow velocity were responsible for the fluctuations in grain size through time. The mode of deposition at these loess islands was probably similar to the process of deposition at the eddy bars which developed at the mouths of pre-flood tributaries to the Channeled Scabland (Baker, 1973a). However, water and sediment entered the loess island re-entrants from divide crossings that were at different angles to one another and from the main channel. The combination of these several flow directions thus created large eddies which accumulated flood debris.

Unsubmerged Remnants

Loess islands whose crests were not topped during the Missoula Flood are characterized by the well-developed steep flanking scarps which truncate and behead the pre-flood drainage systems still evident on their crests. Many of these islands were crossed by small distinct channels during the last flood, but these channels had little effect on the overall morphology and sedimentation. The islands have well-developed quadrilateral shapes; many very similar to the rhombic or diamond shape of longitudinal bars typical of braided streams.

Although the loess islands are dominantly erosional forms, some of their streamlined character is caused by deposition of gravel bars both at the prow and downstream in the lee of the islands (Fig. 6.5C). The bars are easily distinguished from the residual loess by their lower elevation and by their smooth flat surfaces, devoid of any stream network development. An example is the prominent bar which is attached to the prow of the Marengo loess island, perhaps because of the upstream flow stagnation plane. The bar is composed of angular boulders up to .75 m in intermediate diameter. This is a much larger grain size than that of the bar at the lee side of the island which is composed of cobble and granule gravel. No stratification was evident in exposures of either bar.

Discussion

The bars associated with these last two types of loess remnants were originally interpreted as terraces by Flint (1938b). Although Bretz and others (1956) did not make a detailed study of these surfaces, they did hypothesize that the smooth "terrace" surfaces might be the result of gravel bars, buried resistant soil horizons, bedrock ledges or incomplete flood incision. They further noted that eight of Flint's (1938b) examples of terraced loess islands were remnants which had smaller secondary channels cut through their divides. Our study supports the interpretation of Bretz and others (1956). Many of Flint's "terraces" were gravel bars deposited in the reentrants described for the second loess island type. Investigation of several other terraced loess islands showed that bedrock ledges do occasionally form conspicuous terrace-like forms on some loess islands. Finally, the prominent loess island just east of Macall siding (Fig. 6.2, right center) has a prominent flat surface on its downstream end caused by an exhumed resistant petrocalcic horizon. The petrocalcic horizon (caliche) caps a pre-Wisconsin flood deposit (See Baker, Ch. 2, this volume). It is quite probable that here the Palouse Formation was deposited over the gravel bar of a pre-Wisconsin flood.

GRAVEL BARS

Gravel bars in the Cheney-Palouse include pendant bars, expansion bars, and the previously described bars associated with loess islands. In general, the gravel bars in the eastern scablands are smaller than those to the west. The major reason for this is the lateral spreading of Cheney-Palouse flood water, resulting in lower flow depths than those attained in the great coulees of the western Columbia Plateau. The exceptions to this trend are Staircase Rapids Bar at Washtucna and Shoulder and Mid-Canyon Bars on the Snake River. These bars all formed downstream of local flow constrictions.

Pendant Bars

Pendant bars are not restricted to any particular geomorphic setting within the Cheney-Palouse. They occur in channels of all sizes, although they are rare in the deepest scoured channels. They occur most commonly along the margins of the flow and in smaller channels where resistant basalt knobs created the necessary flow obstructions from which the bars could accrete. One of the largest pendant bars in the Cheney-Palouse is adjacent to a scour hole southeast of Rock Lake. An excellent exposure at the toe of the bar demonstrates the large-scale avalanche beds that typically occur within these bars.

Locally, a single resistant basalt flow provided points of flow separation and reattachment, allowing a string of bars to form across the channel. Examples of this can be seen along the Ritzville Macall Road east of the Marengo loess island where numerous bars are attached to the downstream step created by a resistant basalt flow (Fig. 6.2, center). Bar deposition at this location was also favored by the transition at this point from upstream constrictions between loess islands to a major channel expansion that probably reduced the flood velocity. A similar situation occurred northwest of the town of Lamont (Fig. 6.2). Gravel pits show that the pendant bars are made up of large basalt blocks, many of which still have a polygonal columnar structure. Thus, the boulders were probably transported only short distances by the macroturbulent suspension mechanism described by Baker (1973a, 1973b). On the other hand, the bars also contain large granitic boulders that could have been transported only from the Medical Lake area 40 km to the northeast.

The pendant bars in the Cheney-Palouse tract are relatively small when compared to bars in the westward flowing distributaries such as upper Crab Creek and Wilson Creek. Although Cheney-Palouse pendant bars may extend downstream 1.5 km or more from their points of attachment to flow obstructions, the gravel is usually 10 m or less in thickness. This contrasts with the 30 m thick pendant bars reported by Baker (1973a) in Upper Crab Creek. Again, this reflects the relatively shallow flow depths of the Cheney-Palouse scabland tract.

Expansion Bars

Expansion bars are widespread gravel deposits immediately downstream from channel expansions. They often are found where several small channels exit from an assemblage of loess islands. The flow expansion as well as the shadowing effect of the loess islands created a low-velocity zone in the flood flow where deposition was favored. Other expansion bars formed downstream from cataracts or simply where the channel was unusually wide. An example lies south of the Karakul Hills loess islands where several small channels flow out onto a wide plain (Fig. 6.2, upper left). The thin character of this deposit can be seen downstream where basalt crops out only 3 m below the surface of the bar. This bar can be traced downstream to where it is truncated by scabland on the north side of Cow Creek. An example of an expansion bar downstream from a cataract is the gravel accumulation on the east side of the channel which terminates immediately northeast of Benge (Fig. 6.2, bottom). The bar parallels the loess island which forms the eastern side of the channel. At the upstream end of the bar, near the cataract, large slabby basalt boulders up to 3.5 m in intermediate size litter the surface. Three kilometers downstream, the intermediate grain size of the largest boulders has decreased to about one meter, but the roundness of the boulders has only slightly increased. The bar is extremely thin, and, immediately to the west, in the center of the channel, basalt crops out a few meters below the surface of the bar.

Although these bars are thin, they can cover significant areas, up to 50 km^2. The overall extent of these bars is obviously controlled by the channel geometry. Where abrupt constrictions occur, the bars are terminated, such as south of Sprague Lake where a channel converges between two loess islands. Also, where the bars extend into the major channels such as Cow Creek, they are abruptly terminated.

Giant current ripples are fairly common on pendant and expansion bars. In addition, such ripples may form at the downstream ends of scour holes. The ripples in these locations tend to migrate up the adverse slopes of the scour holes. A prominent example is the Macall ripple field (Sec. 18, T.18N., R.38E.) located in a scour hole immediately east of the Marengo loess island. These ripples occur on extremely thin gravel fills. They are really starved ripples, since the troughs of the ripples may be only a meter above bedrock.

Sedimentary Characteristics

A study was made of the distribution of the largest boulders in bars and on bedrock surfaces in the Cheney-Palouse. Unlike the predominant basalt boulders, boulders of granitic composition can be attributed to a known source area at the northern end of the tract, and these boulders were measured wherever they were found. The results indicate no systematic variation in grain size when the Cheney-Palouse is considered as a single unit. This is not surprising because most of the sediment was probably locally derived, and, therefore, the maximum size in any deposit is less a function of distance of transport than it is of local current velocity, turbulence, and joint spacing in the basalt. The largest boulder, 4 x 3 x 3 m in size, was found at midlength in the Cheney-Palouse approximately 12 km north of Benge. The boulder is one of several large basalt blocks deposited on a scarified basalt surface.

Within a single flood bar, there may be a downstream decrease in grain size. Gravel on the previously described expansion bar in the channel north of Benge abruptly decreases in grain size downstream. On the other hand, a large pendant bar in the channel just east of the Marengo loess island has boulders in its downstream tail which are only slightly smaller than those immediately downstream from the basalt flow to which the bar is attached. Expansion bars may be expected to show evidence of hydraulic sorting, since the flow conditions varied along the reach in which they were deposited. Pendant bars, on the other hand, are generally smaller in areal extent, and longitudinal variation in flow conditions is not required for their formation. Therefore, current sorting and large grain size variation would not be evident along their length.

One might hypothesize that granitic boulders derived from the Medical Lake area to the north would decrease in size down the tract as a function of selective sorting and perhaps breakdown of the larger sizes. Our reconnaissance data indicate that the grain size of the granite boulders does not change radically downstream. The largest granite boulder found in a Cheney-Palouse scabland deposit has a long axis of 270 cm and lies south of Rock Lake near Ewan (Bretz and others, 1956). Near Marengo, there is a granite boulder 190 x 120 x 70 cm in size, and Bretz and others (1956) report granite boulders having long axes of 165 cm in the Cow Creek scabland just north of Hooper and in Shoulder Bar in the Snake River. Therefore, from all the available data, it does not appear that there is a rapid downstream decrease in size, in sharp contrast to that reported for the Ephrata Fan in the western scablands (Baker, 1973a). Our hypothesis is that along the deepest channels, the bedrock inner channels, the velocities and competence of the flow were undiminished from one end of the

Cheney-Palouse to the other. This is supported by the nearly uniform water surface profile for the main channels, like Cow Creek (Fig. 6.6). In the western Channeled Scabland, abrupt expansions of major channels, such as the Lower Grand Coulee flowing into the Quincy Basin, caused rapid reduction in stream competence and rapid sediment deposition.

DISCUSSION

When viewed in detail, the Cheney-Palouse appears to be a complex landscape. This is the result of the grouping and superposition of the three primary forms: (1) loess islands, (2) gravel bars, and (3) erosional bedrock scabland (Fig. 6.7). The organization of these primary forms in the Cheney-Palouse can be classified in the same manner as other complex fluvial landscapes such as sandur plains (Church, 1972) and braided streams (Williams and Rust, 1969). The classification demonstrates that the landforms in the Cheney-Palouse form predictable geomorphic assemblages in many aspects similar to modern fluvial systems.

The Karakul Hills loess island assemblage (Fig. 6.8) is analogous to a single spool or diamond bar on a sandur plain (Church, 1972). Isolated from other loess island groups by major scabland channels, the Karakul Hills assemblage is dissected by a sequence of channels, each at higher elevations, smaller and less well developed. The primary zone of deposition is downstream from the loess island group similar to the common pattern for longitudinal bars in braided streams.

By arranging several of these loess island-bar sequences together, more complex geometries can be created such as the assemblage of loess islands northeast of Sprague (Fig. 6.9) or the loess islands and bars at Willow Creek near La Crosse originally described by Bretz (1928b, p. 648).

The above examples are for simple geometries and for fairly limited areas within the Cheney-Palouse. A more complex assemblage comprises the entire center of the map in Fig. 6.2. This particular loess island grouping is in the center of the Cheney-Palouse tract and is bordered on the east by the Cow Creek channel and on the west by the unnamed channel that includes Twelve Mile Lake. The two channels converge and form the southern boundary at Benge. Thus, the loess islands are preserved in a reach where the dominant channels have diverged. The reach covers an area 24 km long and 16 km wide. Directly south of Benge, the convergence of these channels has obliterated all of the loessial topography, and the tract consists entirely of scabland. We consider the two channels to be first-order channel elements (Williams and Rust, 1969). They are almost entirely scabland with only minor sedimentation along their flanks.

Within the area surrounded by the two first-order channels are three distinct levels of smaller

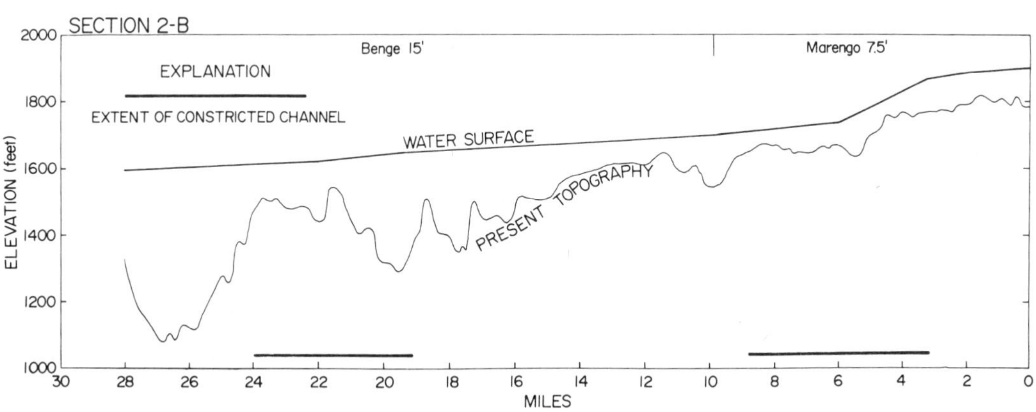

Figure 6.6. Profile 2-B along Cow Creek, a prominent inner channel of the Cheney-Palouse tract (see Fig. 6.1 for location).

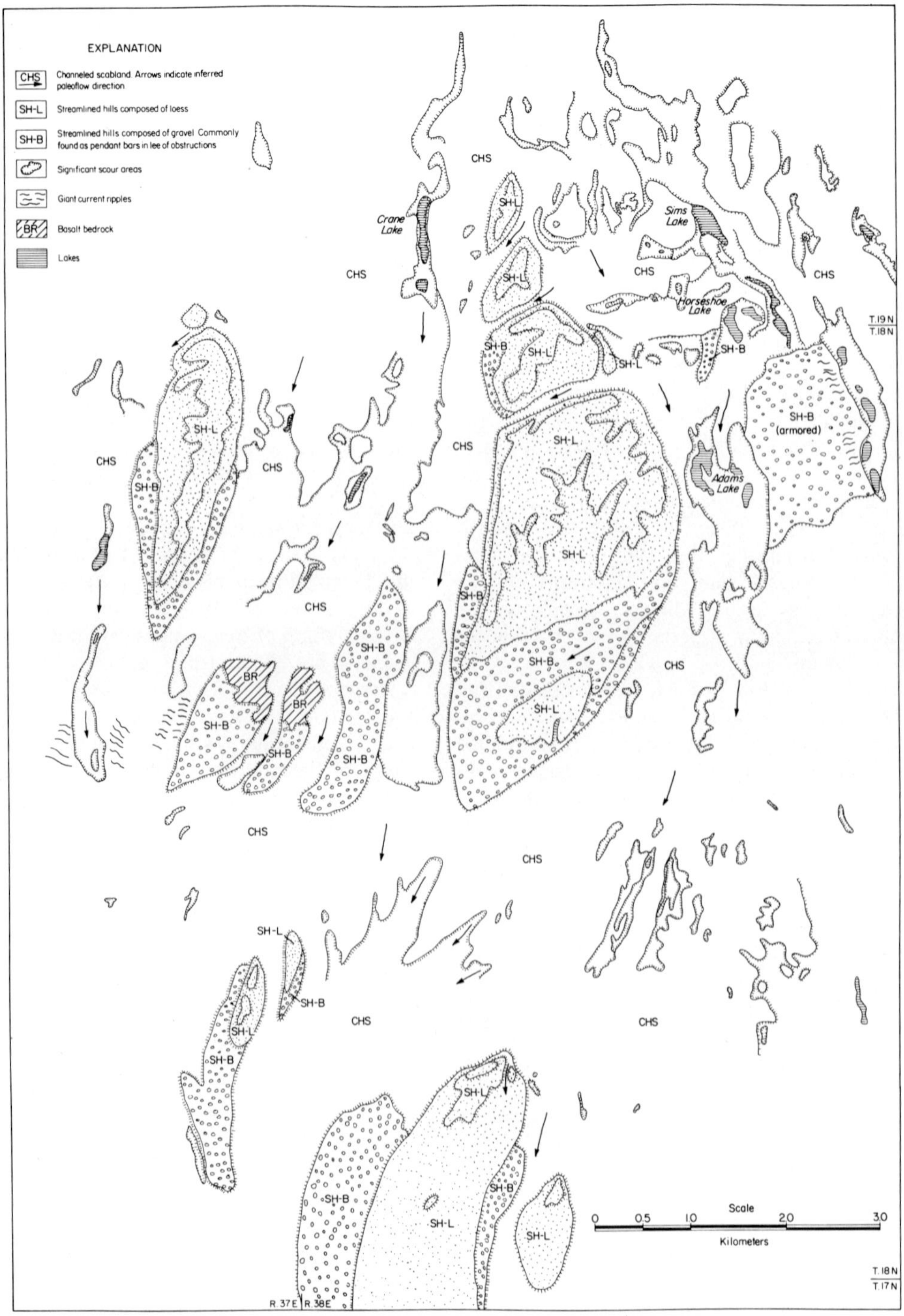

Figure 6.7. Geomorphic map of the Macall area in the central Cheney-Palouse tract.

channels which cut through the basic form. The largest channels have cataracts and weakly developed butte-and-basin topography. Most of the pendant bars and expansion bars downstream from cataracts are associated with these channels (Fig. 6.2). Still smaller channels at higher elevations have been eroded through the loess cover but have only sligthtly scarified the underlying basalt. Finally, the smallest channels are the eroded divide crossings, some of which are filled with slackwater deposits.

The Cheney-Palouse tract contains groupings of erosional residuals (loess islands) that have been modified by three or four levels of stream erosion as indicated by channel size and degree of scabland topographic development. The characteristic arrangement of erosional elements probably represents variations in flow velocities, rates of erosion, and rates of deposition for various elevations in the flood channel way. In modern braided stream environments, Williams and Rust (1969) have noted a decrease in the flow regime, water discharge, rate and mode of sediment transport, and period of activity with increasing elevation or ranking of the channel. We speculate that the same conditions existed during the creation of the Cheney-Palouse. The lowermost channels probably carried the greatest discharges for the longest durations generally with the greatest velocity and turbulence. These channels would logically be expected to exhibit the highest degree of scabland development and the least amount of associated sediment deposition. Sedimentation was greatest in the secondary channels probably because sediment concentration was still high, although velocities were somewhat reduced. These factors, when combined with the numerous channel expansions and large flow obstructions, created numerous zones conducive to deposition.

The analogy between the subaerially exposed loess island complexes and the longitudinal bars of braided streams has already been noted. As in braid bars, the greatest potential for gravel deposition was downstream from the largest channels dissecting individual loess islands. The loess islands created low velocity zones in their lees which localized the deposition of secondary gravel bars. The loess islands also behaved as major elements within a braided stream channel as they forced the main flow against the channel margins. Therefore, the loess islands must be, at least in part, responsible for the width of the Cheney-Palouse tract.

A major question which remains unanswered is exactly what mechanism was responsible for initially allowing the formation of the loess islands. In alluvial braided streams, the formation of a midchannel bar is usually started by deposi-

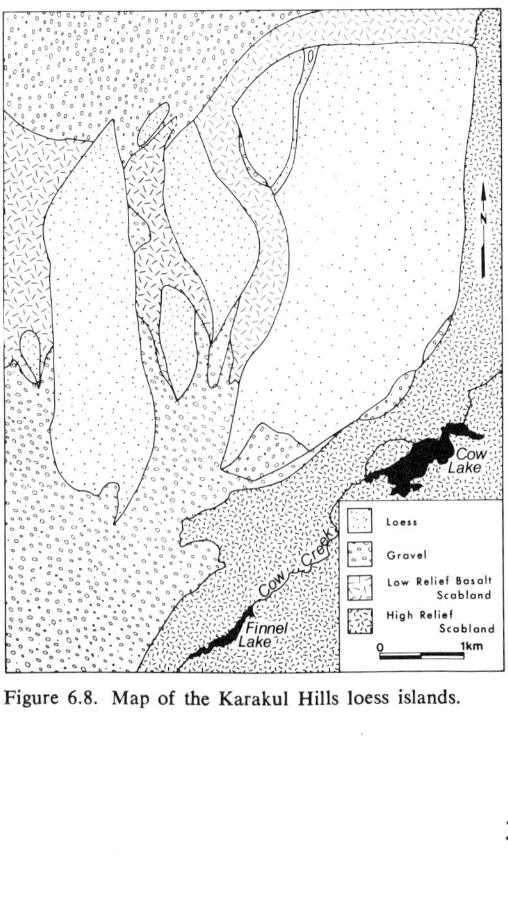

Figure 6.8. Map of the Karakul Hills loess islands.

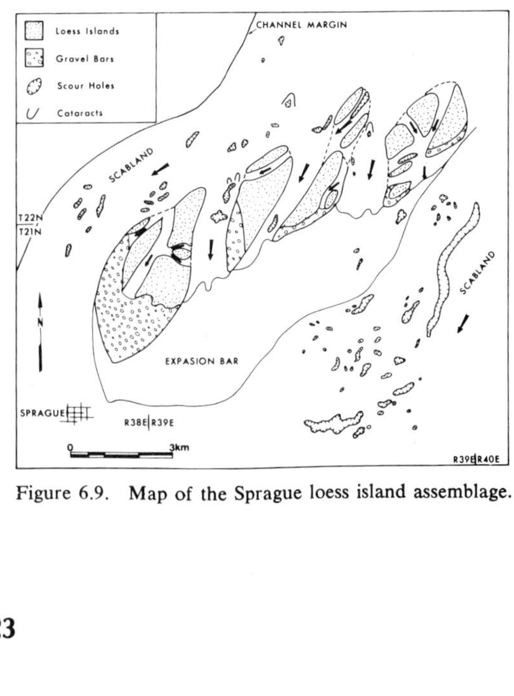

Figure 6.9. Map of the Sprague loess island assemblage.

tion of bedload because of a local incompetence in the flow (Leopold and others, 1964). The braided channels formed in this manner may eventually increase in depth and slope by erosion and cause a temporary increase in the competence of the flow (Leopold and others, 1964).

The Cheney-Palouse loess islands could not have formed in this manner because they are erosional and formed during downcutting. Nevertheless, the end result of the process was the same: increases in relative depth, velocity, and flow competency. Perhaps, as Church (1972, p. 74) suggests, the braiding was partly caused by the increasing boundary resistance that occurred as the channel widened by bank erosion. In order to maintain a great enough velocity for sediment transport, the channel divided, and incision created relatively narrow and deep secondary channels. Therefore, in the Cheney-Palouse, the erodible loess hills which formed the channel margin were probably the ultimate cause of the anastomosing pattern.

[*Editor's Note:* The Patton and Baker article ends here. The following reference list includes the citations for the entire Baker and Nummedal *Channeled Scabland* publication.]

REFERENCES

Alden, W. C., 1927, Discussion; Channeled Scabland and the Spokane flood: Washington Acad. Sci. Jour., v. 17, no. 8, p. 203.

———, 1953, Phyciography and glacial geology of western Montana and adjacent areas: U.S. Geol. Survey Prof. Paper 231, 193 p.

Allen, J. R. L., 1968, Current Ripples, their Relation to Patterns of Water and Sediment Motion: Elsevier, North-Holland Publishing, Amsterdam, 433 p.

———, 1970, Physical Processes of Sedimentation: American Elsevier, N.Y., 248 p.

———, 1971a, Bedforms due to mass transfer in turbulent flows: A kaleidoscope of phenomena: J. Fluid Mech., v. 49, pt. 1, p. 49-63.

———, 1971b, Transverse erosional marks of mud and rock: Their physical basis and geological significance: Sedimentary Geology, v. 5, p. 167-385.

Alexander, H. S., 1932, Pothole erosion: Jour. Geology, v. 40, p. 305-337.

Allison, I. S., 1933, New version of the Spokane flood: Geol. Soc. America Bull., v. 44, p. 675-722.

———, 1941, Flint's fill-hypothesis for channeled scabland: Jour. Geology, v. 49, p. 54-73.

Alwin, J. A., and Scott, W. F., 1970, Clastic dikes of the Touchet beds, southeastern Washington (abstract): Northwest Sci., v. 44, p. 58.

American Society of Civil Engineers, Task Force on Bedforms in Alluvial Channels, 1966, Nomenclature for bedforms in alluvial channels: Am. Soc. Civil Engineers Proc., v. 92, no. HY3, p. 51-64.

Armstrong, J. E., Crandell, D. R. Easterbrook, D. J., and Noble, J. B., 1965, Late Pleistocene stratigraphy and chronology in southwestern British Columbia and northwestern Washington: Geol. Soc. America Bull., v. 76, p. 321-330.

Armstrong, R. L., 1975, Precambrian (1500 m.y. old) rocks of central Idaho—the Salmon River Arch and its role in Cordilleran sedimentation and tectonics: Am. Jour. Sci., v. 275A, p. 437-467.

Atlantic Richfield Hanford Company, 1976, Preliminary feasibility study of storage of radioactive wastes in Columbia River basalts: ERDA Rept. ARH-ST-137, v. 1 (text, 168 p.) and v. 2 (appendices, 264 p.).

Bagnold, R. A., 1966, An approach to the sediment transport problem from general physics: U.S. Geol. Survey Prof. Paper 422-I, 37 p.

Baker, V. R., 1971, Paleohydrology of catastrophic Pleistocene flooding in eastern Washington: Geol. Soc. America, Abstracts with Programs, v. 3, p. 497.

———, 1973a, Paleohydrology and sedimentology of Lake Missoula flooding in eastern Washington: Geol. Soc. America Special Paper 144, 79 p.

———, 1973b, Erosional forms and processes for the catastrophic Pleistocene Missoula floods in eastern Washington: *in* Morisawa, M., *ed.*, Fluvial Geomorphology: Publ. in Geomorphology, State Univ. of N.Y., Binghamton, N.Y., p. 123-148.

———, 1977, Lake Missoula flooding and the Channeled Scabland *in* Brown, E. H., and Ellis, R. C., *eds.*, Geologic Excursions in the Pacific Northwest, Guidebook for the Geol. Soc. America, 1977 Annual Meeting. Western Washington Univ. Press, Bellingham, p. 399-414.

———, and Milton, D. J., 1974, Erosion by catastrophic floods on Mars and Earth: Icarus, v. 23, p. 27-41.

———, and Patton, P. C., 1976, Missoula flooding in the Cheney-Palouse scabland tract: Terrestrial Analogue to the Channeled Terrain of Mars: Geol. Soc. America, Abstracts with Programs, v. 8, no. 3, p. 351-352.

———, and Penteado-Orellana, M. M., 1977, Adjustment to Quaternary climatic change by the Colorado River in central Texas: Jour. Geology, v. 85, p. 395-422.

———, and Ritter, D. F., 1975, Competence of rivers to transport coarse bedload material: Geol. Soc. America Bull., v. 86, p. 975-978.

Barnes, H. L., 1956, Cavitation as a geological agent: Am. Jour. Sci., v. 254, p. 493-505.

Baulig, H., 1913, Le plateaux de lave du Washington Central et la Grand 'Coulee': Annls. Geogr., v. 22, p. 149-159.

Benson, M. A., and Dalrymple, T., 1968, General field and office procedures for indirect discharge measurements: U.S. Geol. Survey, Techniques Water-Resources Inv., Book 3, Chap. A1, 30 p.

Bentley, R. D., 1977, Stratigraphy of the Yakima basalts and structural evolution of the Yakima ridges in the western Columbia Plateau, *in* Brown, E. H., and Ellis, R. C., *eds.*, Geological Excursions in the Pacific Northwest, Western Washington Univ. Bellingham, p. 339-389.

Berggren, W. A., and van Couvering, J. A., 1974, The late Neogene: Biostratigraphy, geochronology, and paleoclimatology of the last 15 million years in marine and continental sequences: Palaeogeography, Palaeoclimatology, Palaeoecology, v. 16, p. 1-216.

Bingham, J. W., 1970, Several probable source vents for the Roza and Priest Rapids type basalts in Whitman and Adams Counties, Washington: *in* Gilmour, E. H., and Stradling, Dale, *eds.*, Proc. Second Columbia River Basalt Symposium: Cheney, Eastern Washington State Coll. Press, p. 171-172.

———, and Grolier, M. J., 1966, The Yakima Basalt and Ellensburg formation of south-central Washington: U.S. Geol. Survey Bull. 1224-G, p. G1-G15.

Bishop, D. T., 1969, Stratigraphy and distribution of

basalt, Benewah Co., Idaho: Idaho Bur. Mines and Geology Pamph. 140, 20 p.

Blissenback, E., 1954, Geology of alluvial fans in semi-arid regions: Geol. Soc. America Bull., v. 65, p. 175-189.

Bond, J. G., 1963, Geology of the Clearwater embayment: Idaho Bur. Mines and Geology Pamph. 128, 83 p.

Boothroyd, J. C., and Ashley, G. M., 1975, Processes, bar morphology and sedimentary structures on braided outwash fans, northeastern Gulf of Alaska: *in* Jopling, A. V., and McDonald, B. C., *eds.*, Glaciofluvial and Glaciolacustrine Sedimentation: Soc. Econ. Paleontologists and Mineralogists Spec. Publ. No. 23, p. 193-222.

———, and Nummedal, D., 1978, Proglacial braided outwash: A model for humid alluvial fan deposits: *in* Miall, A. D. *ed.*, Proceedings, First Internat. Symp. on Fluvial Sedimentology, Calgary, Alberta (in press).

Bradley, W. C., Fahnestock, R. K., and Rowekamp, E. T., 1972, Coarse sediment transport by flood flows on Knick River, Alaska: Geol. Soc. America Bull., v. 83, p. 1261-1284.

Bretz, J H., 1913, Glaciation of the Puget Sound region: Wash. Div. Mines and Geology Bull., no. 8, 244 p.

———, 1919, The late Pleistocene submergence in the Columbia Valley of Oregon and Washington: Jour. Geology, v. 27, p. 489-505.

———, 1923a, Glacial drainage on the Columbia Plateau: Geol. Soc. America Bull., v. 34, p. 573-608.

———, 1923b, The Channeled Scabland of the Columbia Plateau: Jour. Geology, v. 31, p. 617-649.

———, 1924, The Dalles type of river channel: Jour. Geology, v. 32, p. 139-149.

———, 1925, The Spokane flood beyond the Channeled Scabland: Jour. Geology, v. 33, p. 97-115, p. 236-259.

———, 1927a, Channeled Scabland and the Spokane flood: Washington Acad. Sci. Jour., v. 17, no. 8, p. 200-211.

———, 1927b, The Spokane flood; a reply: Jour. Geology, v. 35, p. 461-468.

———, 1928a, Alternative hypotheses for Channeled Scabland: Jour. Geology, v. 36, p. 193-223, p. 312-341.

———, 1928b, Bars of the Channeled Scabland: Geol. Soc. America Bull., v. 39, p. 643-702.

———, 1928c, The Channeled Scabland of eastern Washington: Geog. Rev., v. 18, p. 446-477.

———, 1929, Valley deposits immediately east of the Channeled Scabland of Washington: Jour. Geology, v. 37, p. 393-427, p. 505-541.

———, 1930a, Lake Missoula and the Spokane flood: Geol. Soc. America Bull., v. 41, p. 92-93.

———, 1930b, Valley deposits immediately west of the Channeled Scabland: Jour. Geology, v. 38, p. 385-422.

———, 1932a, The Grand Coulee: American Geog. Soc. Spec. Pub. 15, 89 p.

———, 1932b, The Channeled Scabland: 16th Internat. Geol. Congress, U.S. 1933, Guidebook 22, Excursion C-2, 16 p.

———, 1959, Washington's Channeled Scabland: Wash. Dept. Conserv., Div. Mines and Geology Bull. no. 45, 57 p.

———, 1969, The Lake Missoula floods and the Channeled Scabland: Jour. Geology, v. 77, p. 505-543.

———, 1972, Memories, Part I: Some recollections of a geologist on entering his 90th year: Unpub. manuscript distributed by the University of Chicago Department of Geophysical Sciences and J Harlen Bretz, 64 p.

———, 1973, Memories, Part II: Unpub. manuscript distributed by the University of Chicago Department of Geophysical Sciences and J Harlen Bretz, 136 p.

———, 1974, Memories, Part III: Unpub. manuscript distributed by the University of Chicago Department of Geophysical Sciences and J Harlen Bretz, 136 p.

Bretz, J H., Smith, H. T. U., and Neff, G. E., 1956, Channeled Scabland of Washington; new data and interpretations: Geol. Soc. America Bull., v. 67, p. 957-1049.

Brock, M. R., and Grolier, M. J., 1973, Chemical analyses of 305 basalt samples from the Columbia River Plateau, Washington, Oregon, and Idaho: U.S. Geol. Survey Open-File Rept., 35 p.

Brown, B. H., 1937, The state-line earthquake at Milton and Walla Walla: Seismol. Soc. America Bull., v. 231, p. 205-209.

Brown, R. E., 1971, Volcanic ash beds in the Pasco Basin: Abstracts of the Joint Meetings of the Idaho Academy of Science, Northwest Scientific Association, and Washington State Entomological Society.

———, 1973, The Glacial Lake Missoula floods in the Pasco Basin: Abstracts of the 46 Annual Meeting of the Northwest Scientific Association.

Bryan, K., 1927, The "Palouse soil" problem: U.S. Geol. Survey Bull. 790, p. 21-45.

Byerly, G. R., and Swanson, D. A., 1978, Invasive Columbia River basalt flows along the northwestern margin of the Columbia Plateau, north-central Washington: Geol. Soc. America, Abstracts with Programs, v. 10, no. 3, p. 98.

Calkins, F. C., 1905, Geology and water resources of a portion of east-central Washington: U.S. Geol. Survey Water-Supply Paper 118, 96 p.

Choiniere, S. R., 1977, Magnetostratigraphy and correlation of Miocene basalts of the Oregon coast and Columbia Plateau, southeast Washington: M.S. thesis, Univ. of Rhode Island, Kingston, R.I.

Chorley, R. J., 1959, The shape of drumlins: Jour. Glaciology, v. 3, p. 339-344.

Chow, V. T., 1959, Open-Channel Hydraulics: McGraw Hill, New York, 680 p.

Christiansen, R. L., and Lipman, P. W., 1972, Cenozoic volcanism and plate-tectonic evolution of the western United States, II, Late Cenozoic: Phil. Trans. Royal Soc. London A., v. 271, p. 249-284.

Church, M., 1972, Baffin Island sandurs: A study of arctic fluvial processes: Geol. Survey of Canada, Bull. 215, 208 p.

Clemens, S. (Mark Twain), 1896, Life on the Mississippi: Harper and Row, N.Y., 381 p.

Coleman, J. M., 1969, Brahmaputra River: Channel processes and sedimentation: Sedimentary Geol., v. 3, p. 129-239.

Dahl, R., 1965, Plastically sculptured detail forms on rock surfaces in northern Nordland, Norway: Geogr. Annal., v. 47, p. 83-140.

Davis, W. M., 1913, Meandering valleys and underfit rivers: Annals Am. Assoc. Geographers.

———, 1926, The value of outrageous geological hypotheses: Science, v. 63, p. 463-468.

Dawson, W. L., 1898, Glacial phenomena in Okanogan County, Washington: Amer. Geol., v. 22, p. 203-217.

Dickinson, W. R., 1970, Relations of andesites, granites and derivative sandstones to arc-trench tectonics: Reviews Geophysics and Space Physics, v. 8, no. 4, p. 813-860.

Diery, H. D., and McKee, Bates, 1969, Stratigraphy of the Yakima Basalt in the type area: Northeast Sci., v. 43, p. 47-64.

Dort, Wakefield, Jr., 1967, Late Cenozoic volcanism, St. Joe Valley, Idaho: Northwest Sci., v. 41, p. 141-151.

Dougherty, R. D., 1956, Lind Coulee: Proc. Am. Phil. Soc., v. 100, p. 223-278.

Dury, G. H., 1958, Tests of a general theory of misfit streams: Trans. Institute of British Geographers, Publ. 25, p. 105-118.

———, 1964, Principles of underfit streams: U.S. Geol. Survey Prof. Paper 452-A, 67 p.

———, 1965, Theoretical implications of underfit streams: U.S. Geol. Survey Prof. Paper 452-C, 43 p.

Easterbrook, D. J., 1969, Pleistocene chronology of the Puget Lowland and San Juan Islands, Washington: Geol. Soc. America Bull., v. 80, p. 2273-2286.

———, 1976, Quaternary geology of the Pacific Northwest: *in* Mahaney, W. C., *ed.,* Quaternary Stratigraphy of North America: Dowden, Hutchinson and Ross, Stroudsburg, Pa., p. 441-462.

Einstein, H. A., and Li, H., 1958, Secondary flow in straight channels: Trans. American Geophys. Union, v. 39, p. 1085-1088.

Embleton, C., and King, C. A. M., 1968, Glacial and periglacial geomorphology: New York, St. Martin's Press, 608 p.

Flint, R. F., 1935, Glacial features of the southern Okanogan region: Geol. Soc. America Bull., v. 46, p. 169-194.

———, 1938a, Summary of late-Cenozoic geology of southeastern Washington: Am. Jour. Sci., v. 35, p. 223-230.

———, 1938b, Origin of the Cheney-Palouse Scabland tract: Geol. Soc. America Bull., v. 49, p. 461-524.

Fox, K. R., Jr., Rinehart, C. D., and Engels, J. C., 1977, Plutonism and orogeny in north-central Washington—timing and regional context: U.S. Geol. Survey Prof. Paper 989, 27 p.

Freeman, O. W., Forrester, J. D., and Lupher, R. L., 1945, Physiographic divisions of the Columbia Intermontane Province: Annals Assoc. Am. Geog., v. 35, no. 2, p. 53-75.

Fruchter, J. S., and Baldwin, S. F., 1975, Correlations between dikes of the Monument swarm, central Oregon, and Picture Gorge Basalt flows: Geol. Soc. America Bull., v. 86, p. 514-516.

Fryxell, R., 1962, A radiocarbon limiting date for scabland flooding: Northwest Sci., v. 36, p. 113-119.

———, 1965, Mazama and Glacier Peak volcanic ash layers: Relative ages: Science, v. 147, p. 1288-1290.

———, 1966, Origin and age of Palouse Hills topography, eastern Columbia Plateau: Program, Geol. Soc. America, Annual Meeting, San Francisco.

———, Bielicki, T., Dougherty, R. D., Gustafson, C. E., Irwin, H. T., and Keel, B. C., 1968, A human skeleton from sediments of mid-Pinedale age in southeastern Washington: Am. Antiquity, v. 33, p. 511-514.

———, and Cook, E. F., 1964, A field guide to the loess deposits and channeled scablands of the Palouse area, eastern Washington: Wash. State Univ. Lab. Anthropology Rept. Inv. 27, 32 p.

Fuller, R. E., 1931, The aqueous chilling of basaltic lava on the Columbia River Plateau: Am. Jour. Sci., v. 21, p. 281-300.

Fulton, R. J., 1971, Radiocarbon geochronology of southern British Columbia: Geol. Survey of Canada Paper 71-37, 28 p.

Gilbert, G. K., 1890, Lake Bonneville: U.S. Geol. Survey, Monograph 1, 438p.

Gile, L. H., Peterson, F. F., and Grossman, R. B., 1966, Morphological and genetic sequences of carbonate accumulation in desert soils: Soil Sci., v. 101, p. 347-360.

Gilluly, J., 1927, Discussion: Channeled Scabland and the Spokane flood: Washington Acad. Sci. Jour., v. 17, no. 8, p. 203-205.

Greeley, Ronald, 1977, Basaltic "plains" volcanism: *in* Greeley, Ronald, and King, J. S., *eds.,* Volcanism of the eastern Snake River Plain, Idaho, a Comparative Planetary Geology Guidebook, Office of Planetary Geology, N.A.S.A., Washington, D.C., p. 23-44.

Gresens, R. L., Whetten, J. T., Tabor, R. W., and Frizzell, V. A., Jr., 1977, Tertiary stratigraphy of the Central Cascade Mountains, Washington state: *in* Brown, E. H., and Ellis, R. C., *eds.,* Geological excursions in the Pacific Northwest: Western Washington Univ. Press, Bellingham, p. 84-126.

Griggs, A. B., 1973, Geologic map of the Spokane quadrangle, Washington, Idaho and Montana: U.S. Geol. Survey Misc. Geol. Inv. Map I-768, scale 1:250,000.

———, 1976, The Columbia River Basalt Group in the Spokane quadrangle, Washington, Idaho, and Montana: U.S. Geol. Survey Bull. 1413, 39 p.

Grolier, M. J., and Bingham, J. W., 1971, Geologic map and sections of parts of Grant, Adams, and Franklin Counties, Washington: U.S. Geol. Survey Misc. Geol. Inv. Map I-589, scale 1:62,500.

Gustafson, C. E., 1972, Faunal remains from the Marmes Rockshelter and related archaeological sites in the Columbia Basin: Unpub. Ph.D. dissert., Washington State Univ., 183 p.

———, 1976, An ice age lake in the Columbia Basin:

New Evidence: Geol. Soc. America, Abstracts with Programs, v. 8, no. 3, p. 377.

Gustafson, E., 1973, The vertebrate fauna of the late Pliocene Ringold Formation, south-central Washington: Unpub. M.S. thesis, Univ., Washington, Seattle, 164 p.

———, 1976, A revised chronology for vertebrate fossil faunas in eastern Washington: Geol. Soc. America, Abstracts with Programs, v. 8, no. 3, p. 377-378.

Hammatt, H. H. Foley, L. L., and Leonhardy, F.C., 1976, Late Quaternary stratigraphy in the lower Snake River Canyon: toward a chronology of slack water sediments: Geol. Soc. America, Abstracts with Programs, v. 8, no 3, p. 379.

Hansen, H. P., 1947, Postglacial forest succession, climate and chronology in the Pacific Northwest: Trans. Am. Philos. Soc., v. 37, p. 1-130.

Hanson, L. G., 1970, The origin and development of Moses Coulee and other scabland features on the Waterville Plateau, Washington: Unpub. Ph.D. dissertation, Univ. Washington, Seattle, 137 p.

Harding, H. T., 1929, Possible supply of water for the channeled scablands: Science, v. 69, p. 188-190.

Harms, J. C., 1969, Hydraulic significance of some sand ripples: Geol. Soc. America Bull., v. 80, p. 363-396.

Hjulström, F., 1935, Studies in the morphological activity of rivers as illustrated by the River Fyris: Geol. Inst. Univ. Upsala, Bull., v. 25, p. 221-528.

Hobbs, W. H., 1943, Discovery in eastern Washington of a new lobe of the Pleistocene continental glacier: Science, v. 98, p. 227-230.

———, 1947, The glacial history of the Scabland and Okanogan lobes, Cordilleran continental glacier: Privately printed, J. W. Edwards, Ann Arbor, Michigan, 40 p.

Hodge, E. T., 1934, Origin of the Washington Scabland: Northwest Science, v. 8, p. 4-11.

Hodges, C. A., 1976, Basaltic ring structure of the Columbia Plateau: Geol. Soc. America, Abstracts with Programs, v. 8, no. 3, p. 382.

Holden, G. S., and Hooper, P. R., Petrology and chemistry of a Columbia River basalt section, Rocky Canyon, west-central Idaho: Geol. Soc. America Bull., v. 87, p. 215-225.

Hooper, P. R., 1947, Petrology and chemistry of the Rock Creek flow, Columbia River basalt, Idaho: Geol. Soc. America Bull., v. 85, p. 15-26.

Hooper, P. R., Reidel, S. P., Brown, J. C., Bush, J. H., Holden, G. S. Kleck, W. D., Robinette, M., Sundstrom, C. E., and Taylor, T. L., 1976, Major element analyses of Columbia River Basalt, part 1: Informal Report, Basalt Research Group, Dept. of Geology, Washington State Univ., Pullman.

Hoyt, C. L., 1961, The Hammond Sill—an intrusion in the Yakima Basalt near Wenatchee, Washington: Northwest Science, v. 35, p. 58-64.

Hyndman, D. W., and Williams, L. D., 1977, The Bitterroot lobe of the Idaho Batholith: Northwest Geology, v. 6-1, p. 1-16.

Ippen, A. T., Drinker, P. A., Jobin, W. R., and Shemdin, O. H., 1962, Stream dynamics and boundary shear distribution for curved trapezoidal channels: Mass. Inst. Tech. Report No. 47, 81 p.

Jackson, D. B., 1975, Description of the geoelectric section, Rattlesnake Hills unit 1 well, Washington: Jour. Research U.S. Geol. Survey, v. 3, p. 665-669.

Jackson, R. G., 1975, Hierarchical attributes and a unifying model of bedforms composed of cohesionless material and produced by shearing flow: Geol. Soc. America Bull., v. 86, p. 1523-1533.

———, 1976, Sedimentological and fluid-dynamic implications of the turbulent bursting phenomenon in geophysical flows: Jour. Fluid Mechanics, v. 77, pt. 3, p. 531-560.

———, 1977, Genesis of fluvial bedforms: Program and Abstracts of the First Internat. Symp. on Fluvial Sedimentology, Calgary, Alberta, p. 15-16.

Jaeger, J. C., 1961, The cooling of irregularly shaped igneous bodies: Am. Jour. Sci., v. 259, p. 721-734.

Jenkins, O. P., 1925, Clastic dikes of eastern Washington and their geologic significance: Am. Jour. Sci., 5th ser., v. 10, p. 234-246.

Jones, F. O., and Deacon, R. J., 1966, Geology and tectonic history of the Hanford area and its relation to the geology and tectonic history of the State of Washington and the active seismic zones of western Washington and western Montana: Spokane, Douglas United Nuclear, Inc., 44 p.

Jones, J. C., 1968, Pillow lava and pahoehoe: Jour. Geology, v. 76, p. 485-488.

Jopling, A. V., and Richardson, E. V., 1966, Backset bedding developed in shooting flow in laboratory experiments: Jour. Sed. Petrology, v. 36, p. 821-824.

Karcz, I., 1967, Harrow marks, current aligned sedimentary structures: Jour. Geology, v. 75, p. 113-121.

———, 1968, Fluviatile obstacle marks from the wadis of the Negev (southern Israel): Jour. Sed. Petrology, v. 38, p. 1000-1012.

Keller, E. A., 1971, Areal sorting of bed-load material: The hypothesis of velocity reversal: Geol. Soc. America Bull., v. 82, p. 753-756.

———, and Melhorn, W. N., 1973, Bedforms and fluvial processes in alluvial stream channels: Selected observations: *in* Morisawa, M., *ed.*, Fluvial Geomorphology, State Univ. of N.Y., Binghamton, N.Y., p. 253-283.

King, P. B., and Beikman, H. M., 1974, Geologic map of the United States: U.S. Geol. Survey, scale 1:2,500,-000.

Kittleman, L. R., 1973, Mineralogy, correlation and grain-size distributions of Mazama tephra and other post-glacial pyroclastic layers, Pacific Northwest: Geol. Soc. America Bull., v. 84, p. 2957-2980.

Komar, P. D., 1971, Hydraulic jumps in turbidity currents: Geol. Soc. America Bull., v. 82, p. 1477-1488.

———, 1978, Comparisons of the hydraulics of water flows in Martian outflow channels with flows of similar scale on earth: Icarus (in press).

Krumbein, W. C., 1942, Flood deposits of Arroyo Seco,

Los Angeles County, California: Geol. Soc. America Bull., v. 53, p. 1355-1402.

Kukla, G. J., 1975, Loess stratigraphy of central Europe: *in* Butzer, K. W., and Isaac, G. L., *eds.*, After the Australopithecines: Moulton, The Hague, p. 99-188.

Landye, J. J., 1973, Environmental significance of late Quaternary non-marine mollusks from former Lake Bretz, lower Grand Coulee, Washington: Unpub. M.A. thesis, Washington State Univ., Pullman, 117 p.

Laufer, J., 1975, New trends in experimental turbulence research: Ann. Rev. Fluid Mech., v. 7, p. 307-326.

Laursen, E. M., 1960, Scour at bridge crossings: Jour. Hydraulics Div., Proc. Am. Soc. Civil Engineers, v. 86, no. HY2, p. 39-54.

Lefebvre, R. H., 1970, Columbia River basalts of the Grand Coulee area: *in* Gilmour, E. H., and Stradling, Dale, *eds.*, Proc. Second Columbia River Basalt Symposium: Cheney, Eastern Washington State Coll. Press, p. 1-38.

Lemke, R. W., Mudge, M. R., Wilcox, R. E., and Powers, H. A., 1975, Geologic setting of the Glacier Peak and Mazama ash-bed markers in west-central Montana: U.S. Geol. Survey Bull. 1395-H.

Leopold, L. B., and Maddock, T., Jr., 1953, The hydraulic geometry of stream channels and some physiographic implications: U.S. Geol. Survey Prof. Paper 252, 57 p.

———, Wolman, M. G., and Miller, J. P., 1964, Fluvial Processes in Geomorphology: W. H. Freeman, San Francisco, 522 p.

Lewis, P. F., 1960, Linear topography in the southwestern Palouse, Washington-Oregon: Annals Am. Assoc. Geographers, v. 50, p. 98-111.

Lupher, R. L., 1944, Clastic dikes of the Columbia Basin region, Washington and Idaho: Geol. Soc. America Bull., v. 55, p. 1431-1462.

Mack, R. N., Bryant, V. J., Jr., and Fryxell, R., 1976, Pollen sequence from the Columbia Basin, Washington: Reappraisal of postglacial vegetation: American Midland Naturalist, v. 95, no. 2, p. 390-397.

Mackin, J. H., 1961, A stratigraphic section in the Yakima Basalt and the Ellensburg Formation in south-central Washington: Washington Dept. Conserv., Div. Mines and Geology Rep. Inv. 19, 45 p.

———, 1963, Rational and empirical methods of investigation in geology: *in* Albritton, C. C., *ed.*, The Fabric of Geology: Addison-Wesley Pub. Co., Reading, Mass., p. 135-163.

Malde, H. E., 1968, The catastrophic late Pleistocene Bonneville Flood in the Snake River Plain, Idaho: U.S. Geol. Survey Prof. Paper 596, 52 p.

Marshall, A. G., 1971, An alluvial chronology of the lower Palouse River Canyon and its relation to local archaeological sites: Unpub. M.A. thesis, Washington State Univ., Pullman, 73 p.

Masursky, H., Boyce, J. M., Dial, A. L., Schaber, G. G., and Strobell, M. E., 1977, Classification and time of formation of martian channels based on Viking data: Jour. Geophys. Res., v. 82, no. 28, p. 4016-4032.

Matthes, G. H., 1947, Macroturbulence in natural stream flow: Am. Geophys. Union Trans., v. 28, p. 255-262.

McDougall, Ian, 1976, Geochemistry and origin of basalt of the Columbia River Group, Oregon and Washington: Geol. Soc. America Bull., v. 87, p. 777-792.

McKee, Bates, 1972, Cascadia: McGraw Hill, Inc., 394 p.

———, and Stradling, D., 1970, The sag flowout: A newly described volcanic structure: Geol. Soc. America Bull., v. 81, p. 2035-2044.

McKee, E. H., Swanson, D. A., and Wright, T. L., 1977, Duration and volume of Columbia River basalt volcanism, Washington, Oregon and Idaho: Geol. Soc. America, Abstracts with Programs, v. 9, no. 4, p. 463-464.

McKnight, E. T., 1927, The Spokane flood: A discussion: Jour. Geology, v. 35, p. 453-460.

Mehringer, P. J., Jr., Blinman, E., and Petersen, K. L., 1977, Pollen influx and volcanic ash: Science, v. 198, p. 257-261.

Meier, M. F., 1964, Ice and glaciers: Section 16: *in* Chow, V. T., *ed.*, Handbook of Applied Hydrology: McGraw Hill, N.Y.

Meinzer, O. E., 1918, The glacial history of Columbia River in the Big Bend Region: Jour. Washington Acad. Sci., v. 8, p. 411-412.

———, 1927, Discussion: Channeled scabland and the Spokane flood: Washington Acad. Sci. Jour., v. 17, no. 8, p. 207-208.

Miller, F. K., 1975, Geology of the Chewelah-Loon Lake area, Stevens and Spokane Counties, Washington: U.S. Geol. Survey Prof. Paper 806, 75 p.

Misch, Peter, 1966, Tectonic evolution of Northern Cascades of Washington state—a west-Cordilleran case history: Canadian Inst. Mining and Metallurgy Spec. Pub. Vol. 8, p. 101-148.

———, 1977, Bedrock geology of the North Cascades: *in* Brown, E. H., and Ellis, R. C., *eds.*, Geological excursions in the Pacific Northwest: Western Wash. Univ. Press, Bellingham, p. 1-62.

Moody, U. L., 1977, Correlation of flood deposits containing St. Helens set S ashes and the stratigraphic position of St. Helens set J and Glacier Peak ashes, central Washington: Geol. Soc. America, Abstracts with Programs, v. 9, no. 7, p. 1098-1099.

Moore, J. G., 1975, Mechanism of formation of pillow lava: Am. Scientist, v. 63, p. 269-277.

———, Phillips, R. L., Griggs, R. W., Peterson, D. W., and Swanson, D. A., 1973, Flow of lava into the sea, 1969-1971, Kilauea Volcano, Hawaii, Geol. Soc. America Bull., v. 84, p. 537-546.

Moore, W. L., and Masch, F. D., 1963, Influence of secondary flow on local scour at obstructions in a channel: Federal Inter-Agency Sedimentation Conf. Proc., 1963, U.S. Dept. of Agriculture, Misc. Publ. 970, p. 314-320.

Mullineaux, D. R., Hyde, H. J., and Rubin, M., 1975, Widespread late glacial and postglacial tephra deposits from Mount St. Helens volcano, Washington: U.S. Geol. Survey Jour. Research, v. 3, no. 3, p. 329-335.

———, Wilcox, R. E., Ebaugh, W. F., Fryxell, R., and Rubin, M., 1977, Age of the late major scabland flood of eastern Washington, as inferred from associated ash beds of Mount St. Helens set S: Geol. Soc. America, Abstracts with Programs, v. 9, no. 7, p. 1105.

Nathan, Simon and Fruchter, J. S., 1974, Geochemical and paleomagnetic stratigraphy of the Picture Gorge and Yakima Basalts (Columbia River Group) in central Oregon: Geol. Soc. America Bull., v. 85, p. 63-76.

Nelson, D. O., Swanson, D. A., and Wright, T. L., 1976, Strontium isotopic composition of intracanyon flows of Yakima Basalt, southeast Washington: Geol. Soc. America, Abstracts with Programs, v. 8, no. 3, p. 399.

Newcomb, R. C., 1958, Ringold Formation of Pleistocene age in type locality: Am. Jour. Sci., v. 256, p. 328-340.

———, 1961, Age of the Palouse formation in the Walla Walla and Umatilla River Basins, Oregon and Washington: Northwest Sci., v. 35, p. 122-127.

———, 1962, Hydraulic injection of clastic dikes in the Touchet beds, Washington, Oregon and Idaho (Abstract): Geol. Soc. of the Oregon Country, Geol. Newsletter, v. 28, no. 10, p. 70.

———, 1965, Geology and ground-water resources of the Walla Walla River Basin, Washington-Oregon: Wash. Div. of Water Resources Water Supply Bull. 21, 151 p.

———, 1970, Tectonic structure of the main part of the basalt of the Columbia River Group, Washington, Oregon and Idaho: U.S. Geol. Survey Misc. Geol. Inv. Map 1-587, scale 1:500,000.

Newman, K. R., 1970, Palynology of interflow sediments from the Standard Oil Company of California Rattlesnake Hills No. 1 well, Benton County, Washington: *in* Gilmour, E. H., and Stradling, Dale, *eds.*, Proc. Second Columbia River Basalt Symposium: Cheney, Eastern Washington State Coll. Press. p. 201-207.

Nummedal, D., and Boothroyd, J. C., 1976, Morphology and hydrodynamic characteristics of terrestrial fan environments: Tech. Rept. No. 10-CRD, Coastal Research Division, Dept. of Geology, Univ. of South Carolina, 61 p.

Nummedal, D., Hine, A. C., Ward L. G., Hayes, M. O. Boothroyd, J. C., Stephen, M. F., and Hubbard, D. K., 1974, Recent migrations of the Skeidararsandur shoreline, southeast Iceland: Final report for contract no. N60921-73-C-0258, Naval Ordnance Laboratory, Washington, D.C., 183 p.

Oestreich, K., 1915, Die Grand Coulee: American Geog. Soc. Memorial Volume of the Transcontinental Excursion of 1912 (Am. Geogr. Soc. Spec. Publ. No. 1), p. 259-273.

Offen, G. R., and Kline, S. J., 1975, A proposed model of the bursting process in turbulent boundary layers: Jour. Fluid Mechanics, v. 70, p. 209-228.

Olson, E. C., 1969, Introduction to J Harlen Bretz' paper "The Lake Missoula Floods and the Channeled Scabland": Jour. Geology, v. 77, p. 503-504.

Osawa, Masumi, and Goles, G. G., 1970, Trace-element abundances in Columbia River basalts: *in* Gilmour, E. H., and Stradling, Dale, *eds.*, Proc. Second Columbia River Basalt Symposium: Cheney, Eastern Washington State Coll. Press, p. 55-71.

Pardee, J. T., 1910, The glacial Lake Missoula, Montana: Jour. Geology, v. 18, p. 376-386.

———, 1922, Glaciation in the Cordilleran region: Science, v. 56, p. 686-687.

———, 1942, Unusual currents in glacial Lake Missoula, Montana: Geol. Soc. America Bull., v. 53, p. 1569-1600.

———, and Bryan, Kirk, 1926, Geology of the Latah Formation in relation to the lavas of the Columbia Plateau near Spokane, Washington: U.S. Geol. Survey Prof. Paper 140-A, 17 p.

Parker, S., 1838, Journal of an exploring tour beyond the Rocky Mountains: privately published, Auburn, N.Y.

Pearson, R. C., and Obradovich, J. E., 1977, Eocene rocks in northeast Washington—radiometric ages and correlation: U.S. Geol. Survey Bull. 1433, 41 p.

Pierce, K. L., Obradovich, J. D., and Friedman, I., 1976, Obsidian hydration dating and correlation of Bull Lake and Pinedale Glaciations near West Yellowstone, Montana: Geol. Soc. America Bull., v. 87, p. 703-710.

Prandtl, L., 1926, Über die ausgebildete Turbulenz, Proc. 2nd Internat. Congr. Appl. Mech., Zurich, p. 71.

Raymond, J. R., and Tillson, D. D., 1968, Evaluation of a thick basalt sequence in south-central Washington: A.E.C. Research and Development Rept., BNWL-776, 126 p.

Reynolds, M. W., and Kleinkopf, M. D., 1977, The Lewis and Clark line, Montana-Idaho: A major intraplate tectonic boundary: Geol. Soc. America, Abstract with Programs, v. 9, no. 7, p. 1140-1141.

Rice, D. G., 1972, The Windust phase in lower Snake River region prehistory: unpub. Ph.D. dissertation, Washington State Univ., Pullman, 225 p.

Richardson, L. F., 1920, The supply of energy from and to atmospheric eddies: Proc. Royal Soc., Series A, v. 97, no. 686, p. 354-373.

Richardson, P. D., 1968, The generation of scour marks near obstacles: Jour. Sed. Petrology, v. 38, p. 965-970.

Richmond, G. M., 1965, Glaciation of the Rocky Mountains: *in* Wright, H. E., and Frey, D. G., *eds.*, The Quaternary of the United States: Princeton Univ. Press, Princeton, N.J., p. 217-230.

———, Fryxell, R., Neff, G. E., and Weis, P. L., 1965, The Cordilleran ice sheet of the northern Rocky Mountains and related Quaternary history of the Columbia Plateau: *in* Wright, H. E., and Frey, D. G., *eds.*, The Quaternary of the United States: Princeton Univ. Press, Princeton, N. J., p. 231-242.

Rietman, J. C., 1966, Remnant magnetization of the late Yakima Basalt, Washington State; Ph.D. Diss., Stanford Univ. Stanford, Calif., 87 p.

Rikitake, Tsuneji, 1972, The mechanism of geomagnetic field reversal: Physics of Earth Planetary Interiors, v. 6, p. 340-345.

Rinehart, C. D., and Fox, K. F., Jr., 1976, Bedrock geology of the Conconully quadrangle, Okanogan

County, Washington: U.S. Geol. Survey Bull. 1402, 58 p.
Ringe, D., 1970, Sub-loess basalt topography in the Palouse Hills, southeastern Washington: Geol. Soc. America Bull., v. 81, p. 3049-3060.
Russell, I. C., 1893, Geologic reconnaissance in central Washington: U.S. Geol. Survey Bull. 108, 108 p.
Rust, B. R., 1972, Structure and process in a braided river: Sedimentology, v. 18, p. 221-245.
———, 1975, Fabric and structure in glaciofluvial gravels: *in* Jopling, A. V., and McDonald, B. C., *eds.*, Glaciofluvial and Glaciolacustrine Sedimentation: Soc. Econ. Paleontologists and Mineralogists Spec. Publ. No. 23, p. 238-248.
Salisbury, R. D., 1901, Glacial work in the western mountains in 1901: Jour. Geology, v. 9, p. 721-724.
Schmincke, H. U., 1967a, Stratigraphy and petrography of four upper Yakima Basalt flows in south-central Washington: Geol. Soc. America Bull., v. 78, p. 1385-1422.
———, 1967b, Fused tuff and peperites in south-central Washington: Geol. Soc. America Bull., v. 78, p. 319-330.
———, 1967c, Flow directions in Columbia River basalt flows, and paleocurrents of interbedded sedimentary rocks, south-central Washington: Geol. Rundschau, v. 56, p. 992-1020.
Schumm, S. A., and Shepherd, R. G., 1973, Valley floor morphology; evidence of subglacial erosion: Area, v. 5, p. 5-9.
Schwennessen, A. T., and Meinzer, O. E., 1918, Ground water in Quincy Valley, Washington: U.S. Geol. Survey Water-Supply Paper 425, p. 131-161.
Sharp, R. P., and Malin, M. C., 1975, Channels on Mars: Geol. Soc. America Bull., v. 86, p. 593-609.
Shaw, H. R., 1973, Mantle convection and volcanic periodicity in the Pacific—evidence from Hawaii: Geol. Soc. America Bull., v. 84, p. 1505-1526.
———, and Swanson, D. A., 1970, Eruption and flow rates of flood basalt: *in* Gilmour, E. H., and Stradling, Dale, *eds.*, Proc. Second Columbia River Basalt Symposium: Cheney, Eastern Washington State Coll. Press. p. 271-299.
Shen, H. W., 1971, Scour near piers, *in* Shen, H. W., *ed.*, River Mechanics: Ft. Collins, Colorado, Shen, H. W., Publisher, p. 23.1-23.25.
Shepherd, R. G., 1972, A model study of river incision: Unpub. M.S. thesis, Colorado State Univ., Fort Collins, Colo., 135 p.
———, and Schumm, S. A., 1974, An experimental study of river incision: Geol. Soc. America Bull., v. 85, p. 257-268.
Simons, D. B., and Gessler, J., 1971, Research needs in fluvial processes: *in* Shen, H. W., *ed.*, River Mechanics: Ft. Collins, Colo., Shen, H. W., Publisher, p. 23.1-32.13.
———, and Richardson, E. V., 1966, Resistance to flow in alluvial channels: U.S. Geol. Survey Prof. Paper 422-J, 61 p.
———, Richardson, E. V., and Nordin, C. F., 1965, Sedimentary structures generated by flow in alluvial channels: *in* Middleton, G. V., *ed.*, Primary Sedimentary Structures and Their Hydrodynamic Interpretation: Soc. Econ. Paleontologists and Mineralogists Spec. Publ. 12, p. 34-52.
Smith, N. D., 1971, Transverse bars and braiding in the lower Platte River, Nebraska: Geol. Soc. America Bull., v. 81, p. 2993-3014.
Smith, R. L., 1960a, Ash flows: Geol. Soc. America Bull., v. 71, p. 795-842.
———, 1960b, Zones and zonal variation in welded ash flows: U.S. Geol. Survey Prof. Paper 354-F, p. 149-159.
Spry, A., 1962, The origin of columnar jointing, particularly in basalt flows: Geol. Soc. Australia Jour., v. 8, p. 191-216.
———, and Carr, W. J., 1961, The Michaud delta and
Swanson, D. A., 1967, Yakima Basalt of the Tieton River area, south-central Washington: Geol. Soc. America Bull., v. 78, p. 1077-1110.
———, 1972, Magma supply rate at Kilauea Volcano, 1952-1971: Science, v. 175, p. 169-170.
———, and Wright, T. L., 1976, Guide to field trip between Pasco and Pullman, Washington, emphasizing stratigraphy, vent areas, and intracanyon flows of Yakima Basalt: Geol. Soc. America Cordill. Sect. Mtg., Pullman, Washington, Field Guide No. 1, 33 p.
———, Wright, T. L., Camp, V. E., Gardner, J. N., Helz, R. T., Price, S. A., and Ross, N. E., 1977, Reconnaissance geologic map of the Columbia Basalt Group, Pullman and Walla Walla quadrangles, southeast Washington and adjacent Idaho: U.S. Geol. Survey Open-File Map 77-100, scale 1:250,000.
———, Wright, T. L., and Clem, Richard, 1975a, Intracanyon flows of Yakima Basalt along the Snake River, southeast Washington: Geol. Soc. America, Abstracts with Programs, v. 7, no. 5, p. 645.
———, ———, and Helz, R. T., 1975b, Linear vent systems and estimated rates of magma production and eruption for the Yakima Basalt on the Columbia Plateau: Am. Jour. Sci., v. 275, p. 877-905.
Symons, T. W., 1882, Report of an examination of the upper Columbia River: Forty-Seventh Congress, 1st Session, Sen. Doc. No. 186, 135 p.
Symposium, 1947, Cavitation in hydraulic structures; a symposium: Am. Soc. Civil Engineers, Trans. v. 112, p. 1-124.
Tabor, R. W., Waitt, R. B., Jr., Frizzell, V. A., Jr., Swanson, D. A., and Byerly, G. R., 1977, Preliminary map of the Wenatchee 1:100,000 quadrangle, Washington: U.S. Geol. Survey Open-File Rept. 77-531.
Taubeneck, W. J., 1970, Dikes of Columbia River basalt in northeastern Oregon, western Idaho and southeastern Washington: *in* Gilmour, E. H., and Stradling, Dale, *eds.*, Proc. Second Columbia River Basalt Symposium: Cheney, Eastern Washington State Coll. Press, p. 73-96.
Tomkeieff, S. I., 1940, The basalt lavas of the Giant's Causeway district of northern Ireland: Bull. Volcanologique, v. 6, p. 90-143.

Trimble, D. E., 1950, Joint controlled channeling in the Columbia River basalt: Northwest Sci., v. 24, p. 84-88.

———, 1963, Geology of Portland, Oregon and adjacent areas: U.S. Geol. Survey Bull., 1119, 119 p.

———, and Carr, W. J., 1961, The Michaud delta and Bonneville River near Pocatello, Idaho: U.S. Geol. Survey Prof. Paper 424-B, p. B164-B166.

Waitt, R. B., Jr., 1972a, Geomorphology and glacial geology of the Methow Drainage Basin, eastern North Cascade Range, Washington: Unpub. Ph.D. dissert., Univ. Washington, Seattle, 154 p.

———, 1972b, Revision of Missoula flood history in Columbia Valley between Grand Coulee Dam and Wenatchee, Washington: Geol. Soc. America, Abstracts with Programs, v. 4, p. 255-256.

———, 1977a, Missoula flood *sans* Okanogan lobe: Geol. Soc. America, Abstracts with Programs, v. 9, p. 770.

———, 1977b, Guidebook to Quaternary geology of the Columbia, Wenatchee, Peshastin and upper Yakima Valleys, west central Washington: U.S. Geol. Survey Open-File Rept. 77-753, 25 p.

Walker, G. P. L., 1972, Compound and simple lava flows and flood basalts: Bull. Volcanologique, v. 26, p. 579-590.

Waters, A. C., 1933, Terraces and coulees along Columbia River near Lake Chelan, Washington: Geol. Soc. America Bull., v. 44, p. 763-820.

———, 1955, Geomorphology of south-central Washington, illustrated by the Yakima East quadrangle: Geol. Soc. America Bull., v. 66, p. 663-684.

———, 1960, Determining direction of flow in basalts: Am. Jour. Sci., v. 258-A, p. 350-366.

———, 1961, Stratigraphic and lithologic variations in the Columbia River Basalt: Am. Jour. Sci., v. 259, p. 583-611.

———, 1962, Basalt magma types and their tectonic association—Pacific Northwest of the United States: Am. Geophys. Union, Mon. 6, p. 158-170.

Watkins, N. D., and Baksi, A. K., 1974, Magnetostratigraphy and oroclinal folding of the Columbia River, Steens, and Owyhee basalts in Oregon, Washington and Idaho: Am. Jour. Sci., v. 274, p. 148-189.

Webster, G. D., Baker, V. R., and Gustafson, C., 1976, Channeled Scabland of southeastern Washington, a roadlog via Spokane-Coulee City-Vantage-Washtucna-Lewiston-Pullman: Field Guide No. 2, 72nd Annual Cordill, Sect. Meeting, Geol. Soc. America, Dept. Geology, Washington State Univ., Pullman, 25 p.

Williams, D. T., 1959, A fluid-dynamic mechanism of meteorite pitting: Smithsonian Contr. Astrophys., v. 3, p. 47-67.

Williams, P. F., and Rust, B. R., 1969, The sedimentology of a braided river: Jour. Sed. Petrology, v. 39, p. 649-679.

Wright, T. L., Grolier, M. J., and Swanson, D. A., 1973, Chemical variation related to the stratigraphy of the Columbia River basalt: Geol. Soc. America, Bull., v. 84, p. 371-386.

———, and Hamilton, M. S., 1978, A computer-assisted graphical method for identification and correlation of igneous rock chemistries: Geology, v. 6, p. 16-20.

Yates, R. G., Becraft, G. E., Campbell, A. B., and Pearson, R. C., 1966, Tectonic framework of northeastern Washington, northern Idaho, and northwestern Montana: Canadian Inst. Mining and Metallurgy Spec. Vol. 8, p. 47-59.

Yeats, R. S., 1977, Structure, stratigraphy, plutonism and volcanism of the central Cascades, Washington. Part 1. General geologic setting of the Skykomish Valley: *in* Brown, E. H., and Ellis, R. C., *eds.*, Geological Excursions in the Pacific Northwest: Western Washington Univ. Press, Bellingham, p. 265-275.

15

Reprinted from *Lunar Planet. Sci. Conf. 9th, 1978, Proc.*, pp. 3205–3223 (1978)

A preliminary assessment of the fluid erosional processes that shaped the martian outflow channels

VICTOR R. BAKER

Department of Geological Sciences, The University of Texas at Austin, Austin, Texas 78712

Abstract—Whatever process is invoked to explain the martian outflow channels must also explain a distinctive assemblage of channel landforms: regional and local anastomosing patterns, expanding and contracting reaches associated with flow constrictions, streamlined uplands, inner channels with recessional headcuts, pendant forms (bars or erosional residuals) on the downcurrent sides of flow obstacles, longitudinal grooves, irregular "etched" zones on channel floors, and scour marks around obstacles. Maja Vallis shows longitudinal grooves converging on constricted reaches cut through ridges oriented transverse to the fluid flow. Elevation-controlled scour features imply an upper flow boundary in the responsible fluid. Candidate fluids that can create morphological forms similar to one or several of the martain examples include extremely low-viscosity lava, glacial ice, density flows of sediment in water or similar liquid, wind, some varieties of mass movement, and catastrophic floods of water. Only one of these candidates, catastrophic floods, acting alone, can explain all the observed features. The only single terrestrial landscape that contains analogs to all the martain outflow features is the Channeled Scabland of eastern Washington, created by catastrophic flood flows across differentially resistant basalt, loess, and sedimentary rocks. Lack of knowledge concerning large-scale fluvial erosion by catastrophic floods has led to misconceptions in previous studies of martian channels. Braided and meandering rivers have generally evolved to quasi-equilibrium with the terrestrial environment, and they are not appropriate analogs to the martian outflow channels. Catastrophic flood erosion is a decidedly disequilibrium process. However, certain bed forms develop rapidly enough to reflect the character of the generating flows. The martian channel bed forms develop as a consequence of both the hydrodynamic character of the responsible fluid (passive bed theory) and the properties of the deformed bounday (defect theory). The influence of defect theory appears as obstacle marks. Eroded impact craters, chaotic terrain remnants, and other residual terrain features probably served as obstacles in the path of deep flood outflows on Mars. These obstacles induced flow separations and vorticity that localized both deposition and erosion. Typical forms resulting from these processes are pendant bars, scour holes, and streamlined residual uplands. Hydrodynamic influences (passive bed theory) are best expressed on exposed rock surfaces. Longitudinal grooves and inner channels indicate the probable paleoflood streamline patterns.

INTRODUCTION

One of the most significant results of the Mariner 9 and Viking missions was the realization that fluvial processes may have shaped portions of the martian surface. Persuasive evidence of past water on Mars is provided by a variety of trough-like features that are considered to be relict fluvial channels (Masursky, 1973; Milton, 1973; Sagan *et al.*, 1973; Baker, in Thomsen, 1974; Baker and Milton, 1974; Milton, 1974; Hartmann, 1974; Sharp and Malin, 1975; Malin, 1976; Pieri, 1976; Masursky, 1976; Baker, 1977; Masursky *et al.*, 1977). Moreover, the case for water as an erosive agent on Mars has been immensely strengthened by the Viking discovery that the residual summer ice cap at the martian north pole is water ice, not frozen carbon dioxide (Farmer *et al.*, 1976; Kieffer *et al.*, 1976). Indeed the early appraisals of the Viking mission designated the three most important mission results as (1) the fluvial origin of the channels, (2) the nature of the polar ice caps, and (3) the size of the ancient atmosphere (Metz, 1976).

The martian channels occur in several major varieties. The outflow channels considered in this paper are distinguished as very large forms displaying evidence of scour on an immense scale and related to very localized probable source points (Sharp and Malin, 1975). Masursky *et al.*, (1977) refer to these channels as "broad channels" and distinguish them from sinuous channels of intermediate size and dendritic (runoff) channels. The outflow channels appear fully developed on the landscape with somewhat indistinct source relationships. Some channels clearly do head at collapse zones (chaotic terrain), but others do not.

Based on the Mariner 9 studies, a variety of morphological features in the outflow channels were used to support a clear analogy to the great Pleistocene flood channels scoured through loess and basalt bedrock in the Channeled Scabland of eastern Washington (Baker and Milton, 1974; Sharp and Malin, 1975; Nummedal, 1976). The following are some important properties of the martian outflow channels that were recognized in previous studies:

1) Taken together the martian outflow channels form great composite anastomosing patterns that appear to be eroded across diverse types of terrain, as is apparent in the Channeled Scabland.
2) Separating the anastomosing channel elements are various quadrilateral-shaped or diamond-shaped forms. Many of these are probably erosional residuals of prechannel topography. Similar relationships are common in the Channeled Scabland (Baker and Nummedal, 1978).
3) The channels emanate from potential fluid-source regions (chaotic terrain) and flow down the regional slope of the martian surface from heavily cratered uplands toward the relatively smooth northern plains.
4) Streamlined forms, called "teardrop-shaped" islands and "bars" by some investigators, occur in several channels. In all cases, the streamlining conforms to fluid outflow directions dictated by gradient and source constraints.
5) Unlike the channels of most terrestrial rivers, both martian and Scabland channel width is quite irregular. Expansion and contraction occurs abruptly along the flow path.
6) Several associated bedrock erosional bed forms that are unique to terrestrial scabland erosion also occur in several martian channels. These include longitudinal grooves, inner channels, and cataracts.
7) Distinctive large-scale scour marks occur in association with flow obstructions as would be expected for highly turbulent flows.
8) Both martian and Scabland channels have relatively high width-depth ratios and low sinuosity, characteristics of very high discharge.
9) Features of differential resistance to erosion, such as buried craters and joint patterns, were differentially eroded and exhumed.

Despite the morphological evidence indicating a flood-outflow origin for several of the large martian channels, a number of theoretical difficulties persist for the hypothesis. Sharp and Malin (1975) have summarized these:

(1) A disparity exists between the dimensions of some channels and the volume of fluid that could be released from probable source areas.
(2) The current martian environment poses severe problems for the persistence of water on the planet's surface.
(3) The source of the water, lithospheric or atmospheric, remains an enigma.

The Viking project has provided superior resolution of many outflow channels, increasing the information concerning their extent, source relations, and detailed enchannel features. For some investigators the new data has led to alternative genetic schemes involving erosion by low-viscosity turbulent-flow lava (Schonfeld, 1976, 1977; Cutts *et al.*, 1978a; Cutts and Blasius, 1978). Alternatively, the large channels that occur in obvious non-volcanic settings have been ascribed to wind erosion (Cutts *et al.*, 1976; Cutts, 1977; Cutts *et al.*, 1978b). Schonfeld (1977), Blasius *et al.* (1978), and other investigators have listed the following features and phenomena not apparent on Mars, which they would expect to be apparent if a fluvial origin were hypothesized: deltas at river mouths, obvious sediment accumulations in "sink" regions that for water flows should be depositional basins (e.g., Chryse Planitia), great differences in age between channel (valley) systems and the volcanic terrains into which they are cut, proper relationships in scale between the potential volume of fluid released in a source region and the magnitude of downstream channel erosion, favorable environmental conditions for the persistence of water in liquid form, and the tendency for erosional scale to decrease downstream along major river systems.

This brief paper will consider some relationships for outflow channels related to the above problems. It will begin by describing some of the large-scale erosional and depositional forms associated with the martian channels. It will be argued that this entire assemblage is uniquely associated with scabland flood processes on earth. Moreover, the paper will question the assumption of some investigators that the well-documented qualities of relatively small-scale terrestrial rivers can be used as a guide to martian outflow channel processes.

Large-scale Erosional and Depositional Forms

A bed form can be considered to be any deviation from a flat bed, generated by the flow of a fluid along the bed of its confining channel. Table 1 lists the channel and bed forms recognized from Viking imagery as characterizing the martian outflow channels. Any fluid invoked to explain the channels must also account for this enchannel detail.

Channel anastomosis and streamlining are especially well displayed in the Chryse outflow area (Baker and Kochel, 1978a). The distinct upper elevational limit to the eroded terrain implies that the responsible fluid must have had a free, defined surface. This criterion allows a dense fluid such as water, glacial ice, or lava, but disallows a diffuse media such as atmospheric flows.

Table 1. Characteristic morphological features of the martian outflow channels. All features on this list also characterize the Missoula Flood path from Montana to the Pacific Ocean, especially the Channeled Scabland.

A. Macroscale: Channel Pattern
1. Discrete source areas that present a disparity in volume between probable fluid volume and downstream trough volume showing fluid-erosional forms.
2. Indistinct sink relationships, including the lack of an obvious delta.
3. Anastomosing pattern of channels.
4. Erosion of diverse rock types thousands of kilometers from probable fluid source areas.
5. Residual uplands separating channels.
6. Many uplands streamlined by fluid flow.
7. Pronounced flow constrictions and expansions.
8. High width-depth ratio.
9. Low sinuosity.
10. Differential erosion controlled by structure and lithology; exhumation of resistant pre-flow topography.

B. Mesoscale: Bed Forms
1. Longitudinal grooves.
2. Inner channels.
3. Transverse headcuts (cataracts).
4. Scabland.
5. Scour marks around flow obstacles.
6. Pendant forms (small-scale streamlined features that may be depositional bars).
7. Expansion bar complexes.

Streamlined uplands are perhaps the most striking residual forms between and within the martian outflow channels. Residual uplands separating the various anastomotic channel branches include several varieties. The simplest forms have shapes shapes that are approximated in map view by the mathematical form of a single lemniscate loop (Baker and Kochel, 1978b). These appear to display rather precise streamlined shape characteristics that can develop rapidly from pressure drag and skin resistance factors in the responsible fluid (Baker and Kochel, 1978b). Many are developed on the downflow margins of "defects" that occurred on the channel floors. Often these defects were pre-channel craters or residual blocks of chaotic terrain. Other lemniscate forms lack obvious flow obstructions for their isolation.

Another variety of residual channel upland is characterized by secondary channels that cut through the residual form, but do not erode as deeply as the primary channel floor. Such uplands usually possess a quadrilateral shape, often very similar to the rhombic or diamond shape of eroded braid bars in terrestrial rivers. These channeled uplands are usually larger than nearby lemniscate forms (Fig. 1).

A third class of streamlined upland has steep flanking scarps but shows no modification by secondary channels across the upland crest. These are often the largest residual uplands within the channels. The extensive zones of unmodified prechannel terrain preserved on their crests give these uplands the character of "islands" in an overall channeled region.

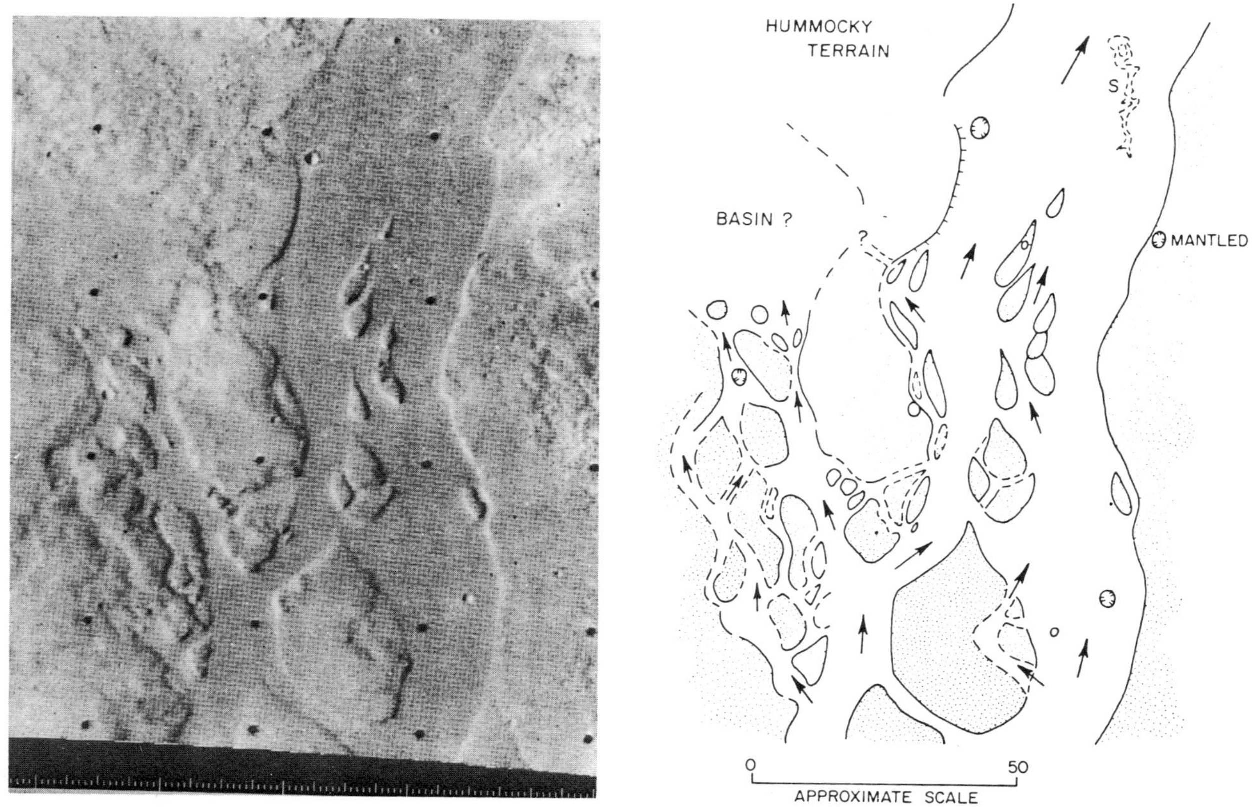

Fig. 1. Sketch map (right) of unnamed channels in Elysium (Lat. 30°N., Long. 224°W.) derived from a high-resolution B frame image (left) of Mariner 9 (DAS 8910729). Arrows show the postulated fluid flow directions around large quadrilateral residual forms (with secondary channels) and smaller streamlined forms. Some of the channel floor relief may be scabland (S). Malin (1976) attributed the highly subdued appearance of these channels to mantling by a blanket of weakly consolidated material currently under erosional attack by eolian processes.

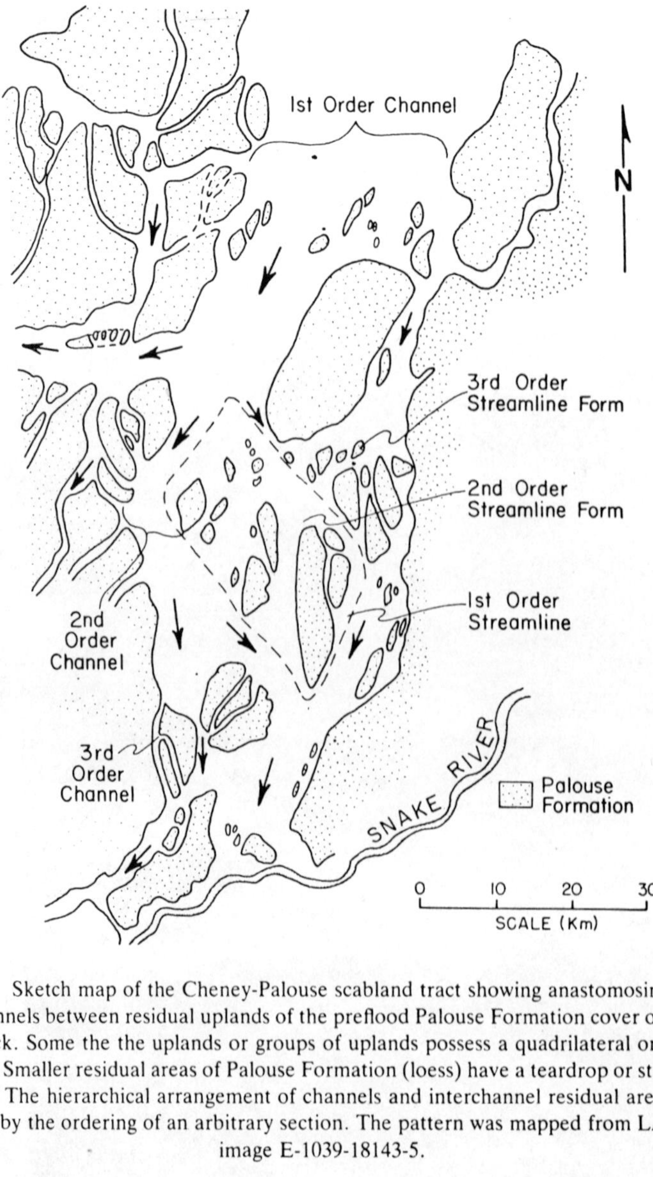

Fig. 2. Sketch map of the Cheney-Palouse scabland tract showing anastomosing pattern of channels between residual uplands of the preflood Palouse Formation cover over basalt bedrock. Some the the uplands or groups of uplands possess a quadrilateral or diamond shape. Smaller residual areas of Palouse Formation (loess) have a teardrop or streamlined shape. The hierarchical arrangement of channels and interchannel residual areas is illustrated by the ordering of an arbitrary section. The pattern was mapped from LANDSAT image E-1039-18143-5.

The three classes of streamlined upland in martian channels all have close analogs in the Channeled Scabland. Studies of the Cheney-Palouse scabland tract suggest that the characteristic arrangements of the residual elements probably are due to variations in flow velocities, rates of erosion, and rates of deposition for various elevations in a hierarchy of flood channelways (Patton and Baker, 1978). With continued flooding and erosion large residual uplands can be cut by channels and eventually reduced to smaller streamlined hills of minimum flow resistance. Figure 2 illustrates the resulting hierarchical arrangement.

Patton and Baker (1978) suggest an analogy between the residual hills of catastrophic flood channels and the longitudinal bars of braided streams. Initially, the flooding spread widely over the rough planetary surface in relatively wide-shallow channels. As these widened by bank erosion, they encountered increasing boundary resistance. Then, in order to maintain a sufficient velocity for sediment transport, the channel divided, and incision created relatively narrow and deep secondary channels. Thus the residual hills that separate the anastomosing channels are part of a pattern of stream adjustment to achieve over-all equilibrium between an erosive fluid and a resistant boundary. It must be emphasized, however, that the anastomosing pattern of equilibrium braided streams generally results from depositional processes with secondary erosion. Catastrophic flood anastomosis is produced by *erosion* into rock with secondary deposition associated mainly with flow obstructions.

As described by Baker and Kochel (1978a), some of the streamlined uplands of martian channel floors may be in part depositional. Residual uplands in the Channeled Scabland, much as longitudinal bars in a braided stream, created low velocity zones in their lees which localized the deposition of fluid-entrained debris. This debris then could form a pendant extending downchannel from the flow defect. A typical scabland pendant bar is shown in Fig. 3. Of course, similar pendant forms can result from a variety of fluids *that transport sediment*. In the Channeled Scabland pendant bars of flood gravel resulted from flow conditions that locally reduced the fantastic competency of the peak flood discharges (Fig. 3). The transport rate of flood gravel into the separation zone downstream from the bar was simply greater than the transport rate out of that zone. The relationship of the bars to the flow conditions is evinced by the fact that they are streamlined to present minimum resistance to the flood water.

The Channeled Scabland is most famous for the erosive stripping of basalt bedrock to produce the terrain called "scabland" (Fig. 4). An important result of Viking orbiter imagery of the southern Chryse Basin was the depiction of erosively stripped zones on the channel floors in the outflow region of Tiu and Ares Valles (Carr *et al.*, 1976; Greeley *et al.*, 1977). Two aspects of the stripped zones are remarkably similar to relationships in the Channeled Scabland. First, commonly the stripping occurs between residual uplands in locations where flows would have been somewhat constricted. Second, some areas show organization of the stripped surface into a distinct inner channel that follows the presumed flow streamlines around the streamlined residual uplands (Baker and Kochel, 1978a). On the other hand, the angular margins of the martian depressions are unlike the characteristic rounded pothole margins of scabland butte-and-basin topography.

An important association in the Channeled Scabland is that of erosional longitudinal grooves and inner channels that head at impressive cataracts (Baker, 1973b, 1978b). Figure 5 shows the Viking orbital imagery of a scene in lower Kasei Vallis that was first described in a study of Mariner 9 imagery by Baker and Milton (1974). Note the prominence of erosional grooves paralleling the presumed paleoflow streamlines around the large erosional residual. Two transverse headcuts occur at the proximal ends of inner channels. The grooves lead right to the lips of

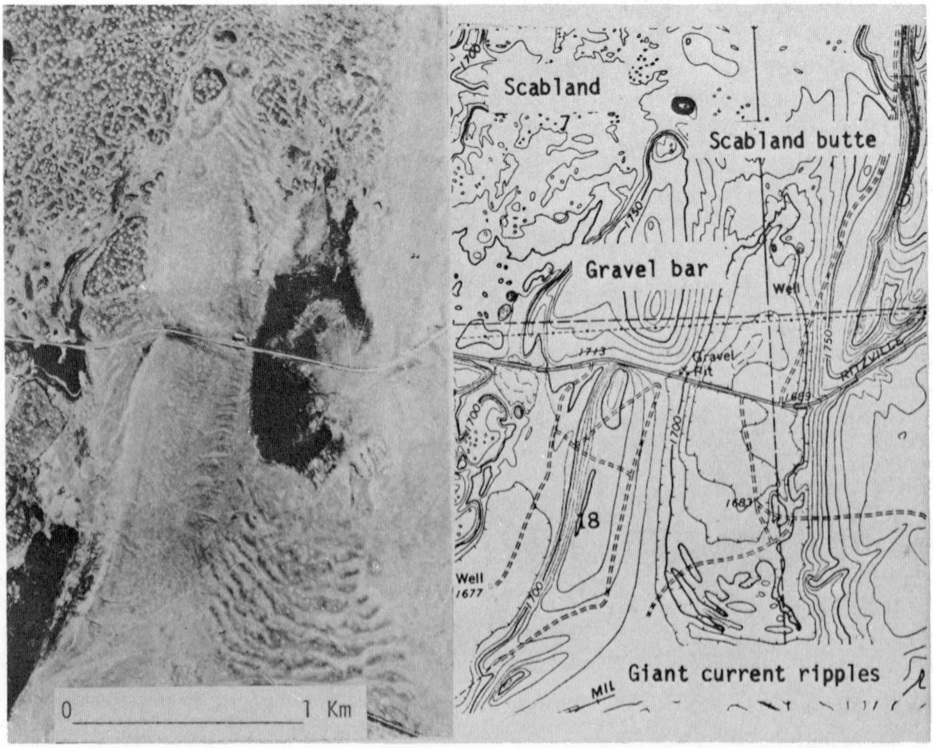

Fig. 3. Vertical aerial photograph and topographic map of a small pendant bar in the Cheney-Palouse scabland tract near Macall, Washington. Contour interval is approximately 3 m (10 feet). The bar accumulated downstream from a residual knob of basalt. Water velocities in this area averaged 12 m/sec for flow depths of 30–40 m (Baker, 1973a) during the flood maximum.

these horseshoe-shaped headcuts. It would appear to require an unusual coincidence for this assemblage of typical scabland landforms to have been produced by processes other than catastrophic erosion by a highly turbulent liquid.

In the Channeled Scabland the inner channels form by the headward recession of subfluvial cataracts (headcuts). The cataract lips and inner channel walls are held up by relatively resistant rock layers that form the general channel floor. The flow separation developed in the water passing over the cataract lip induces plucking erosion from beneath the relatively resistant layer. In the Channeled Scabland this undermining and recession process proceeded rapidly enough that the 250 m deep inner channel of the upper Grand Coulee receded headward over 30 km during the course of one or two short-lived floods (Bretz, 1932).

Scour marks are often generated on an erodible bed by a turbulent fluid boundary layer as it approaches a blunt-nosed obstacle. Crescentic scour holes develop on the upcurrent sides of positive relief defects on the channel floor. In the

Fig. 4. Oblique aerial·photograph typical scabland topography in the Hartline Basin. The foreground relief 100–150 m was generated by plucking erosion of basalt flows. The general channel floor level is preserved because of the relatively resistant entablature of a single basalt flow. The higher cliffs in the background have a relief of 300–500 m and were undercut by flood water along a zone of structural weakness (the Coulee Monocline). The maximum depth of flood water over the lowest channel floor at this location was approximately 200 m. The initial flood flows entered this part of the basaltic Columbia Plateau because it was a preflood trough on the land surface.

martian outflow channels these defects include bedrock projections, craters, and streamlined residual hills (Fig. 6). The scour probably results from a horseshoe-vortex system developed at the front of the obstacle (Richardson, 1968).

In the Channeled Scabland scour marks, longitudinal grooves, and transverse headcuts all appear to require high turbulence in the responsible fluid. Moreover, this turbulence apparently was organized into macroturbulent flow structures that are considered to be a characteristic quality of very deep, high gradient flood flows (Baker, 1973a, 1973b, 1978a).

Catastrophic Flood Processes

A lack of knowledge concerning large-scale fluvial erosion processes in catastrophic floods has led to misconceptions in some previous studies of the martian channels. Milton (1973) compared an anastomosing reach of Mangala Vallis to depositional processes in a braided river. Weinhaupt (1974) applied terrestrial fluvial regimen theory to the sinuous rill-like portion of the complex Nirgal Vallis

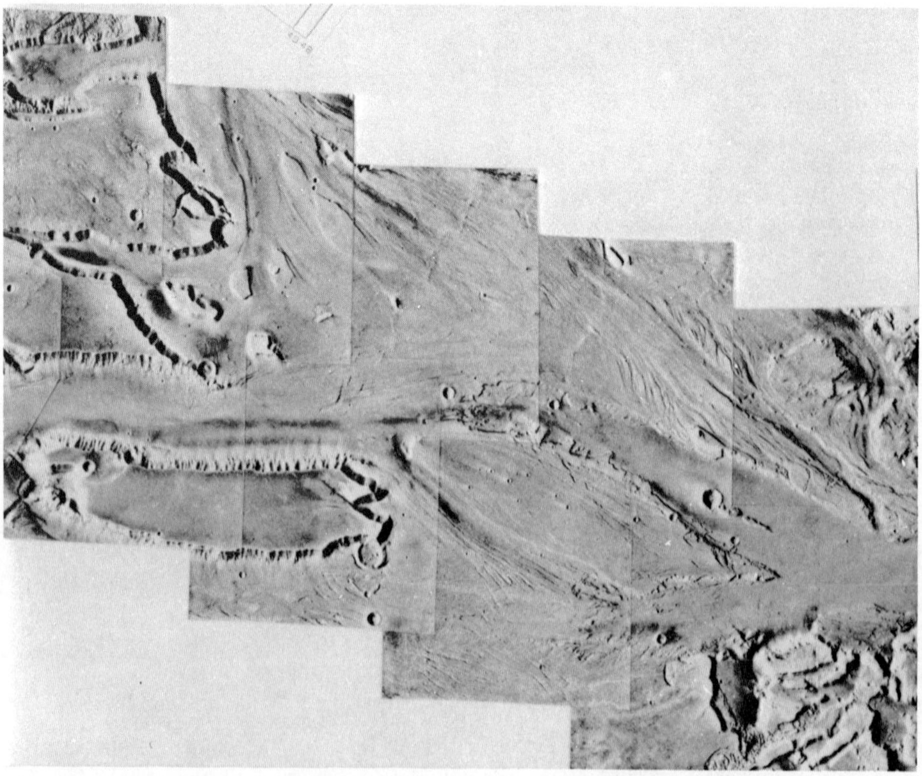

Fig. 5. Part of lower Kasei Vallis near 62°W, 25°N with landforms indicating catastrophic flood erosion. The relationships were first described from a study of Mariner 9 imagery by Baker and Milton (1974). A detailed interpretive map of this reach is given by Baker and Kochel (1978a, this volume). This mosaic of Viking orbiter imagery was prepared from photographs A06 through A13 (part of JPL Mosaic 211-5371).

system. These apparent analogs appear reasonable given comparison to "normal" terrestrial rivers. The martian outflow channels, however, have probably evolved from high-magnitude conditions that are extremely rare on earth. The requisite conditions have only been achieved in a few rather spectacular terrestrial settings.

An important point that is often not appreciated in attempting to envision a fluvial origin of the martian channels is that the channel floors display mostly large-scale bed forms eroded into rock. Most terrestrial rivers show small-scale forms developed on their own alluvium. Catastrophic flood channels on earth do not display the same adjustments of form to process that are observed in most equilibrium river channels; we should not expect them to do so on Mars. Floods imposed on an ungraded planetary surface deposit their load at every flow expansion and then erode and streamline their own deposits as they drain through the downstream constriction that was responsible for the ponding. This observation is

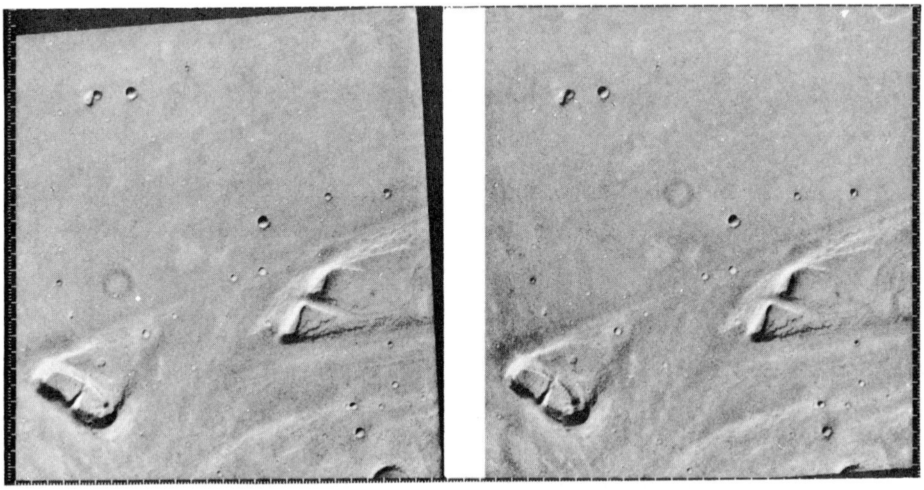

Fig. 6. Obstacle marks associated with residual uplands at the outflow Ares Vallis into the southern Chryse Basin (18.0°N., 33.9°W.). These two Viking photographs form a stero pair (left 04A23; right 04A83). A prominent horseshoe-shaped depression is wrapped around the upflow side of the upland at left. Fluid flows entered from lower left and possessed sufficient turbulence to generate this scour mark in the channel floor sediment. Computer stereophotogrammetry by the JPL Image Processing Laboratory shows that the uplands rise 200–400 m above the adjacent channel floor and that the scour depression is less than 100 meters deep. The photographs show a scene approximately 40 × 40 km.

especially pertinent for the debouchment of Maja Vallis into the western Chryse Planitia (Fig. 7).

The following is a somewhat speculative outline of events. Converging and diverging flow patterns are typical of a river's attempt to create an alternating pattern of energy dissipation. Given sufficient time terrestrial rivers will eventually scour at the convergences to create pools. With continued discharge the pool-and-riffle pattern gradually will evolve to the equilibrium morphology of a meandering or braided river (Keller and Melhorn, 1973). Catastrophic flood channel-ways only display the incipient stages of this process (Baker, 1978b). Channel equilibrium is rarely achieved. As martian flows passed from various constricted canyons into the Chryse Basin they were no longer constrained by narrow channel walls. To move as a sheetflood across an irregular preflood surface would have been highly inefficient relative to energy dissipation. Therefore, the fluid eroded into the existing troughs of the topography, between the larger craters and along preexisting structural lows. This produced an anastomosing system of secondary channels separated by residual hills of the preflood topography. The individual channels were now relatively efficient compared to the original land surface, but the complete development was inhibited by the short duration of the flow event (or events).

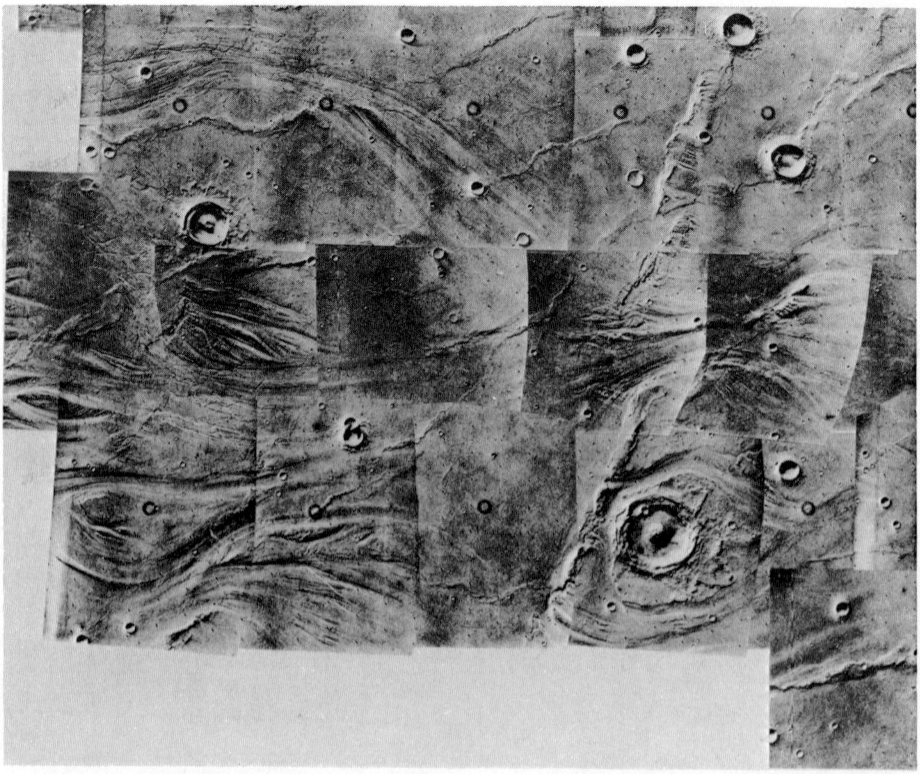

Fig. 7. Anastomosing channel patterns in the western Chryse Planitia. Ridges transverse to the fluid flow direction ponded water on their upflow margins (left) until fluid could flow through low points on the ridge crests. The convergent and divergent patterns of grooves and streamlined forms is typical of relatively wide, shallow fluid flow. Some of smaller "wrinkle" ridges appear to postdate the fluid-flow event. Scale for this scene is indicated by the largest impact crater at lower right which is 15 km in diameter.

The movement of martian fluid flows across surfaces that were unprepared by previous fluvial activity is well-displayed in Ares Vallis. There fluid even spilled into one side of a crater, filled it, and then spilled out the other side (Fig. 8). These relationships, described by Sharp and Malin (1975), show that the fluid possessed a well-defined free surface and that it was capable of eroding a variety of materials over rather immense distances from the channel heads.

The enormous flow velocities achieved in many scabland channels caused considerable reduction of absolute pressure, perhaps to the fluid vapor pressure. The resulting cavitation processes can be extremely erosive. Engineering experiments have shown that bubble collapse near materials such as steel or concrete produce hammer-like blows with local pressures as high as 30,000 atmospheres (Barnes, 1956). These very intense local pressures are capable of shattering the surface of nearly any solid material. The fractured surface layers are quickly carried away by turbulent water and fracturing continues.

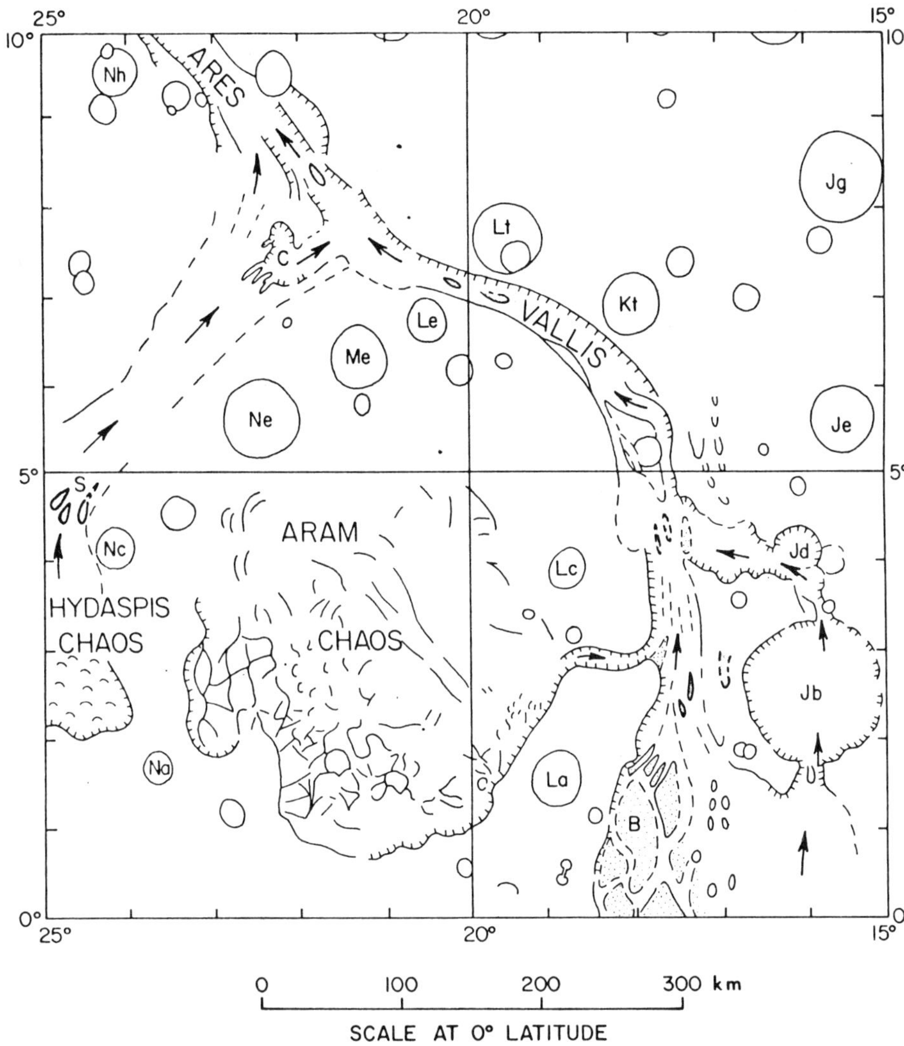

Fig. 8. Sketch map of Ares Vallis showing the effects of fluid flows on ancient cratered terrains and the relation of channel heads to zones of chaotic terrain. Craters Jb and Jd were breached by fluid flows (arrows) as discussed by Malin (1976). Streamlined uplands (S) are indicated by shading. A possible expansion bar complex that was later streamlined by waning flows occurs at B. Other possible flood bars are indicated by the dotted pattern at locations where deposition of flood load would be expected. A possible inner channel cataract complex (C) occurs where relatively shallow flows spilled into Ares Vallis after crossing the interchannel divide from Hydaspis Chaos (which mainly drained to the west). The complete terrain relationship is geomorphically unique to catastrophic flood erosion.

Another important process for rapid erosion in catastrophic flood flows is macroturbulent plucking associated with the intense energy dissipation of upward vortex action. The pressure and velocity gradients of vortex action can produce a phenomenal hydraulic lift force along the filament of the vortex. The required vortices would be expected along the extremely irregular floors of scabland channels and the martian outflow channels. The lips of potholes, canyon walls, and cataracts are but a few examples.

The combination of cavitation and macroturbulent plucking appears to have generated a sequence of erosional forms in time and space for rock surfaces in the Channeled Scabland (Baker, 1973b). The initial channels had relatively smooth floors. These are marked by longitudinal grooves, which mimic the longitudinal vorticity of the streaming fluid. Deeper erosion created irregular surfaces that generated flow separation and/or kolks. As erosive activity concentrated at these sites, the result was greater accentuation of the surface irregularities. A critical threshold had to be crossed to achieve this change from longitudinal forms to the production of irregular scabland. Indeed the entire sequence could only begin to form above a critical threshold of flow depth and water-surface gradient that is only rarely achieved in terrestrial water flows.

Whereas the mechanics of equilibrium fluvial processes are generally well described (e.g., Leopold *et al.*, 1964), catastrophic flood mechanics have not been adequately investigated. The poor understanding of flood erosion may account for the fact that it took over thirty years for Bretz' (1923) pronouncement of a catastrophic flood origin of the Channeled Scabland to gain general acceptance by the geological community (Baker, 1978c). The debates over the origin of the Channeled Scabland centered on the erosive capability of running water in a short time period, the phenomenal magnitude of the required erosion, the lack of an apparent source sufficient to supply enough water at the required rate, the relative scarcity and localized nature of depositional features in a landscape dominated by massive erosion, the presence of spectacular erosion far downstream from any implied source ($>10^3$ km), and the presence of obvious erosional forms in great basins where the flood water would not have been expected to have been erosive. Many of the above relationships can now be easily explained for the Channeled Scabland by the disequilibrium character of massive flood flows.

DISCUSSION

Table 2 summarizes the compatibility of various fluid-flow processes with certain morphological features of the martian outflow channels. Whereas every flow system can explain some of the evidence (marked by "X" in the table), all the flow systems save one are either questionable generators of such morphology (question marks) or unlikely causes (dash). The only single terrestrial landscape that contains clear analogs to all the martian outflow features is the Channeled Scabland of eastern Washington, created by catastrophic flood flows across differentially resistant basalt, loess, and sedimentary rocks.

A complete discussion of the various alternative fluids is beyond the scope of this paper. Consider only the volcanic hypothesis. It is consistent with certain relationships such as the apparent penecontemporaneity (based on crater size-frequency relationships) that exists between the channels and the volcanic terrain

Table 2. Compatibility of various fluid-flow processes with certain martian channel features.

POSSIBLE TERRESTRIAL ANALOGS

Martian Channel Forms	Lava	Glacier	Mudflow	Density Flow	Wind	Cata-strophic Flood
Anastomosis	?	?	X	?	?	X
Streamlined Uplands	?	X	X	X	X	X
Longitudinal Grooves	?	X	X	X	X	X
Scour Marks	?	–	?	X	X	X
Scabland	–	?	–	?	?	X
Inner Channels	?	?	?	X	?	X
Headcuts	–	X	?	?	–	X
Constrictions	X	X	X	X	X	X
Pendant Forms	?	X	X	X	X	X

into which they are eroded. Models of martian volcanism (Smyth *et al.*, 1978) indicate that martian primary lavas should be of unusually low viscosity. Certainly lava flows can reach very high velocities; a highly fluid melilite nephelinite lava from Nyiragongo lava lake reached record speeds of 60 km/hr (16.7 m/sec) (Simkin and McClelland, 1978). However, it is far from clear that the detailed morphological assemblage noted above can be produced by lava which would subsequently cool, leaving immense zones of solidified rock. Could the martian lavas mimic all the detailed depositional and erosional features known to be characteristic of terrestrial catastrophic flood flows?

The Channeled Scabland may be an excellent testing ground for this question. The Columbia Plateau lavas into which the flood channels are carved represent the earth's most impressive example of catastrophic flooding by lava (Swanson and Wright, 1978). Many of the lava flows follow pre-flow river canyons. The highly fluid Pamona Flow advanced 500 km from its suspected source in Idaho to the Pacific Ocean. Catastrophic flooding of water followed nearly the same route as the catastrophic lava flows which preceded it by 6–14 m.y. The structural properties of the basalt were a critical factor in producing the distinctive erosional forms created by the water floods (Baker, 1973b). The remarkable coincidence in time and space for these greatest of terrestrial fluid-flow phenomena may not be so unusual for Mars.

The most perplexing problem for martian outflow channels involves flow volumes and rate of release. How could the relatively small source areas (chaotic terrain) have contributed the amounts of water required to generate the observed scabland erosion? An initial suggestion was that ground ice was locally melted by geothermal processes. If the liquified subsurface slurry of water and debris were contained by pressure, it might eventually burst on to the surface as a short-lived mudflow followed by collapse (McCauley *et al.*, 1972). Milton (1974) suggested

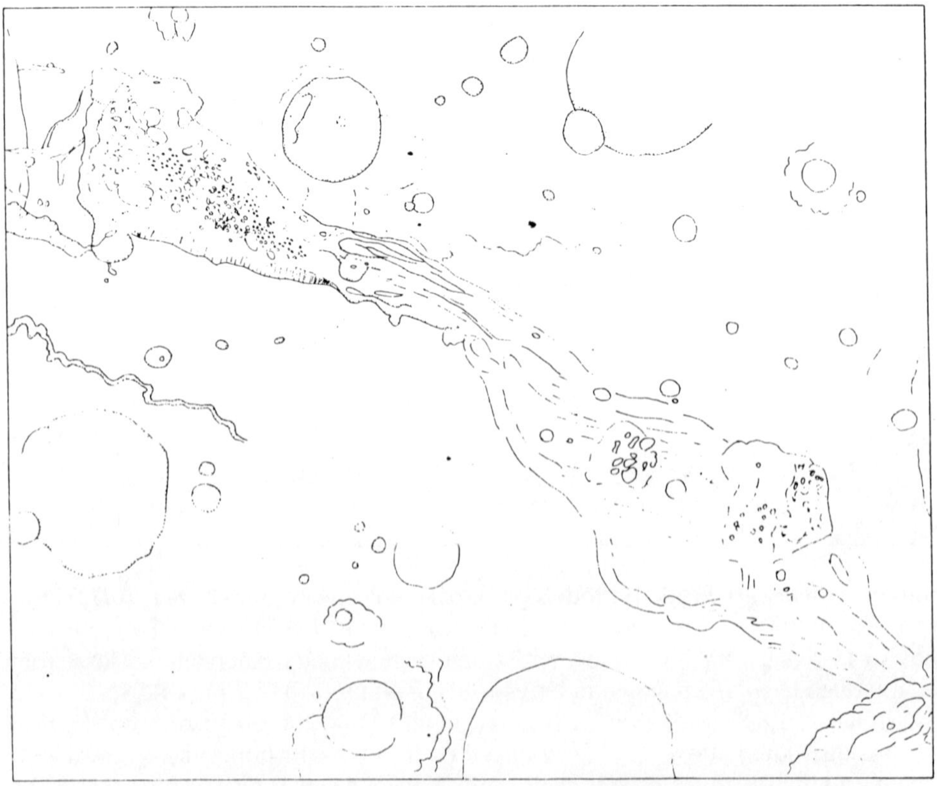

Fig. 9. Sketch map of a channel system (42°W, 0°N) tributary to Hydraotes Chaos (lower right). The map shows an area measuring about 250 × 200 km. Zones of chaotic terrain mark the channel head and probably provided catastrophic outbreak flooding. Other chaotic terrain appears to truncate the grooved and streamlined channel topography.

that ground ice on Mars might be bound in a clathrate compound. Heat for the clathrate melting can be initiated by climatic change and further generated by mechanical work as collapse occurs. Peal *et al.* (1975) argue that liquid water was probably already present beneath the martian permafrost layer. A variety of processes, including meteorite impact, faulting, and volcanic activity, would then release the water. The formation of a relatively thin skin of ice on the surface of the flowing water would then prevent boiling in the martian atmosphere by processes discussed by Schubert *et al.* (1970). Masursky *et al.* (1977) suggest that water is released after subterranean emplacement of local volcanic complexes. This volcanism melts the overlying permafrost, generating an outward spreading front of meltwater that bursts to the surface when it intersects a cliff face.

The morphological evidence that can test the above hypotheses is elusive, and serious questions can be raised about required fluid volumes versus the areas of preserved source terrains. Figure 9 illustrates some important relationships in a source region. Cross-cutting relationships can be discerned in this channel which suggest that several chaotic collapse zones downstream from the channel head occurred after the main channelling event. In other outflow areas one observes chaotic remnants eroded by the channel fluids. Fluid release on Mars is apparently not instantaneous, but spread through time. Such relationships are produced on a much smaller scale by several terrestrial processes that concentrate water in the subsurface and release it as surface flows.

The piping process (Parker, 1963; Parker *et al.*, 1964) occurs in weak, incoherent sediments, such as loess and tuff. Fluid concentrates in conduits, often along structural lines. These conduits draw in all subsurface fluid flows and enlarge, growing by headward saping. Eventually the roof collapses over this pipe forming a chaotic collapse zone. As the point of fluid release migrates headward, the downstream zones of collapsed terrain are scoured by the most recent flows. Thus the channels are collapse troughs which were subsequently modified by floods emanating from upstream migrating collapse zones. Deeper and later piping can produce collapse within the zones that were scoured by earlier floods. Piping is facilitated by sporadic fluid inputs as characteristic of semiarid regions.

A related process that has the added advantage of terrestrial association with permafrost is suffosion, a bursting out on the surface of erupted highly mobile or water-saturated material. The process is common in peri-glacial areas where underground water resulting from partial melting of ground ice, exerts upward pressure and bursts through a hard dried upper skin (Muller, 1947). The abundant physical evidence for permafrost with high percentages of ground ice in the old cratered terrain of Mars (Carr and Schaber, 1977) suggests a possible setting for suffosion. Outburst could be initiated by various critical events, such as faulting, cliff retreat, and volcanism.

Both suffosion and piping are consistent with a working hypothesis that the martian outflow channels extended headward by progressive transformation of cratered terrain underlain by water-saturated, fractured rock into chaotic terrain. With continued headward extension, the downstream zones of chaotic terrain were scoured by the most recent flooding. Thus, the preserved zones of chaotic terrain would have no relation to channel volume that is partly produced by collapse over earlier fluid-release zones and subsequently modified by flows produced at new collapse zones upstream. Thus downstream zones could have been preflood lows on the topography or collapse-generated troughs that subsequently were modified by flooding to produce the characteristic assemblage of flood bed forms on their floors.

Acknowledgments—The research on which this preliminary report is based was performed under NASA Grant NSG-7326 from the Planetary Geology Program, Office of Space Science and under grant NAS 1-15164 for the NASA Viking Guest Investigator Program. R. Craig Kochel and Peter C. Patton contributed to the development of these ideas. Publication support was provided by the Geology Foundation, University of Texas at Austin. I thank Paul D. Komar and David E. Thompson for their helpful comments on the manuscript.

REFERENCES

Baker V. R. (1973a) Paleohydrology and sedimentology of Lake Missoula flooding in eastern Washington. *Geol. Soc. Amer. Spec. Paper 144.* 79 pp.

Baker V. R. (1973b) Erosional forms and processes for the catastrophic Pleistocene Missoula floods in eastern Washington. In *Fluvial Geomorphology* (M. Morisawa, ed.), p. 123–148. State Univ. N.Y., Binghamton.

Baker V. R. (1977) Viking–Slashing at the martian scabland problem (abstract). NASA TMX-3511, p. 169–172.

Baker V. R. (1978a) Paleohydraulics and hydrodynamics of scabland floods. Large-scale erosional and depositional features of the Channeled Scabland. In *The Channeled Scabland* (V. R. Baker and D. Nummedal, eds.), p. 59–79. NASA Office of Space Science, Planetary Geology Program, Washington, D.C.

Baker V. R. (1978b) Large-scale erosional and depositional features of the Channeled Scabland. In *The Channeled Scabland* (V. R. Baker and D. Nummedal, eds.), p. 81–115. NASA Office of Space Science, Planetary Geology Program, Washington, D.C.

Baker V. R. (1978c) The Spokane Flood controversy. In *The Channeled Scabland* (V. R. Baker and D. Nummedal, eds.), p. 3–15. NASA Office of Space Science, Planetary Geology Program, Washington, D.C.

Baker V. R. and Kochel R. C. (1978a) Morphological mapping of martian outflow channels. *Proc. Lunar Planet. Sci. Conf. 9th.* This volume.

Baker V. R. and Kochel R. C. (1978b) Morphometry of streamlined forms in terrestrial and martian channels: A preliminary assessment. *Proc. Lunar Planet. Sci. Conf. 9th.* This volume

Baker V. R. and Milton D. J. (1974) Erosion by catastrophic floods on Mars and earth. *Icarus* **23,** 27–41.

Baker V. R. and Nummedal D. (1978) *The Channeled Scabland.* NASA Office of Space Science, Planetary Geology Program, Washington, D.C. 186 pp.

Barnes H. L. (1956) Cavitation as a geological agent. *Amer. J. Sci.* **254,** 493–505.

Blasius K. R., Cutts J. A. and Roberts W. J. (1978) Large scale erosive flows associated with Chryse Planitia, Mars: Source and sink relationships (abstract). In *Lunar and Planetary Science IX*, p. 109–110. Lunar and Planetary Institute, Houston.

Bretz J H. (1923) The Channeled Scabland of the Columbia Plateau. *J. Geol.* **31** 617–649.

Bretz J H. (1932) The Grand Coulee. *Amer. Geog. Soc. Spec. Publ. 15*, 89 pp.

Carr M. H., Masursky H., Baum W. A., Blasius K. R., Briggs G. A., Cutts J. A., Duxbury T., Greeley R., Guest J. E., Smith B. A., Soderblom L. A., Veverka J. and Wellman J. B. (1976) Preliminary results from Viking orbiter imaging experiments. *Science* **193,** 766–776.

Carr M. H. and Schaber G. G. (1977) Martian permafrost features. *J. Geophys. Res.* **82,** 4039–4054.

Cutts J. A. (1977) Chryse Basin channels (abstract). In *Abstracts for Special Session on Mars.* Lunar and Planetary Institute, Houston.

Cutts J. A. and Blasius K. R. (1978) Models of lava erosion relevant to channel formation on Mars. *Icarus.* In press.

Cutts J. A., Blasius K. R. and Farrell K. W. (1976) Mars: New data on Chryse basin land forms (abstract). In *Bull. Amer. Astron. Soc.* **8,** 480.

Cutts J. A., Blasius K. R. and Roberts W. J. (1978b) Chaotic terrain and channels associated with Chryse Planitia, Mars (abstract). In *Lunar and Planetary Science IX*, p. 206–208. Lunar and Planetary Institute, Houston.

Cutts J. A., Roberts W. J. and Blasius K. R. (1978a) Martian channels formed by lava erosion (abstract). In *Lunar and Planetary Science IX*, p. 209. Lunar and Planetary Institute, Houston.

Farmer C. B., Davies D. W. and LaPorte (1976) Mars: Northern summer ice cap—water vapor observations from Viking 2. *Science* **194,** 1339-1341.

Greeley R., Theilig E., Guest J. E., Carr M. H., Masursky H. and Cutts J. A. (1977) Geology of Chryse Planitia. *J. Geophys. Res.* **82,** 4093–4109.

Hartmann W. K. (1974) Geological observations of martian arroyos. *J. Geophys. Res.* **79,** 3951–3957.

Keller E. A. and Melhorn W. N. (1973) Bedforms and fluvial processes in alluvial stream channels. In *Fluvial Geomorphology* (M. Morisawa, ed.), p. 253–283. State Univ. N.Y., Binghamton.

Kieffer H. H., Chase S. C., Martin T. Z., Miner E. D. and Palluconi F. D. (1976) Martian north polar summer temperature: Dirty water ice. *Science* **194,** 1341–1343.

Leopold L. B., Wolman M. G. and Miller J. P. (1964) *Fluvial Processes in Geomorphology*. W. H. Freeman, San Francisco. 522 pp.

Malin J. C. (1976) Age of martian channels. *J. Geophys. Res.* **81,** 4825–4845.

Masursky H. (1973) An overview of geological results from Mariner 9. *J. Geophys. Res.* **78,** 4009–4030.

Masursky H. (1976) Martian channels (abstract). NASA TMV-3364, p. 169–171.

Masursky H., Boyce J. M., Dial A. L., Schaber G. G. and Strobell M. E. (1977) Classification and time of formation of martian channels based on Viking data. *J. Geophys. Res.* **82,** 4016–4038.

McCauley J. F., Carr M. H., Cutts J. A., Hartmann W. K., Masursky H., Milton D. J., Sharp R. P. and Wilhelms D. E. (1972) Preliminary Mariner 9 report on the geology of Mars. *Icarus* **17,** 289–327.

Metz W. D. (1976) Viking (II): water, ice, and argon—three puzzles resolved. *Science* **194,** 924–927.

Milton D. J. (1973) Water and processes of degradation in the martian landscape. *J. Geophys. Res.* **78,** 4037–4047.

Milton D. J. (1974) Carbon dioxide hydrate and floods on Mars. *Science* **183,** 654–656.

Muller S. W. (1947) Permafrost or permanently frozen ground and related engineering problems. Edwards, Ann Arbor, Michigan. 231 pp.

Nummedal D. (1976) Fluvial erosion on Mars: A review. In *Colloquium on Water in Planetary Regoliths*, p. 47–54. U.S. Army Cold Regions Research and Engineering Lab., Hanover, New Hampshire.

Parker G. G. (1963) Piping, a geomorphic agent in landform development of the drylands. *Inter. Assoc. Sci. Hydrol. Publ.* **65,** 103–113.

Parker G. G., Shown L. M. and Ratzlaff K. W. (1964) Officer's Cave, a pseudokarst feature in altered tuff and volcanic ash of the John Day Formation in eastern Oregon. *Geol. Soc. Amer. Bull.* **75,** 393–402.

Patton P. C. and Baker V. R. (1978) Origin of the Cheney-Palouse scabland tract. In *The Channeled Scabland* (V. R. Baker and D. Nummedal, eds.), p. 117–130. NASA Office of Space Science, Planetary Geology Program, Washington, D.C.

Peale S. J., Schubert G. and Lingenfelter R. E. (1975) Origin of martian channels: Clathrates and water. *Science* **187,** 273–274.

Pieri D. (1976) Martian channels: Distribution of small channels on the martian surface. *Icarus* **27,** 25–50.

Richardson P. D. (1968) The generation of scour marks near obstacles. *J. Sediment Petrol.* **38,** 965–970.

Sagan C., Toon O. B. and Gierasch P. J. (1973) Climatic change on Mars. *Science* **181,** 1045–1049.

Schonfeld E. (1976) On the origin of the martian channels (abstract). *EOS (Trans. Amer. Geophys. Union)* **57,** 1948.

Schonfeld E. (1977) Martian volcanism (abstract). In *Lunar Science VIII*, p. 843–845. The Lunar Science Institute, Houston.

Schubert G., Lingenfelter R. E. and Peale S. J. (1970) The morphology, distribution, and origin of lunar sinuous rilles. *Rev. Geophys. Space Phys.* **8,** 199–224.

Sharp R. P. and Malin M. C. (1975) Channels on Mars. *Bull. Geol. Soc. Amer.* **86,** 593–609.

Simkin T. and McClelland L. (1978) Volcanology. *Geotimes* **28,** 49–50.

Smyth J. R., Huguenin R. and McGetchin T. R. (1978) Composition of martian primary lavas—Convergence of model results. In *Lunar and Planetary Science IX*, p. 1077–1079. Lunar and Planetary Institute, Houston.

Swanson D. A. and Wright T. L. (1978) Bedrock geology of the northern Columbia Plateau and adjacent areas. In *The Channeled Scabland* (V. R. Baker and D. Nummedal, eds.), p. 37–57. NASA Office of Space Science, Planetary Geology Program. Washington, D.C.

Thomsen D. E. (1974) The day the dam burst. *Science News* **106,** 250–251.

Weinhaupt J. G. (1974) Possible origin and probable discharges of meandering channels on the planet Mars. *J. Geophys. Res.* **79,** 2073–2076.

AUTHOR CITATION INDEX

SUBJECT INDEX

About the Editor

VICTOR R. BAKER is an associate professor of geological sciences at The University of Texas at Austin. His research interests include geomorphic processes, paleohydrology, remote sensing, Quaternary geology, geologic hazards, and surficial processes on Mars.

Professor Baker received the B.S. degree from Rensselaer Polytechnic Institute (1967) and the Ph.D. degree from the University of Colorado (1971). During 1977–1979 he served as a Guest Investigator on the NASA Viking program for the study of Mars. During 1979–1980 he was named Visiting Senior Research Scholar (Fulbright-Hays Program) at the North Australian Research Unit, Australian National University, for hydrogeomorphic studies of floods in northern Australia. He is the author of over 70 research papers and monographs concerning geomorphology, hydrology, remote sensing applications, and the geology of Mars.